Unsolved Problems in Astrophysics

John N. Bahcall
and
Jeremiah P. Ostriker, Editors

UNSOLVED PROBLEMS

IN ASTROPHYSICS

PRINCETON UNIVERSITY PRESS

Princeton, New Jersey

ISBN 0-691-01607-0 (cl)
ISBN 0-691-01606-2 (pa)

The publisher would like to acknowledge the editors of this volume for
providing the camera-ready copy from which this book was printed

Princeton University Press books are printed on acid-free paper and meet the
guidelines for permanence and durability of the Committee on Production
Guidelines for Book Longevity of the Council on Library Resources

Printed in the United States of America
by Princeton Academic Press

1 3 5 7 9 10 8 6 4 2

OVERVIEW

Contents

PREFACE

The articles in this volume were written in response to the following hypothetical situation. A second year graduate student walks into the author's office and says: "I am thinking of doing a thesis in your area. Are there any good problems for me to work on?"

Leading astrophysical researchers answer the student's question in this collection by providing their views as to what are the most important problems on which major progress may be expected in the next decade. The authors summarize the current state of knowledge, observational and theoretical, in their areas. They also suggest the style of work that is likely to be necessary in order to make progress. The bibliographical notes at the end of each paper are answers to the parting question by the hypothetical graduate student: "Is there anything I should read to help me make up mind about a thesis?"

As everyone knows who reads a newspaper or listens to the daily news, astrophysics is in the midst of a technologically driven renaissance; fundamental discoveries are being made with astonishing frequency. Measured by the number of professional researchers, astrophysics is a small field. But, astronomical scientists have the entire universe outside planet earth as their exclusive laboratory. In the last decade, new detectors in space, on earth, and deep underground have, when coupled with the computational power of modern computers, revolutionized our knowledge and understanding of the astronomical world. This is a great time for a student of any age to become acquainted with the remarkable universe in which we live.

In order to make the texts more useful to students, each of the papers was "refereed" by cooperative graduate students and colleagues. On average, each paper was refereed four times. We would like to express our gratitude to the referees; their work made the papers clearer, more accessible, and in some cases, more correct. We are grateful to each of the authors for wonderful manuscripts and for their cooperation in what must have at times seemed like an endless series of iterations. The excellent quality of the final texts justifies their hard work.

In looking over the material as it now appears, we believe that these papers may have a wider readership than we originally anticipated. Most of the articles are accessible to junior or senior undergraduate students with a good science background. The book can therefore be useful as an undergraduate introduction to some of the important topics in modern astrophysics. We hope that readers who are graduate students now or in the future will solve many of the problems listed here as unsolved. Anyone, from an undergraduate science major to a senior science faculty member, who would like to know more about some of the active areas of contemporary astrophysics can profit by reading about what these researchers think are the most important solvable problems.

The articles collected here were originally presented as invited talks at a conference entitled 'Some Unsolved Problems in Astrophysics" that was held at the Institute for Advanced Study on April 27-29, 1995. This conference was sponsored in part by the Sloan Foundation, to whom we express our gratitude. The dates for the conference were related to the 60th birthday (and 25th year at the Institute) of one of us (JNB), but nearly every effort was made to focus the meeting on science, not anniversaries. However, a large fraction of the attendees and speakers were alumni of the Institute's postdoctoral program in astronomy and astrophysics.

The manuscript for this book was expertly prepared by Margaret Best. All of us are grateful to Maggie for her exceptional editorial and TeX skills and for her constant good nature.

John N. Bahcall, Jeremiah P. Ostriker
Princeton, June 1996

Unsolved Problems in Astrophysics

CHAPTER 1

THE COSMOLOGICAL PARAMETERS

P. J. E. PEEBLES

Joseph Henry Laboratories, Princeton University, Princeton, NJ

ABSTRACT

The tests of the relativistic cosmological model are well understood; what is new is the development of means of applying them in a broad variety of ways capable of producing a network of consistency checks. When the network is tight enough we will have learned either that general relativity theory passes a highly nontrivial test or that something is wrong with classical physics. I see no reason to look for the latter but we should keep it in mind. In the former case we will have gained the boundary conditions for a deeper cosmology, and a new set of puzzles to study.

1.1 INTRODUCTION

The first thing to know about the measurement of the parameters of the standard relativistic cosmological model is that the problem has been with us for a long time. By the 1930s people understood the physics of the evolving relativistic cosmology and how astronomical observations might be used to test and constrain the values of its parameters. The first large-scale application of the astronomical tests, the count of galaxies as a function of apparent magnitude (Hubble 1936), had already reached redshift $z \sim 0.4$ (inferred from the values of Hubble's counts and more recent measurements of the mean redshift-count relation). The application of the cosmological tests was a "key project" for the 200 inch telescope when it was under construction in the 1930s; now it is a key project for the Hubble Space Telescope and the Keck Telescope.

There has been ample time for the development of strong opinions on what the results of the measurements are likely to be, and for a tendency to lose sight of

1

the reasons for measuring the parameters. To my mind there are two main goals: extend the tests of the physics of the standard model, and seek clues to a cosmology based on a deeper level of physics. Despite the sobering record there is reason to believe we may actually be witnessing the end game in finding useful measurements of the parameters.

1.2   WHY MEASURE THE PARAMETERS?

1.2.1   *Testing the Physics*

In the standard cosmological model the universe is close to homogeneous and isotropic in the large-scale average, and it is homogeneously expanding and cooling. The evidence, which most cosmologists agree is quite strong, is summarized for example in Peebles, Schramm, Turner, and Kron (1991) and Peebles (1993). (I refer the reader to these references for details of the following comments.) The evolution of the standard model is described by general relativity theory. This part is not as closely probed, and one goal of the cosmological tests is to broaden the constraints on the underlying gravity physics. Here is the situation.

To begin, I assume the geometry of our spacetime is described by a single metric tensor, that is, a single line element which determines the relations among measured distance or time intervals between events in spacetime. The evidence is that our universe is close to homogeneous and isotropic in the large-scale average, and we know the line element of a homogeneous and isotropic spacetime is unique up to coordinate transformations; the Robertson-Walker form is[1]

$$ds^2 = dt^2 - a(t)^2 dl^2 \,, \qquad dl^2 = \frac{dr^2}{1 \pm r^2/R^2} + r^2 d\theta^2 + r^2 \sin^2 \theta d\phi^2. \quad (1.1)$$

In this equation a comoving observer, who moves so the universe is seen to be isotropic in the large-scale mean, has fixed position r, θ, ϕ. The proper time kept by such comoving observers is t (if the observers' clocks are synchronized so all see the same mean mass density ρ_i at a given time t_i). The parameter R^{-2} in the expression dl^2 for the spatial part of the line element measures the radius of curvature of the three-dimensional space sections of fixed world time t. If the prefactor of R^{-2} is negative space is closed, as in the surface of a balloon. The expansion factor $a(t)$ in the expression for ds^2 means the balloon in general may be expanding or contracting; the evidence is that our universe is expanding. If the prefactor of R^{-2} is negative space sections are open, the circumference of a circle of radius r being larger than $2\pi r$. In this case the nearly homogeneous

---

[1] Corrections for the departures from homogeneity are important for some of the cosmological tests, and will have to be reconsidered as the tests improve. The issues are discussed in Peebles (1993).

space we observe may extend to indefinitely large distances, or space might be periodic, or conditions beyond the distance we can observe might be very different. If $R^{-2} = 0$ then dl^2 is the familiar Cartesian form for flat space, and spacetime is said to be cosmologically flat.

The proper physical distance between comoving observers, measured at given world time, is $D = la(t)$, where the coordinate distance l is the result of integrating the second part of equation (1.1) along the geodesic connecting the observers at fixed t. The rate of change of the proper distance is

$$v = \frac{dD}{dt} = HD, \qquad H = \frac{1}{a}\frac{da}{dt}. \tag{1.2}$$

If v is much less than the velocity of light this gives a good picture for the predicted linear relation between distance and relative velocity in a homogeneous expanding universe.[2] Thus a galaxy at distance D is moving moving away at recession velocity $v = HD$, and the resulting Doppler effect stretches the wavelength of the radiation received from the galaxy by the amount

$$z \equiv \frac{\delta\lambda}{\lambda} = HD/c. \tag{1.3}$$

The constant of proportionality H is Hubble's constant; its present value usually is written as H_o. The linear relation between redshift and distance, which is Hubble's law, has been tested to redshifts on the order of unity; for an example see Figure 7 in McCarthy (1993).

Hubble's law was one of the first pieces of evidence leading to the discovery of the relativistic cosmology, but we see that the functional form follows more generally from the observed large-scale homogeneity of the universe.

When the redshift is comparable to or larger than unity equation (1.2) does not directly apply (because D is measured along a surface of fixed cosmic time t, which is not how light from a distant galaxy travels to us). The easy way to analyze the redshift in this case is to imagine that the electromagnetic field is decomposed into normal modes of oscillation with fixed boundary conditions in the space coordinates of equation (1.1) The boundary conditions mean the physical wavelength λ of a mode stretches as the universe expands:

$$\lambda \propto a(t). \tag{1.4}$$

If the interaction of the radiation with other matter and fields is weak then adiabaticity tells us the number of photons in each mode is conserved, which is to

---

[2]As in equation (1.1), this assumes perfect homogeneity. In the real world the galaxies are moving at peculiar velocities ~ 500 km s^{-1} relative to the ideal Hubble flow in equation (1.2).

say that equation (1.4) gives the time evolution of the wavelength of freely propagating radiation as measured by comoving observers placed along the path of the radiation. The cosmological redshift factor z is defined by the equation

$$1 + z \equiv \frac{\lambda_{\mathrm{obs}}}{\lambda_{\mathrm{em}}} = \frac{a(t_{\mathrm{obs}})}{a(t_{\mathrm{em}})}. \tag{1.5}$$

The wavelength at emission from a source at epoch t_{em} is λ_{em}, as measured by a comoving observer at the source, and the radiation is detected at time t_{obs} at wavelength λ_{obs}. If the time between emission and detection is small we can expand $a(t)$ in a Taylor series to get equation (1.3). It has become customary to label epochs in the early universe by the redshift factor z considered as an expansion factor even when there is no chance of detection of radiation freely propagating to us from this epoch.

The thermal cosmic background radiation (the CBR) detected at wavelengths of millimeters to centimeters is very close to homogeneous — the surface brightness departs from isotropy by only about one part in 10^5 — and the spectrum is very close to blackbody at temperature $T = 2.73$ K. To analyze the behavior of this radiation in an expanding universe suppose the homogeneous space of equation (1.1) at time t contains a homogeneous sea of thermal radiation at temperature T. The photon occupation number of a mode with wavelength λ is given by Planck's equation,

$$N = \frac{1}{e^{hc/kT\lambda} - 1}. \tag{1.6}$$

If the radiation is freely propagating the occupation number N is conserved, so we see from equation (1.4) that the mode temperature scales with time as

$$T \propto 1/a(t). \tag{1.7}$$

Since this is independent of the mode wavelength an initially thermal sea of radiation remains thermal, even in the absence of the traditional thermalizing grain of dust.

We know the universe now is transparent to the CBR, at least along some lines of sight, because distant galaxies are observed at CBR wavelengths. At some earlier epoch the universe could have been dense and hot enough to have been opaque and therefore capable of relaxing the radiation to the observed thermal spectrum. When this was happening the interaction of matter and radiation cannot be neglected, of course, but since the heat capacity of the radiation is much larger than that of the matter[3] equation (1.7) still applies to the coupled matter and radiation when sources or sinks of the radiation may be neglected.

---

[3]The energy density in the radiation is aT^4, where a is Stefan's constant. In a plasma of n protons per unit volume and a like number of free electrons the energy density is $3nkT$. The ratio is $aT^3/(3nk) \sim 10^9$, nearly independent of redshift.

The conclusion is that the CBR is a fossil of a time when our expanding universe was hotter and denser than it is now. This argument does not require general relativity theory, only conventional local physics, a nearly homogeneous and isotropic expansion described by a single line element (as in eq. [1.1]), and an expansion factor large enough to lead back to a time when the universe was opaque enough to have been capable of relaxing to equilibrium. I think all who have given the matter serious thought would agree with this; the issue is the minimum expansion factor needed to account for the observations. Hoyle, Burbidge, and Narlikar (1993) propose a Quasi-Steady State scenario in which the present expansion phase traces back to a redshift only slightly greater than the largest observed for galaxies. Others doubt that the properties of the postulated thermalizing dust grains can be chosen to relax the radiation to blackbody at such low redshifts while still allowing the observed visibility of high redshift galaxies at CBR wavelengths, though the issue certainly could be analyzed in more detail than has been done by either side so far. Most cosmologists accept the evidence for the origin of the light elements as remnants of the rapid expansion and cooling of the universe through temperatures on the order of 1 MeV, at redshift $z \sim 10^{10}$. This model for element formation depends on the rate of expansion through $z \sim 10^{10}$ and thus tests the gravity theory, as follows.

In general relativity the expansion factor $a(t)$ in equation (1.1) satisfies

$$H^2 = \left(\frac{1}{a}\frac{da}{dt}\right)^2 = \frac{8}{3}\pi G\rho \pm \frac{1}{a^2 R^2} + \frac{1}{3}\Lambda. \qquad (1.8)$$

The mean mass density is ρ, Λ is Einstein's cosmological constant, and the constant R^2 appears with the same algebraic sign as in equation (1.1). (I simplify the equations by choosing units so the velocity of light is unity.) The equation of local energy conservation is $\dot{\rho}/\rho = -3(\dot{a}/a)(\rho + P)$, where P is the pressure. If the pressure is not negative the mass density varies with the expansion parameter at least as rapidly as a^{-3}, meaning it is the most rapidly varying term in equation (1.8), and hence the dominant term at high redshift. Thus the predicted expansion rate through the epoch of light element production is very well approximated as

$$\left(\frac{1}{a}\frac{da}{dt}\right)^2 = \frac{8}{3}\pi G\rho. \qquad (1.9)$$

For a reasonable value of the baryon number density, and assuming the baryon distribution at high redshift is close to homogeneous and the lepton numbers are small, the predicted values of the light element abundances left over from the rapid expansion and cooling of the early universe are close to the observed abundances (with modest and reasonable corrections for the effects of nuclear burning in stars).

This is a non-trivial test of general relativity theory, through equation (1.9). It is not difficult to arrive at equation (1.9) in a quasi-Newtonian picture, however, so the test arguably is not very specific.

The program of cosmological tests may extend the checks of the gravity theory. One part of the program uses galaxies or other markers to measure the curvature of spacetime. In the description of our universe by the homogeneous and isotropic line element in equation (1.1) this amounts to measuring the value and sign of the parameter R^2 and the expansion parameter a as a function of world time t. These results with equation (1.8) predict the present value of the mean mass density ρ_o, which we can hope to compare to what is deduced from studies of the dynamics of systems of galaxies, and the expansion time $t(z)$ as a function of redshift, which we can hope to compare to radioactive decay ages and stellar evolution ages applied at the present epoch and, from stellar evolution times derived from the spectra of distant galaxies, at earlier times. In equation (1.8) the parameter R^{-2} plays the role of a constant of integration or a conserved energy for the effective value $\dot{a}^2/2$ of the kinetic energy of expansion per unit mass, while in equation (1.1) it is a measure of the curvature of space sections. It will be fascinating to see a check that the same parameter plays both roles, as predicted in general relativity.

It may be useful to note in a little more detail the theory behind the use of astronomical observations to constrain $a(t)$ and R^2. We can define two distance functions, the proper rate of radial displacement of a light packet with respect to redshift,

$$s(z) = \frac{dt}{dz},\tag{1.10}$$

and the angular size distance,

$$r(z) = R \sinh\left[\frac{1}{R} \int \frac{dt}{a(t)}\right],\tag{1.11}$$

which is the coordinate distance r in equation (1.1). This expression assumes an open model; in a closed model the hyperbolic sine is replaced by a trigonometric sine. Here are examples of the uses of these functions.

The integral of $s(z)$ in equation (1.10) over redshift is the expansion time, to be compared to other measures of time. The function $s(z)$ also enters the analysis of the absorption lines in quasar spectra that give such a remarkably detailed picture of the distribution of gas along the line of sight. If a class of clouds has proper number density $n(z)$ and cross section $\sigma(z)$ at redshift z then the density of absorption lines produced by these clouds, measured as the probability of finding a line in a quasar spectrum at redshift z in the range dz, is

$$dP = \sigma(z)\, n(z)\, s(z)\, dz.\tag{1.12}$$

A galaxy with angular diameter θ appearing at redshift z has proper diameter

$$d = a(z)r(z)\theta, \tag{1.13}$$

which is why r is called the angular size distance.

In a metric theory the surface brightness of a galaxy integrated over wavelength varies as $(1 + z)^{-4}$. It is an interesting exercise for the student to derive this from Liouville's theorem, and to check that the angular size distance in equation (1.13) thus predicts the detected energy flux density f from a galaxy with luminosity L isotropically radiating at redshift z:

$$f = \frac{L}{4\pi a_o^2 r(z)^2 (1 + z)^2}. \tag{1.14}$$

The count of objects per steradian depends on the radial and angular size distances:

$$\frac{dN}{dz} = [a(z)r(z)]^2 s(z)n(z). \tag{1.15}$$

The first factor is the proper area per steradian subtended at redshift z, the second factor is the proper radial displacement of a light packet per increment of redshift, and $n(z)$ is the number of objects per unit proper volume.

Similar expressions follow for other measurements that might be done at least in principle. A real measurement is quite another matter, of course, but there has been impressive progress, a few examples of which will be considered in the next section.

1.2.2   *How Will It All End?*

It is fascinating to think we can discover how the world ends; the standard cosmological model offers a few definite possibilities. One is a violent collapse back to a "Big Crunch," a sort of time-reversed "Big Bang." Another is expansion into the indefinitely remote future, a "Big Chill" marked by the eventual deaths of all the stars and by the eventual relativistic collapse of gravitationally bound islands of matter and the slow evaporation and dissipation of the mass of the resulting black holes. All this is romantic, but the science is debatable: why should we trust an extrapolation into the remote future of a theory we know can only be an approximation to reality? Surely the more likely outcome of a successful application of the cosmological tests would be the discovery of the boundary conditions for some deeper physical theory which most of us hope will not be blemished by the singular initial and final states of the universe or parts of it that follow within general relativity theory. According to this way of thinking an examination of the history of ideas on how the universe ought to end is valuable for the hints it might offer to a deeper theory, and we hope a knowledge of how the world ends within the present model will focus our minds on the search for a deeper picture. While we wait to see whether anything comes of this we can contemplate some interesting examples in the sociology of science.

If the curvature term does not stop the expansion then within the standard model in equation (1.8) the long-term future of the universe depends on the value of Λ. A negative value, however small, eventually returns the universe to a Big Crunch. If Λ is identically zero and the curvature term does not stop the expansion then spacetime becomes arbitrarily close to Minkowski in the remote future. Dyson (1979) points out that this allows interesting things to continue happening for a very long time, albeit operating at ever increasing timescales. If Λ is positive and the curvature term does not stop the expansion then the universe approaches the static de Sitter spacetime (where the cosmological constant term dominates the right-hand side of eq. [1.8]). In this limit spacetime returns to initial conditions like those of the cosmological inflation scenario described in these proceedings by Paul Steinhardt, who was one of the leaders in its development. Ome might imagine that this signals another round of inflation that will be followed by a phase transition that converts the present Λ into entropy. The characteristic mass corresponding to an astronomically interesting Λ is only a milli-electron volt, but it does not seem inconceivable that a complex new world could develop out of what is to us an exceedingly small energy density (Misner 1992). If this does seem inconceivable one might consider the idea that the Big Chill ends in a burst of creation of new material, as in the Quasi-Steady State scenario (Hoyle, Burbidge, & Narlikar 1993), the material being created at densities and temperatures large enough to satisfy the observational constraints (assuming that what comes after resembles what we see).

Lemaître (1933) considered the idea that the collapse of a universe to a Big Crunch might be followed by a bounce and a new expanding phase. To Lemaître this picture has (in my loose translation) "an incontestable poetic charm, bringing to mind the Phoenix of the legend," and others certainly have agreed. For example, Dicke was led to the idea that the universe might contain an observable thermal cosmic radiation background (the CBR) through the idea that starlight from the previous phase of an oscillating universe would be thermalized during a deep enough bounce and would be capable of evaporating the stars and heavy elements to provide fresh hydrogen for the next cycle (Dicke, Peebles, Roll, & Wilkinson 1965). There are no generally accepted ideas on the physics of a bounce, but evidence that there will be a Big Crunch might be expected to concentrate our thoughts.

Einstein's original static cosmological model requires nonzero values for space curvature and the cosmological constant, and it was natural therefore that the first discussions of the evolving case, by Friedmann and by Lemaître, included these terms. Einstein and de Sitter (1932) noted that neither is required to account for the expansion of the universe, and neither was needed fit the available constraints on the values of the expansion rate, the mean mass density, and space curvature. They proposed therefore that one might pay particular attention to the case in which the mass density dominates the right-hand side of the expansion equation (1.8), as in equation (1.9). This has come to be called the Einstein-de Sitter model. Einstein and de Sitter concluded by remarking that the "curvature is, however, essentially determinable and an increase in the precision of the data derived from observations will enable us in the future to fix its sign and to determine its value." I do not get the impression from this that the authors felt the only reasonable possibility is the Einstein-de Sitter model.

Prior to the discovery of the inflation scenario others were similarly cautious. Robertson (1955) stated that the Einstein-de Sitter case is "of some passing interest." In the addendum to the second edition of his book, *Cosmology*, Bondi (1960) was a little more enthusiastic: he wrote that, "in addition to its outstanding simplicity, this model has the remarkable property (unique amongst relativistic models) that $\gamma\rho R^2/\dot{R}^2$," which is the density parameter defined in equation (1.16) up to a numerical factor (and $\gamma = G$ is Newton's constant), "is constant. It may well be argued that as important a simple pure number as $\gamma\rho R^2/\dot{R}^2$ should be constant during the evolution of the universe so as to provide in some sense a constant background to the application of the theory."

Bondi's remark is close to the coincidences argument I believe originated with Dicke. He noted that if the universe did expand from a dense state then there would have been an earliest time in the past when the physics of the expanding universe can be well approximated by classical field theory. Whatever had happened earlier would have set the initial conditions for the subsequent classical evolution, and in particular would have set the values of the time t_1 at which the mass density ceases to be the dominant term in the right-hand side of the expansion equation (1.8) and the time t_2 at which we came on the scene and measured the parameters in this equation. We know t_2 is not much larger than t_1, because the known masses of the galaxies contribute at least 10% to the present value of H^2 in equation (1.8). It would be a curious coincidence if t_1 and t_2 had similar values. The likely possibility therefore is that t_1 is much larger than t_2, meaning we still are in the Einstein-de Sitter phase. I think I first learned this argument from Dicke in about 1960, when he led me into research in gravity physics and cosmology; he got around to publishing it a decade later (Dicke 1970).

If others knew the Dicke coincidences argument before the discovery of the inflation scenario it did not prevent discussions of alternatives. For example, Petrosian, Salpeter, and Szerkes (1967) and Shklovsky (1967) discussed the possible role of a nonzero cosmological constant Λ in interpreting the quasar redshift distribution, and Gott et al. (1974) argued that the estimates of the mean mass density and the expansion rate H_o, and the theory of light element production in the early universe, seemed to be most readily understandable if the mean mass density is less than the Einstein-de Sitter value.

During the mid-1980s there was a sharp swing toward the opinion that the Einstein-de Sitter model is the only reasonable possibility. This was driven by an idea, cosmological inflation (Guth 1981).

The main point of inflation for our purpose is the concept that the universe might have evolved through an epoch of rapid expansion that would have stretched the length scales of the wrinkles of the primeval ooze into values much larger than those we can explore: the universe may be disorganized and without form in the large yet very close to homogeneous and isotropic in the bit we can see. The stretching that ironed out the wrinkles would have ironed out any mean space curvature too, so in this picture the curvature term in equation (1.8) for the present value of H^2 is negligibly small.

This leaves the cosmological constant Λ, a term for which the particle physics community demonstrates a love-hate attitude. In the modern and successful theories for the weak, electromagnetic, and strong interactions there very naturally appear a set of contributions to the stress-energy tensor that act like a time-variable Λ. This was one of the element that led Guth to inflation: he postulated that a large effective Λ from particle physics drives the rapid expansion during inflation. The decay of this term into ordinary matter and radiation would end inflation and begin evolution according to the standard cosmological model. A residual nonzero Λ could be left over after inflation, and present now, but the expected value of a Λ term coming out of standard particle theory is ridiculously large compared to what is acceptable for cosmology. The natural presumption is that the cosmological constant has settled down to the only reasonable and observationally acceptable value, $\Lambda = 0$. It would follow from all this that inflation predicts the Einstein-de Sitter model.

During the past decade many people, including respected and thoughtful observational astronomers, were led to conclude that the observable part of the universe likely is well described by the Einstein-de Sitter model. I have mentioned the three main drivers: the Dicke coincidences argument, the belief that an astronomically interesting value for a cosmological cosmological is not likely to come out of particle physics, and the inflation scenario for the early universe. All three are worth serious consideration, and inflation in particular has been very influential in the development of ideas of what our universe might have been like before it was expanding. Theory was the driver, however: there was little observational evidence for the Einstein-de Sitter case and a nontrivial case against it. I can say this without excessive resort to hindsight because as inflation was becoming popular my reading of the evidence was leading me to abandon my earlier enthusiasm for the Einstein-de Sitter model: it seemed difficult to see where the large mass required by this model might be located (for the reasons discussed in Peebles 1986). As described in the next section, where some of the observational arguments pro and con $\Omega = 1$ are discussed, I have seen no reason to change my mind, but the case certainly is not yet closed.

The observational pressures on the Einstein-de Sitter model have led people to explore alternatives. Inflation's original explanation for the near homogeneity of the observable space requires that the curvature term R^{-2} be negligibly small, but it certainly would allow a nonzero Λ. N. Bahcall and Ostriker describe in these proceedings the benefits this parameter offers in interpreting the observations. Another possibility is inflation in an open universe (Ratra and Peebles 1994; Bucher, Goldhaber, & Turok 1995). Here we need another explanation for homogeneity; the best bet seems to be Gott's (1982) picture for the growth of an open universe out of an event in a de Sitter spacetime.

My impression is that research on inflation is in a healthy state: the science is being driven by advances in the observational evidence that are leading people to consider new ideas. It is too soon to decide whether the rich flow of ideas on how the world begins and ends within inflation and other scenarios is leading us toward a deeper understanding of physical reality; perhaps that will depend on the outcome of the measurements of the cosmological parameters.

1.3   THE STATE OF THE MEASUREMENTS

The first thing to understand is that we have no measurement presently capable of unambiguously distinguishing a Big Crunch from a Big Chill. We do have some promising lines of evidence, however, and the reasonable hope that research in progress will show us how this evidence can be tied together in a concordance tight enough to be believable.

The situation is illustrated in Table 1.1. The density parameter Ω is the fractional contribution of the mass density to the present value of the expansion rate in equation (1.8):

$$\Omega = \frac{8\pi G\rho}{3H_o{}^2}. \tag{1.16}$$

The case in the second column of the table, labeled $\Omega = 1$, is meant to be the Einstein-de Sitter model. It is possible that the curvature and Λ terms cancel each other at the present epoch, leaving $\Omega = 1$ with nonzero values of R^{-2} and Λ, but until we are driven from it it is sensible to operate under the assumption the Nature would not have been so unkind. By the same hopeful reasoning the third column assumes there are just two significant terms in the expansion rate equation, the dominant one being space curvature or else a cosmological constant, with almost all the rest in the mass density.

The mark $\sqrt{}$ means there is significant evidence in favor of the case, the mark X significant evidence against. I have attempted to give some indication of the degree of significance by assigning question marks to the more debatable evidence.

Table 1.1: Scorecard 1995

Observation	$\Omega = 1$	$\Omega \sim 0.1$
Dynamics and biasing on scales $\lesssim 3$ Mpc	X	$\sqrt{}$
Dynamics on scales $\gtrsim 10$ Mpc	$\sqrt{}$	$\sqrt{}$
Expansion time $H_o t_o$	?	$\sqrt{}$
Radial and angular size distances	$X?$	$\sqrt{}?$
Plasma mass fraction in clusters	X	$\sqrt{}$
Models for structure formation	$\sqrt{}?$	$\sqrt{}?$

The first entry in the table is based on the dynamical studies of groups and the central parts of clusters of galaxies, on scales less than a few megaparsecs. The derived mass per galaxy multiplied by the mean galaxy number density yields the contribution to the mean mass density from the material that is concentrated around galaxies; the result is equivalent to the density parameter

$$0.1 \lesssim \Omega \lesssim 0.2. \tag{1.17}$$

Another way to put it is that the small-scale dynamical measurements of the mass per galaxy agree with the mass found in the dark halos of spiral galaxies within radii

$$r_{\text{halo}} \sim 300 \text{ kpc}. \tag{1.18}$$

This is discussed further in these proceedings by N. Bahcall.

The measurement in equation (1.17) seems to be secure; the open issue is whether there might be a good deal more mass in a more broadly distributed component. The galaxy masses derived from the relative gravitational accelerations in samples of close pairs of galaxies are measures of the mass concentrated around the pairs on scales less than or comparable to the separation, and miss mass in common more extended envelopes. This is the mass biasing picture, meaning the galaxies are not fair measures of where the mass is. I think it is pretty clearly established that mass biasing is required if $\Omega = 1$.

The biasing concept is most commonly discussed in terms of the cold dark matter (CDM) cosmogony. The key idea here is that galaxies form at the peaks of primeval initially Gaussian mass density fluctuations, and that the familiar giant spiral and elliptical galaxies form at unusually high peaks that tend to appear in groups and clusters, leaving most of the mass on the outskirts of these concentrations. Branchini and Carlberg (1994) present an instructive example of the effect within a CDM cosmogony with $\Omega = 1$; they find that the mass concentrated around pairs of galaxies increases markedly between separations of 1 Mpc and 3 Mpc. We have a test, from samples of the radial velocities of the galaxies and groups of galaxies around the Local Group, at distances less than about 4 Mpc. The CDM prediction is not observed. Rather, a reasonable fit to the radial velocities follows if the mass per galaxy is consistent with equation (1.17), in halos with sizes given by equation (1.18) (Zaritsky et al. 1989; Peebles 1995).

We have another more qualitative but I think very significant test of the biasing concept within the CDM model. The cosmogony assumes galaxy formation is seeded by Gaussian primeval density fluctuations, which means that all matter has been seeded for galaxy formation. The mass biasing assumption here is that seeds in the matter outside the halos of the bright galaxies failed to germinate or else produced unobtrusive galaxies. Since the failure of germination could hardly be complete it would follow that space outside the concentrations of the bright galaxies, at $r > r_{\text{halo}}$, contains dwarf irregular galaxies marked by the low fertility of the ground on which their seeds were cast. The effect is seen in numerical N-body studies, of course (an example is given by Brainerd and Villumsen 1992), but it is not observed: known dwarf and irregular galaxies avoid the voids defined by the giant galaxies. I offer one example.

Absorption lines in quasar spectra reveal gas clouds that happen to intersect the lines of sight. For our purpose a useful sample is the clouds with column densities $\Sigma \gtrsim 10^{17}$ neutral hydrogen atoms cm^{-2}, three or four orders of magnitude below that typical of the bright parts a spiral galaxy, and at redshifts in the range $0.5 \lesssim z \lesssim 1.5$, close enough that ordinary large galaxies are observable and deep enough to give good statistics. The remarkable Bergeron effect[4] is that with high probability the clouds at this column density are within about 100 kpc of a luminous galaxy (Steidel, Dickinson, & Persson 1994; Lanzetta et al. 1995). That is, at

---

[4]I name the effect after the discovery that one can identify the galaxy whose gaseous halo is responsible for a specific absorption line (Bergeron 1986). The effect was anticipated well before that (Bahcall & Spitzer 1969).

redshift $z \sim 1$ this diffuse and generally unobtrusive gas occupies only about one part in 10^5 of space: it is within the dark halo radii $r_{halo} \sim 300$ kpc (eq. [1.18]) of the large galaxies. Under the biasing picture most of the mass of the universe has to be outside r_{halo}, is seeded for galaxy formation, yet contains insignificant concentrations of neutral gas at column densities $\gtrsim 10^{17}$ cm^{-2}. Is it reasonable to think germination could have been so efficiently suppressed?

My conclusion is that the biasing concept of the $\Omega = 1$ CDM model is unpromising. The story is different if $\Omega \sim 0.1$, of course, for then we can imagine that at $z \lesssim 1.5$ there is little matter of any kind further than r_{halo} from a giant galaxy.

The story is different also if $\Omega = 1$ and the dominant mass component at $r > r_{halo}$ was not seeded for galaxies, or the dominant mass component is not capable of clustering on the scale of individual galaxies. We have a test from the relative peculiar velocities on larger scales, and here there are conflicting indications. Analyses of large-scale peculiar velocities by the method of Bertschinger and Dekel (1989) are consistent with $\Omega = 1$ (Dekel et al. 1993; Dekel & Rees 1994). Another approach to the interpretation of the large-scale peculiar velocities using independent data and a different method of analysis fails to reproduce the evidence for a high mass density, however; it indicates $\Omega \sim 0.1$ (Shaya, Peebles, & Tully 1995). A galaxy-by-galaxy comparison of predictions of peculiar velocities by the two camps is under study, and we may hope a resolution of the discrepancy is within reach. But meanwhile each side can claim observational support, as indicated in the second entry in the table.

The relative contributions of space curvature and the cosmological constant to the expansion rate equation (1.8) determine the predicted value of the dimensionless product $H_o t_o$ of Hubble's constant and the expansion time from high redshift to the present. In the Einstein-de Sitter model $H_o t_o = 2/3$. Lowering Ω lowers the predicted gravitational deceleration of the expansion rate and so increases the time back to high redshift. A positive Λ has a larger effect than positive R^{-2} at the same Ω, because the former tends to accelerate the expansion rate. If $H_o = 50$ km s^{-1} Mpc^{-1}, close to the smallest value currently under discussion, the Einstein-de Sitter model predicts $t_o = 13$ Gyr, about the smallest value under discussion for stellar evolution ages. If, as many but not all experts are arguing, H_o is not likely to be less than about 60 in the above units (e.g., Freedman 1994), the expansion time in an Einstein-de Sitter model would seem to be too small. I have accordingly entered the result of the timescale test as tentative bad news for this model, good news for the low density cases.

Most of the current work on the extragalactic distance scale follows Hubble's path: use the inverse square law to estimate distances by the comparison of the apparent brightnesses and intrinsic luminosities of stars that are luminous enough to be observationally interesting at interesting distances, and maybe also periodic or exploding. New directions have been found. One involves the measurement of time delays in the gravitational lensing events Blandford discusses in these proceedings. Another uses the physics of plasmas in clusters of galaxies, as discussed recently by Herbig et al. (1995). Yet another uses the remarkably precise measurements of the velocities and accelerations of massing interstellar clouds close to the nuclei of galaxies (Miyoshi et al. 1995). Perhaps out of all of this we may hope to know H_o to 10% by the turn of the millennium.

The angular size distance (eq. [1.11]) enters the analysis of other cosmological tests, including galaxy counts, the rate of lensing of quasar images by the mass concentrations in foreground galaxies (e.g., Fukugita & Turner 1991), and the magnitude-redshift relation applied to distant supernovae (e.g., Goobar & Perlmutter 1995).

A recent example of the first of these tests is the deep K-band (2.2 micron) counts of Djorgovski et al. (1995). Many experts advise against interpretation until the counts as a function of apparent magnitude are better established and we have a better understanding of the time evolution of galaxy luminosities and colors, but we can take note of the following indications. If the giant galaxies that tend to dominate a sample selected by apparent magnitude are evolving only through the stellar evolution of populations formed at high redshifts, the predicted counts at the faint end of the Djorgovski et al. sample differ by nearly an order of magnitude in the high and low density cases with $\Lambda = 0$. That is, the observations seem to be reaching redshifts where the effects of the cosmology are large. The effects of galaxy evolution can be large too, but we do have checks. For example, the galaxies selected by the Bergeron effect at redshift $z \sim 1$ show little evolution from the present (Steidel, Dickinson, & Persson 1994). If evolution in galaxies selected in the K-band is modest then the counts favor low Ω, with $\Lambda = 0$, and I have accordingly entered this result as tentative good news for the low density case.

The gravitational lensing of quasar images by foreground galaxies also samples large redshifts, and again the effect of the cosmological parameters on the predicted lensing rate is large (Fukugita & Turner 1991). The analysis by Maoz and Rix (1993) indicates that, if space curvature vanishes, Ω is no less than about 0.3. An interesting problem in this analysis is that it assumes elliptical galaxies have massive dark halos, in analogy with the dark halos needed in Newtonian mechanics to account for the rotation curves in spiral galaxies. Massive dark halos are needed to gravitationally contain the pools of plasma observed around giant ellipticals in clusters of galaxies. At issue here are massive halos around the more numerous less luminous ellipticals that would be responsible for lensing cases where the angular separations of the lensed images is less than about one second of arc. Tests of the mass distributions in such ellipticals based on the distribution and motions of the stars are difficult to apply, and the constraints on massive halos in these galaxies are still subject to debate. There is one case, the galaxy M 105, where a detailed study of the motions of planetary nebulae (Ciardullo, Jacoby, & Dejonghe 1993) and of a conveniently in placed ring of atomic hydrogen (Schneider 1991) yield a well-motivated mass model that requires little dark matter. If this were a common situation among the less luminous ellipticals, it would reduce the bound on Ω.

The fifth entry in the table is an elegant new test based on the standard model for the origin of the light elements and the measurements of the baryonic mass fraction in clusters of galaxies (White et al. 1993; White & Fabian 1995). A reasonable fit to the observed light element abundances follows out of the computed production of elements as the young universe expands and cools, if the density parameter in baryons is $\Omega_B \sim 0.01h^{-2}$, where $H_o = 100h$ km s^{-1} Mpc^{-1} (Walker et al. 1991). If the Hubble parameter is no less than about $h = 0.5$ and $\Omega = 1$ this would say the cosmic baryonic mass fraction is no more than about 0.05, the rest being some kind of nonbaryonic dark matter. In the central parts of rich clusters the mass fraction in observable baryons — stars and intracluster plasma — exceeds this number if $h \gtrsim 0.5$. Numerical studies of cluster formation indicate that the baryons are not likely to have been able to settle relative to the dark matter. The straightforward interpretation is that the cosmic baryonic mass fraction is significantly larger than 0.05 because the mean mass density is less than the Einstein-de Sitter value.

The last entry in the table refers to theories of the origin of structure: intergalactic gas clouds, galaxies, and the large-scale galaxy distribution. The theories depend on the cosmological model, because the cosmology determines the relation between the mass distribution, the peculiar velocities relative to the general expansion, and the perturbations to the angular distribution of the CBR (as discussed in these proceedings by N. Bahcall, Ostriker, and Steinhardt). The unraveling of clues to structure formation thus will guide our assessment of the cosmological parameters. Of particular interest now is the relation between the parameters of the cosmological model and the angular irregularities in the thermal cosmic background radiation. As Steinhardt describes, the measurements of the spectrum of angular fluctuations as a function of angular scale already significantly constrain the cosmological parameters. We should bear in mind that the constraints depend on the model for structure formation, however, and theorists have shown great ingenuity in tuning models to fit the measurements. Perhaps the improved precision of measurement of the angular fluctuation spectrum to be expected in the next few years will allow only model that will prove to be consistent with all the other observational constraints. But we should not underestimate the devious natures of astronomy and theorists.

It certainly is premature to settle bets on the value of Ω, whether consistent with the Einstein-de Sitter model or significantly lower, but we do have quite a few promising lines of evidence and good reason to expect that improvements from measurements in progress will yield enough cross checks with sufficient consistency to make a convincing case. If so, we are at last near the end of the trail Hubble (1936) pioneered six decades ago.

1.4 COSMOLOGY FOR THE NEXT GENERATION

When (or if) we arrive at a convincing interpretation of all the lines of evidence indicated in Table 1.1, along with the related tests people may invent, in terms of a set of values of the cosmological parameters, it will not mean cosmology has come to an end; we will turn to other problems. I expect we will still be debating the issue of what really happened in the remote past and what is really going to happen in the distant future. We may hope the debate will be sharper because we will know what the present standard model predicts, but I would not count on any more direct help from the observational evidence. I discuss here examples of what I expect will be observationally-driven problems for the next decade or so.

Finding a believable set of values of the cosmological parameters is likely to depend on a demonstration that the cosmology admits a successful cosmogony, a theory of the origin of galaxies. This successful cosmogony will include a description of the dominant mass components and the nature of their initial distributions at high redshift, but, unless we are lucky, there need not be an unambiguous picture of where these initial conditions came from in terms of some deeper theory. If not, we will continue to sort through the observational evidence for clues to the origin of the initial conditions for the standard model. Present research along these lines concentrates on large-scale structure, because that is close to linear and therefore relatively easily interpreted. A next step will have to be the study of the origin of the small-scale structure of the universe, which in hierarchical cosmogonies ties to the early history of structure formation. The analyses will be a good deal more difficult than for large-scale structure, but there are some openings to explore.

The drag of the CBR inhibits structure formation out of diffuse baryons at $z \gtrsim 1000$. At $z \sim 1000$, when the mean baryon density is about $100 \, \text{protons} \, \text{cm}^{-3}$, the CBR is cool enough to allow the plasma to combine to neutral atomic hydrogen that can slip past the CBR to form gravitationally bound structures. Galaxies as we know them could not have formed then, because the density is at least two orders of magnitude too large, but there could have been star clusters. Could objects from this epoch have survived more or less intact as parts of present-day galaxies? My favorite candidate has been the globular star clusters, but other ideas would be welcome.

At $1000 \gtrsim z \gtrsim 100$ matter reionized by massive stars is recoupled to the CBR and radiation drag tends to force the plasma to rejoin the general expansion of the universe. That is, very early baryonic structure formation tends to be self-limiting. A closer analysis of this effect would be feasible and welcome.

If the mass of the universe were dominated by nonbaryonic pressureless matter, as in the CDM model, then mass clustering at high redshift would have grown independent of what the baryons were doing, maybe producing potential wells into which the baryons fell when radiation drag or matter pressure at last allowed it. This effect is taken into account in numerical simulations of the cosmic evolution of the distributions of baryons and dark matter, if the spatial resolution allows an adequate representation of the early development of small-scale mass clustering. I am looking forward to seeing analytic studies of what happens to the early small-scale evolution of the distributions of mass and baryons. An example of what might be found is the presence of dark massive halos around globular star clusters, at least those which formed at the baryonic Jeans mass and escaped disruption.

Galactic winds from supernovae are capable of stripping the gas from a young galaxy, leaving the dark matter halo, if it had one, and whatever stars had already formed. It has been suggested that such failed galaxies may be present now in the voids between the giants, along with most of the mass of an Einstein-de Sitter universe. Failed galaxies could be visible by the starlight from the generations produced before the gas was expelled. Perhaps evidence for this picture comes from the weaker clustering of galaxies selected for strong emission lines (Salzer 1989), or for low surface brightness (Mo, McGaugh, & Bothun 1994). In the former case one sees from maps of the galaxy distributions that the clustering is weaker because the emission line galaxies avoid dense regions. This has a natural interpretation in terms of environment: collisions and the ram pressure of the intergalactic plasma in dense regions tend to strip the galaxies of the interstellar gas responsible for the emission. Missing so far is evidence of a population that lives in the voids between the concentrations of bright galaxies. I expect to see continued discussion of this observational issue and the related theoretical one: would failed galaxies be expected to have accreted enough gas to be detectable at low redshift in quasar absorption line studies?

The baryon densities of objects starting to form at $z \sim 100$, when drag by the CBR becomes unimportant, would be about 1 proton cm^{-3}. Might it be significant that this is the density characteristic of the bright parts of a giant galaxy?

The massive dark halos are less dense, but they could have been added at lower redshifts.

In the past decade there has been considerable interest the idea that even the bright parts of the giant galaxies were assembled much later, at $z \lesssim 2$. This grew out of early numerical studies of the biased $\Omega = 1$ CDM model, and it was reenforced by the observations of substantial galaxy evolution at $z \lesssim 0.5$.[5] Quasars at $z \sim 5$ might be accommodated in this late assembly picture, but it certainly looks like a tight fit. The same is true of the observations of galaxies at redshifts as high as $z \sim 3$ with relaxed-looking morphologies and spectra characteristic of star populations. There have been dramatic recent improvements in such observations, as Ellis and Sargent describe in these proceedings, but it is worth noting that the subject is not entirely new: consider Oke's (1984) remark, "When one looks at the spectra of first-ranked cluster galaxies over the whole range of z covered", to redshift $z \sim 0.8$, "one is impressed by the fact that the vast majority are very similar to each other and to nearby ellipticals." Nature certainly has been inviting us to consider the idea that galaxy assembly occurred well before $z = 1$. It will be interesting to see whether the current evidence for the assembly of the giant elliptical galaxies $z \gtrsim 3$ holds up, and if so whether any of the variants of the CDM model now under discussion can account for this early formation.

There is a plateau in the abundance of quasars at $2 \lesssim z \lesssim 3$. Is this telling us this is when the L_* galaxies were assembled? Can we imagine instead that the galaxies were assembled earlier and the time needed to produce a central engine brought the age of the quasars to $z \sim 2.5$?

At $z \sim 4$ there is about as much diffuse baryonic mass in the intergalactic medium in the Lymanα forest clouds as there is in the young galaxies. If the galaxies had been assembled at $z \gg 4$ they would have tended to gravitationally accrete the intergalactic matter, removing the Lyα forest. Because the forest clearly is present at $z = 4$ this line of argument suggests galaxies were not assembled much earlier. A counter argument uses the abundant evidence of mass exchange between present-day galaxies and the intergalactic medium (e.g., Irwin 1995). Could it be that galactic winds in gas-rich young galaxies born at $z \gg 4$ fed the intergalactic medium, accounting for the presence of the Lymanα forest without overly perturbing the CBR or overly polluting the forest clouds with heavy elements?

Which formed first, galaxies or clusters of galaxies? Hubble (1936) argued for the latter.[6] Lemaître (1934) argued that galaxies and clusters of galaxies may have been produced by the gravitational instability of the expanding universe. The

---

[5] A good sample of the debate is in Frenk et al. (1989)

[6] The argument is mainly of historical interest. Following Jeans, Hubble considered the idea that elliptical galaxies evolve into spirals. Since spirals prefer less dense surroundings than ellipticals, this suggested the field galaxies might have evolved out of ellipticals that escaped from clusters.

density $\bar{\rho}$ of a newly virialized mass concentration would be some characteristic multiple of the cosmic mean $\langle\rho\rangle$ at the time of virialization. Since galaxies are a good deal denser than clusters it would follow that galaxies formed a good deal earlier. I consider the Local Group of galaxies a good example of Lemaître's picture: the evidence is that this system is forming now by the mutual gravitational attraction of old galaxies.[7]

An important number in Lemaître's picture is the ratio $\bar{\rho}/\langle\rho\rangle$ of the mean density within a newly virialized system and the cosmic mean density at this time. The spherical collapse model indicates $\bar{\rho}/\langle\rho\rangle \sim 200$ (e.g., Efstathiou & Rees 1988). In rich clusters of galaxies the ratio of the galaxy number density within the Abell radius $r_A = 1.5h^{-1}$ Mpc to the present cosmic mean is $\bar{n}/\langle n\rangle \sim 200$, and the mass contrast is similar. Does this mean the great clusters are being assembled now at the Abell radius? I have trouble believing it, because this seems to be an extreme application of spherical symmetry, and because the Local Group seems to be a counterexample, having been assembled and virialized at a much smaller density contrast. We may hope to see a key test in the reasonably near future: the analysis of the typical mass distribution within clusters at redshift $z \sim 1$. The spherical model predicts the mean density at $r_A = 1.5h^{-1}$ Mpc and $z \sim 1$ is well below the value observed in low redshift clusters. If the density collapse factor at formation is an order of magnitude less than the spherical prediction then the densities at $r_A = 1.5h^{-1}$ Mpc will about the same at $z \sim 1$ and $z \sim 0$.

The collapse factor issue applies also to the epoch of gravitational assembly of the galaxies. The dark halo in a spiral galaxy such as the Milky Way dominates the mass at radius $r \sim 10$ kpc, where the circular velocity of rotation is $v \sim 200$ km s^{-1}. The enclosed mass is $M = v^2r/G$, and the ratio of the enclosed mean mass density $\bar{\rho}$ to the large-scale mean $\langle\rho\rangle$ evaluated at redshift z is

$$\frac{\bar{\rho}}{\langle\rho\rangle} = \frac{2}{\Omega(1+z)^3}\left(\frac{v}{Hr}\right)^2 \sim \frac{10^5}{\Omega(1+z)^3}. \tag{1.19}$$

If the mass concentration were assembled at $z = 2$ the density contrast at assembly would have been $\bar{\rho}/\langle\rho\rangle \sim 3000/\Omega$, which requires a collapse factor even larger than that of the spherical model. This is why the very old discussions emphasized galaxy assembly at $z \gtrsim 10$ (Partridge & Peebles 1967). In the past decade many have favored more recent formation, but the trend of recent discussions has swung back to higher redshifts. We must expect still more swings of opinion, but observations back to $z \sim 5$ show us a lot about what galaxies actually were doing and perhaps out of this will come evidence that will make believers of us all.

---

[7]Details and references are in Peebles (1995).

ACKNOWLEDGMENTS

This paper was improved by helpful comments from Jonn Bahcall, Karl Fisher, Wayne Hu, and David Spergel. The work was supported in part at Princeton University by the National Science Foundation.

BIBLIOGRAPHIC NOTES

- **Dicke, R. H. 1970, Gravitation and the Universe (Philadelphia: American Philosophical Society).** Three lectures on gravitational physics by one of the master physicists and astrophysicists. The third lecture is an illuminating account of the discovery of the cosmic microwave background radiation.

- **Hubble, E. 1936, The Realm of the Nebulae (New Haven: Yale University Press).** A thrilling and charming account of early cosmological research.

- **Kolb, E. W. and Turner, M. S. 1990, The Early Universe (Redwood City: Addison-Wesley).** A standard compendium of the physics of the behaviour of the universe at very high redshift.

- **Ohanian, H. C. and Ruffini, R. 1994, Gravitation and Spacetime (New York: Norton).** There are many textbooks on general relativity theory; in my experience this presents the subject at about the right level.

- **Peebles, P. J. E. 1980, The Large-Scale Structure of the Universe; 1993, Principles of Physical Cosmology (Princeton: Princeton University Press).** These monographs provide introductions and standard treatments of the subjects discussed in this lecture.

BIBLIOGRAPHY

[1] Bahcall, J. N., & Spitzer, L. 1969, ApJ, 156, L63.

[2] Bergeron, J. 1986, A&A, 155, L8.

[3] Bertschinger, E., & Dekel, A. 1989, ApJ, 336, L5.

[4] Bondi, H. 1960, Cosmology, 2nd edition (Cambridge: Cambridge Univ. Press), 166.

[5] Brainerd, T. G., & Villumsen, J. V. 1992, ApJ, 394, 409.

[6] Branchini, E., & Carlberg, R. G. 1994, ApJ, 434, 37.

[7] Butcher, M., Goldhaber, A. S., & Turok, N. 1995, Phys. Rev. B, in press.

[8] Ciardullo, R., Jacoby, G. H., & Dejonghe, H. B. 1993, ApJ, 414, 454.

[9] Dekel, A., et al. 1993, ApJ, 412, 1.

[10] Dekel, A., & Rees, M. J. 1994, ApJ, 422, L1.

[11] Dicke, R. H. 1970, Gravitation and the Universe (Philadelphia: American Philosophical Society).

[12] Dicke, R. H., Peebles, P. J. E., Roll, P. G., & Wilkinson, D. T. 1965, ApJ, 142, 414.

[13] Djorgovski, S., et al. 1995, ApJ, 438, L13.

[14] Dyson, F. 1979, Rev. Mod. Phys., 51, 447.

[15] Efstathiou, G., & Rees, M. J. 1988, MNRAS, 230, 5P.

[16] Einstein, A., & de Sitter, W. 1932, Proc. NAS, 18, 213.

[17] Freedman, W. L., et al. 1994, Nature, 371, 757.

[18] Frenk, C. S., et al. 1989, The Epoch of Galaxy Formation (Dordrecht: Kluwer).

[19] Fukugita, M., & Turner, E. L. 1991, MNRAS, 253, 99.

[20] Goobar, A., & Perlmutter, S. 1995, ApJ, 450, 14.

[21] Gott, J. R. 1982, Nature, 295, 304.

[22] Gott, J. R., Gunn, J. E., Schramm, D. N., & Tinsley, B. M. 1974, ApJ, 194, 543.

[23] Guth, A. H. 1981, Phys. Rev. D, 23, 347.

[24] Herbig, T., Lawrence, C. R., Readhead, A. C. S., & Gulkis, S. 1995, ApJ, 449, L5.

[25] Hoyle, F., Burbidge, G., & Narlikar, J. V. 1993, ApJ, 410, 437.

[26] Hubble, E. 1936, Realm of the Nebulae (New Haven: Yale Univ. Press).

[27] Irwin, J. I. 1995, PASP, 107, 715.

[28] Lanzetta, K. M., Bowen, D. V., Tytler, D., & Webb, J. K. 1995, ApJ, 442, 538.

[29] Lemaître, G. 1933, Ann. Soc. Sci. Brussels, A53, 51.

[30] Lemaître, G. 1934, Proc. NAS, 20, 12.

[31] Maoz, D., & Rix, H.-W. 1993, ApJ, 416, 425.

[32] McCarthy, P. J. 1993, ARA&A, 31, 639.

[33] Misner, C. W. 1992, in the 13th International Conference on General Relativity and Gravitation, Univ. Nacional de Cordoba, Argentina, ed. P. W. Lamberti, & O. E. Ortiz, p. 417.

[34] Miyoshi, M., et al. 1995, Nature, 373, 127.

[35] Mo, H. J., McGaugh, S. S., & Bothun, G. D. 1994, MNRAS, 167, 129.

[36] Oke, J. B. 1984, in Clusters and Groups of Galaxies, ed. F. Mardirossian, et al. (Dordrecht: Reidel), 99.

[37] Partridge, R. B., & Peebles, P. J. E. 1967, ApJ, 147, 868.

[38] Peebles, P. J. E. 1986, Nature, 321, 27.

[39] Peebles, P. J. E. 1993, Principles of Physical Cosmology (Princeton: Princeton Univ. Press).

[40] Peebles, P. J. E. 1995, ApJ, 449, 52.

[41] Peebles, P. J. E., Schramm, D. N., Turner, E. L., & Kron, R. G. 1991, Nature, 352, 769.

[42] Petrosian, V., Salpeter, E. E., & Szekeres, P. 1967, ApJ, 147, 1222.

[43] Ratra, B., & Peebles, P. J. E. 1994, ApJ, 432, L5.

[44] Robertson, H. P. 1955, PASP, 67, 82.

[45] Salzer, J. J. 1989, ApJ, 347, 152.

[46] Schneider, S. E. 1991, in Warped Discs and Inclined Rings Around Galaxies, ed. S. Casertano, P. D. Sackett, & F. H. Briggs (Cambridge: Cambridge Univ. Press), 25.

[47] Shaya, E. J., Peebles, P. J. E., & Tully, R. B. 1995, ApJ, 454, 15.

[48] Shklovsky, I. S. 1967, Astr. Tsirk., No. 429.

[49] Steidel, C. C., Dickinson, M., & Persson, S. E. 1994, ApJ, 437, L75.

[50] Walker, T. P., et al. 1991, ApJ, 376, 51.

[51] White, D. A., & Fabian, A. C. 1995, MNRAS, 273, 73.

[52] White, S. D. M., Navarro, J. F., Evrard, A. E., & Frenk, C. S. 1993, Nature, 366, 429.

[53] Zaritsky, D., Olszewski, E. W., Schommer, R. A., Peterson, R. C., & Aaronson, M. A. 1989, ApJ, 345, 759.

Chapter 2

In the Beginning ...

Paul J. Steinhardt

Department of Physics and Astronomy,
University of Pennsylvania, Philadelphia, PA

Abstract

New observational technologies are transforming cosmology from speculative inquiry to rigorous science. In the next decade, observations will be made which will provide redundant, quantitative tests of theoretical models for the origin and evolution of large-scale structure in the universe. Using the inflationary theory as an example, this paper discusses some of the crucial tests and how they can be combined to discriminate among different models and measure essential parameters.

2.1 The Future Fate of Cosmology

In the beginning ...

... long before there was science, there was cosmology. As soon as the first intelligent eyes were capable of peering up at the heavens, questions began to arise concerning the origin and evolution of the Universe: How big is the Universe? How old is the Universe? How did it begin and how did it develop into what we see today? Many answers have been proposed over the centuries, yet these remain the fundamental, unsolved problems of cosmology today (see the contribution by P. J. E. Peebles to these proceedings).

What has changed is that there is now the technology needed to test our proposed answers to these questions. The transformation began near the turn of this century with the construction of the first, giant, optical telescopes. It was first discovered that the Universe is composed of galaxies and that the galaxies are re-

25

ceding from another. New observational technologies in successive decades have brought further breakthroughs.

Now, as the beginning of a new millennium approaches, the number and diversity of new technologies is increasing at an incredible pace. The forthcoming decade will bring the first galaxy surveys with red shifts of millions of galaxies allowing the first extensive three-dimensional reconstruction of large-scale structure in the Universe; the first high-resolution maps of the cosmic microwave background, providing a snapshot of the universe in its infancy; detections of cosmic sources of gamma-rays, x-rays, infra-red radiation, and perhaps even gravitational waves; and, the first extensive catalogs of supernovae and gravitational lenses (see, e.g., the contributions by N. Bahcall and R. Blandford to these proceedings). Each of these new measurements represents a fundamentally new probe of the cosmos. Together, they form a powerful, quantitative, and redundant suite of tests that will make it possible to decisively discriminate among competing cosmological models.

It appears that, for the first time in human history, cosmology will be as much observation-driven as theory-driven, finally achieving the balance required for a true, healthy science. Young scientists selecting directions for their research careers should be aware of the lessons of history which suggest that, once a field reaches this epochal stage, a heroic age will ensue in which some of the fundamental problems of the ages may be resolved.

What is discovered in the next decade may determine whether cosmology ever achieves its grand ambition. The nominal purpose of cosmology is to weave together a story, a history of the evolution of the Universe. In this sense, the science of cosmology is most similar to archaeology or paleontology. Its approach is similar, too. The basic methods entail gathering fossil relics from different epochs and tracing the evolutionary links between them. In addition, though, cosmology also has a not-so-hidden, grand ambition: to reduce this history by explaining it as a natural consequence of some simple, predictive and explanatory model based on symmetries, general physical principles, and known physical processes. It has been, to a large degree, an article of faith that there exists such a simple predictive and explanatory theory. This faith has remained resolute as long as one has found that the Universe is a relatively simple place. But, with the torrent of new observations, it is conceivable that the Universe will be found to be a more complex place than has been supposed.

Only a tiny patch of the Universe can be observed. Causality forbids seeing beyond 15 billion light years or so. Most likely, the total Universe is much larger. The hope has been that the Universe is simple enough that, by observing just our one, tiny patch, one can understand the Universe entire. Instead, it may be found that the properties of our patch depend on the chance initial conditions within one patch; or on local choices of parameters, which cannot be fathomed; or on physics outside the Hubble horizon. In this case, a fundamental explanation of the Universe lies beyond our grasp. Cosmologists will have to be content with an archaeological reconstruction of the history of the Universe.

2.2   TESTING INFLATION

The fate of cosmology, as either explanatory science or archaeology, rests on the discoveries of the coming decade. I believe that the critical observations will be efforts to test inflationary cosmology. In part, this is because inflationary cosmology is our best hope for a simple explanatory and predictive theory [1]–[4]. In a single stroke, inflation explains the homogeneity and isotropy of the Universe; the flatness and energy density of the Universe; the absence of magnetic monopoles and other particle monsters that would have been produced at Planckian temperatures; and the spectrum of inhomogeneities [5] that perturbed the microwave background and seeded large-scale structure formation. But, more generally, the tests of inflation are important because they will reveal the simplicity or complexity of the Universe and thereby determine if any explanatory model, inflation or otherwise, is feasible.

Inflation makes three predictions which can be decisively tested using current technology [6]:

(1) **Spatial flatness** [1, 7]: $\Omega_{\text{total}} = 1$, where Ω_{total} includes radiation energy density, matter density, and any cosmological constant contribution. Models can be constructed in which $\Omega_{\text{total}} < 1$, but these require very special adjustments of parameters and initial conditions.

(2) **Nearly scale-invariant spectrum of fluctuations, $0.7 < n < 1.2$** [5]: The spectrum of fluctuations originates through quantum fluctuations which run rampant when the Universe occupies a microphysical size ($< 10^{-25}$ cm) prior to inflation. Inflation stretches the quantum inhomogeneities to cosmological length scales. The resulting spectrum is *nearly* scale-invariant, tilted slightly compared to a perfectly scale-invariant (Harrison-Zel'dovich) spectrum. If the expansion of the Universe were purely exponential, say, then one would find a strictly scale invariant spectrum. However, in all inflationary models, the expansion rate must slow down when you come near the end of inflation, which leads to small corrections to scale invariance. In addition, most models of inflation don't require exponential expansion at any stage. Power-law expansion or other forms will do just fine in terms of solving the cosmological conundrums. Non-exponential potentials produce spectra that are nearly, but not precisely, scale-invariant on all wavelengths

[8]. Instead of spectral index $n = 1$ for a strictly scale invariant spectrum, a range of tilt between 0.7 and 1.2, roughly, is achievable [6].

It should be emphasized that observing a tilt away from scale invariance is good news for inflation. Although it was not fully appreciated in the literature, the fact that the inflationary perturbation spectrum is "almost" but not quite scale-invariant has been known from the first papers on the subject [9]. As a typical feature of inflation, tilt ought to be incorporated as a well-motivated degree-of-freedom in models of large-scale structure formation inspired by inflationary cosmology (e.g., cold dark matter models).

(3) **Two Species of Fluctuations: Scalar and Tensor**: Inflation generates both scalar and tensor perturbations of the metric. Both contribute to the cosmic microwave background anisotropy, and the scalar perturbations may also act as gravitational potential wells that seed large-scale structure formation. Scalar fluctuations in the space-time metric are caused by quantum fluctuations in the inflation field which drives inflation [5]. Since the inflation field dominates the energy density of the universe during inflation, the scalar fluctuations result in inhomogeneities in the energy density distribution. Tensor fluctuations in the space time metric caused by the same quantum effects correspond to gravitational waves [10]–[12]. The scalar fluctuations are perturbations in the gravitational potential which can seed structure formation and perturb the distribution of microwave background photons on the last-scattering surface. The gravitational waves are fluctuations in the space-time metric that also induce microwave background perturbations, but they have no effect on large-scale structure formation.

In addition to these three observational predictions, there are other detailed quantitative predictions of inflation, such as relations between the degree of tilt and the ratio of scalar-to-tensor fluctuation effects on the microwave background [13, 14]. But the three features listed above will likely be the first targets since they are accessible using current technology and distinguish inflation from competing cosmological models.

2.3   THE POWER OF THE COSMIC MICROWAVE BACKGROUND

Cosmology has been transformed by the COBE discovery of anisotropy in the cosmic microwave background [15]. The anisotropy may provide a snapshot of the Universe in its infancy, displaying fluctuations initiated by events that took place during the first instants after creation, such as those predicted by inflation. A pressing challenge for cosmologists in the coming decade is to determine how to use the snapshot to decisively discriminate among competing cosmological models. Each cosmological model produces a distinguishing anisotropy fingerprint. The fingerprint entails a broad-band spectrum of information that can be redundantly tested by diverse experiments. An important advantage is that the anisotropy is a linear response to well-understood physical processes. Consequently, the physical interpretation of anisotropy measurements is less subject to model-dependent assumptions than other cosmological tests. For these reasons, there is real hope that microwave background measurements in the next decade will make a historic contribution to our understanding of the universe.

The cosmic fingerprint [6] is obtained from a temperature difference map (Fig. 2.1) which displays $\Delta T(\mathbf{x})/T_\gamma$ as a function of sky direction $\mathbf{x}$. The map represents the deviations in temperature from the mean value, $T_\gamma = 2.726 \pm .010$. The root-mean-square deviation from average in the COBE map is of order $\Delta T/T_\gamma = 0.001\%$. The temperature autocorrelation function, $C(\theta)$, compares the temperature at points in the sky separated by angle θ:

$$C(\theta) = \left\langle \frac{\Delta T}{T_\gamma}(\mathbf{x}) \frac{\Delta T}{T_\gamma}(\mathbf{x}') \right\rangle$$

$$= \frac{1}{4\pi} \sum_\ell (2\ell + 1) C_\ell P_\ell(\cos\theta), \tag{2.1}$$

where $\langle\rangle$ represents an average over the sky and $\mathbf{x} \cdot \mathbf{x}' = \cos\theta$. The coefficients, C_ℓ, are the *multipole moments* (for example, C_2 is the quadrupole, C_3 is the octopole, etc.). Roughly speaking, the value of C_ℓ is determined by fluctuations

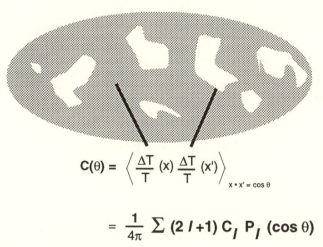

Figure 2.1: The temperature autocorrelation function, $C(\theta)$, is obtained from a map of the sky (here represented by the oval) displaying the difference in the microwave background temperature from the average value, $\Delta T/T$. $C(\theta)$ is computed by taking the map-average of the product of $\Delta T/T$ measured from any two points in the sky separated by angle θ. If $C(\theta)$ is expanded in Legendre polynomials, $P_\ell(\cos\theta)$, the coefficients C_ℓ are the *multipole moments*.

on angular scales $\theta \sim \pi/\ell$. A plot of $\ell(\ell + 1)C_\ell$ is referred to as the *cosmic microwave background (CMB) power spectrum*. An exactly scale-invariant spectrum of primordial energy density fluctuations, if there were no evolution when the fluctuations pass inside the Hubble horizon, produces a flat CMB power spectrum curve (*i.e.*, $\ell(\ell + 1)C_\ell$ is independent of ℓ).

There is valuable information in the cosmic microwave background anisotropy in addition to the CMB power spectrum that will be extracted some day. Higher-point temperature correlation functions (entailing three or more factors of $\Delta T/T_\gamma$) could be obtained from the temperature difference map and be used to test if the fluctuation spectrum is gaussian, as predicted by inflation. However, the fact that statistical and systematic errors in $\Delta T/T_\gamma$ compound for higher-point correlations makes precise measurements very challenging. Polarization of the microwave background by the last scattering of photons from the anisotropic electron distribution is another sky signal that provides quantitative data that can be used to test models. However, for known models, the predicted polarization requires more than two orders of magnitude better accuracy than anisotropy measurements alone in order to discriminate models [16]. For the coming decade, the most reliable information will be the CMB power spectrum. Fortunately, the CMB power spectrum is packed with information that can be used, by itself, to discriminate the leading cosmological models.

Here we will focus on how measurements of the CMB power spectrum can be used to test the inflationary paradigm and measure cosmological parameters. For alternative theories of large-scale structure formation, the model tests and criteria for constraining parameters are similar, but differ in detail. The CMB power spectrum for inflationary models is a direct reflection of the key, generic features of inflation outlined in the previous section. These can be seen by studying Figure 2.2, which displays a representative power spectrum for inflation [17]. For this example, the spectral index is $n = 0.85$ with equal scalar and tensor contributions to the quadrupole moment, within the range allowed by inflation. Left to right in the figure spans large- to small-angular scale fluctuations.

It is useful to imagine a vertical line at $\ell \approx 100$. To the left of $\ell \approx 100$ are multipoles dominated by fluctuations with wavelengths much larger than the size of the Hubble horizon at the time of last scattering, spanning angles $> 1° - 2°$ on the sky. According to the inflationary model, these wavelengths did not have a chance to evolve before last scattering and the beginning of the photon trek towards our detectors. Hence, these fluctuations preserve the imprint of whatever fluctuations were there in the first place. If the fluctuations are remnants of a nearly scale-invariant inflationary spectrum, then the low-ℓ part of the CMB power spectrum should be featureless, just as shown in the figure.

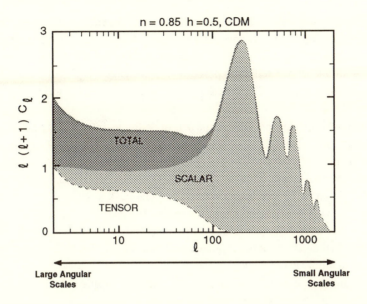

Figure 2.2: A plot of $\ell(\ell + 1)C_\ell$ vs. multipole moment number ℓ is the cosmic microwave background power spectrum. For a given ℓ, C_ℓ is dominated by fluctuations on angular scale $\theta \sim \pi/\ell$. In inflation, the power spectrum represents the sum of two independent, scalar and tensor contributions.

The spectrum includes, in general, both scalar and tensor contributions, as indicated in the example of Figure 2.2. For inflationary models, the two contributions are predicted to be statistically independent. Consequently, the total power C_ℓ is simply the sum of the scalar and tensor contributions, $C_\ell^{(S)}$ and $C_\ell^{(T)}$, respectively. The fluctuations in $\Delta T/T$ are also predicted to be Gaussian-distributed for inflationary models. The C_ℓ's, which are an average over $2\ell + 1$ Gaussian-distributed variables, have a χ^2-distribution.

To the right of $\ell \approx 100$ are multipoles dominated by fluctuations with wavelengths smaller than the horizon at last scattering. This side of the power spectrum appears different from the left because there are additional physical effects once inhomogeneities enter the Hubble horizon and begin to evolve. Gravitational waves inside the horizon red-shift away. For energy density fluctuations, the baryon and photon begin to collapse and oscillate acoustically about the centers of high and low energy density, adding to the net microwave background perturbation.

Each wavelength laid down by inflation begins its acoustic oscillation shortly after entering the Hubble horizon. Hence, there is a well-defined phase-relation between the acoustic oscillations on different wavelengths. Waves just entering the horizon and smaller-wavelength waves which have completed a half-integral number of oscillations by last scattering will be at maximum amplitude. Wavelengths in-between are mid-phase and will have smaller amplitudes. In a plot of C_ℓ's, increasing ℓ corresponds to multipoles dominated by decreasing wavelengths. The variations of the oscillation phase with wavelength results in a sequence of peaks as a function of ℓ. These peaks are sometimes referred to as Doppler peaks or acoustic peaks.

The position of the first Doppler peak is of particular interest. Its position along the ℓ-axis, left or right, is most sensitive to the value of Ω_{total}. The peak moves to the right in proportion to $1/\sqrt{\Omega_{\text{total}}}$ [18], for large Ω_{total}. There is only weak dependence on the Hubble constant and other cosmological parameters [19]. Decreasing Ω_{total} to 0.2, say, causes the first Doppler peak to shift to $\ell \approx 600$ instead of $\ell \approx 200$, a dramatic and decisive difference. As a test of Ω_{total}, the first Doppler peak has the advantage that it is relatively insensitive to the form of the energy density, whether it be radiation, matter, or cosmological constant and it is relatively difficult to mimic using other physical effects.

In sum, Figure 2.2 illustrates how all three key features of inflation can be

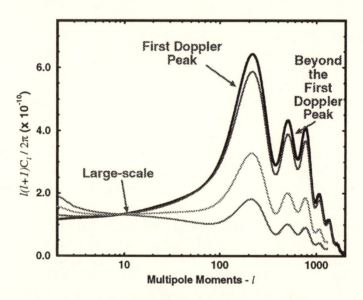

Figure 2.3: A band of microwave background power spectra allowed by inflation. The uppermost curve is a pure-scalar, scale-invariant spectrum, and the lower curves have tilt ($n < 1$) and gravitational waves. Inflationary models with spectra somewhat higher than the uppermost curve are also possible. The common features among these curves—the prime targets for experimental tests of inflation—are a plateau at large angular scales, a prominent first Doppler (or acoustic) peak, and subsequent acoustic oscillation peaks at small angular scales.

tested by the microwave background power spectrum. Large-angular scale fluctuations are consequences of scale-invariance and the combination of scalar and tensor perturbations. A nearly flat CMB power spectrum at small ℓ is the signature of being nearly scale-invariant. Small-angular scale fluctuations, especially the position of the first Doppler peak, are consequences of Ω_{total} being unity. The presence of a combination of scalar and tensor fluctuations can be determined from more subtle features, such as the ratios of the Doppler peaks to the plateau at small ℓ.

The inflationary prediction for the CMB power spectrum is not unique, since there are undetermined, free parameters. Figure 2.3 is a representative band of predicted curves for different values of the spectral index. Each example lies within the parameter space achievable in inflationary models. Although the band is wide, there are common features among the curves which can be used as the critical tests of inflation. All have a plateau at small ℓ, a large first Doppler peak at $\ell \approx 200$, and then smaller Doppler hills at larger ℓ.

The key microwave background tests of inflation for the next decade break down into two key battles. The first is the "Battle of the Primordial Plain," illustrated in Figure 2.4. Displayed is a blow-up of the power spectrum over the

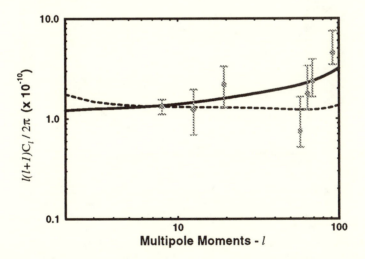

Figure 2.4: Blow-up of the power spectrum focusing on large angular scales. The upper curve corresponds to a pure-scalar $n_s = 1$ (scale-invariant) spectrum, and the lower curve has $n_s = 0.85$ and equal scalar and tensor contributions to the quadrupole. Superimposed are the experimental flat band power detections with one-sigma error bars. Left-to-right, these correspond to: (a) COBE; (b) FIRS; (c) Tenerife; (d) South Pole 1991; (e) South Pole 1994; (f) Big Plate 1993-4; and (g) PYTHON.

range $\ell = 2$ to $\ell = 100$, where the featureless plateau predicted by inflation can be tested. The predictions of two representative inflation models are compared to the present measurements from various space-borne, air-borne, and land-based cosmic microwave background anisotropy experiments. Experimental results are from Ref. [15] and Refs. [20]–[28]. What this battle will determine is whether the spectrum really is featureless or whether there are bumps and wiggles; and, if it is featureless, the battle will determine the slope or spectral index and whether it lies within the inflationary range.

The second battle, taking place concurrently, is the "Battle of the Doppler Peak." Figure 2.5 shows a blow-up with multipoles $\ell = 2$ through $\ell = 350$, through the first Doppler peak. The presence of a peak near $\ell = 200$ is decisive evidence for discriminating models. Inflation, for example, is a model which predicts a peak at $\ell = 200$, while other models, including open universe models [18] and cosmic defect models [29]–[31] predict peaks at larger values of ℓ. As noted

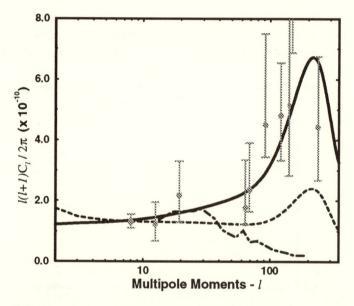

Figure 2.5: Power spectra through intermediate scales (1/2 degree), including the first Doppler (or acoustic) peak. The upper two curves are the same as in Figure 2.4. The dot-dashed curve is the prediction for cosmic texture models, assuming late reionization of the intergalactic medium. For the data, error bars represent one-sigma. Left-to-right, the experiments correspond to: (a) COBE; (b) FIRS; (c) Tenerife; (d) South Pole 1991; (e) South Pole 1994; (f) Big Plate 1993-4; (g) PYTHON; (i) MSAM (2-beam); (j) MAX4 (ι-Draconis region); (k) MAX3 (GUM region); and (l) MSAM (3-beam).

above, finding a peak near $\ell \approx 200$ would also be a strong indication that Ω_{total} is close to unity, a historic breakthrough in cosmology parameter measurement as well as a triumph for inflation.

If inflation wins both battles, then near-scale invariance and $\Omega_{\text{total}} \approx 1$ will have been confirmed on the basis of more-or-less qualitative tests. There remains detailed quantitative information to be extracted from the spectral shape which can discriminate scalar and tensor contributions, test other predictions of inflation, and measure cosmological parameters. To do all of these, one must push on to smaller angular scales, $1'$-$30'$, where cosmic microwave background anisotropy measurements can be combined with other astrophysical observations to form a suite of precision tests.

2.4 COSMIC CONCORDANCE

Inflation, like every cosmological model to date, includes some parameters that are not fixed by the theory: the Hubble constant, the cosmological constant, the baryon density, the spectral tilt, the ratio of gravitational waves to energy density fluctuations, etc. The ultimate validity of a model will rest on whether there is a plausible choice of the parameters which, joined with the theoretical framework, explains all available astrophysical and cosmological observations.

Several groups have been studying whether any models survive the current astrophysical and cosmological constraints [32]–[34] (see also the discussion of Table 1 in the contribution of P. J. E. Peebles to these proceedings). It is somewhat premature since both astrophysical and cosmological observations have large uncertainties. Nevertheless, the game is irresistible and perhaps instructive. It may be good practice for the real cosmic concordance game to come.

As an example, I will focus on results for special subset of models which seem most promising, based on work done in collaboration with J. Ostriker [33]. The primary focus is on spatially flat models because these are motivated by inflation. However, the analysis has been extended to cosmic defect models and open universe models as well. The most promising models, by the objective comparison with observations, are those with cold dark matter (CDM) plus non-negligible cosmological constant (CDM + Λ).

Because CDM + Λ models are spatially flat, they are consistent with inflationary cosmology. The inclusion of a non-negligible cosmological constant does require a fine-tuning problem to explain why the decreasing matter density and the time-independent cosmological constant happen to be comparable today. However, this unattractive aspect is not directly attributable to inflation, which only guarantees a spatially flat universe, whether $\Lambda = 0$ or not. Perhaps it is better to suspend theoretical bias until after comparison with observations.

The concordance tests can be divided into astrophysical measures, which constrain Ω_Λ and $H_0 = 100h$ km/sec/Mpc, and cosmic microwave background measurements, which fix the spectral index n. Together, these tests determine the range of the (Ω_Λ, h, n) parameter space which is allowed by observations. Here, for simplicity, we assume that the contribution of energy density fluctuations to the quadrupole anisotropy is linked to the spectral tilt as in most chaotic or new or supersymmetric inflationary models [6, 13, 14].

In Figure 2.6, a plot of the Ω_Λ vs. h plane is shown which illustrates the key astrophysical constraints. The constraints are:

(1) bounds on the Hubble constant [35]–[38] ($h = 0.7 \pm 0.15$) based on recent Hubble Space Telescope studies using classic Cepheid variables ($H_0 = 82 \pm 17$ km/sec/Mpc) and studies using Type I supernovae ($H_0 = 67 \pm 7$ km/sec/Mpc) as standard candles;

(2) the age of the Universe ($t_0 > 11.5$ Gyr or 13.7 Gyr), based on recent globular cluster age estimates from extrapolation of red-giant [39] and main-sequence stars [40], respectively. The first limit, which suggests $t_0 > 13.7$ Gyrs at the one-sigma level, is more stringent and, since it virtually

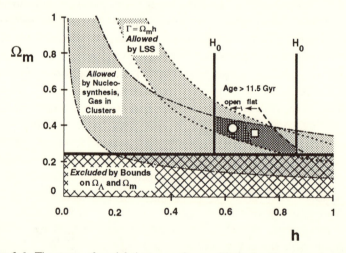

Figure 2.6: The range of models in concordance with the best known astronomical observations. The entire, heavily shaded region (with and without stripes) is the concordance domain for flat CDM + Λ models; the vertically striped subregion alone applies to open models with $\Omega_\Lambda = 0$. (The age constraint—see dashed lines—differs in the two cases.) Square indicates a representative flat model with $h = 0.7$ and $\Omega_\Lambda = 0.65$. Circle represents a representative open model, $h = .625$, $\Omega_m = 0.375$.

excludes open models, is stronger support for our conclusions. However, to be conservative, we adopt the latter lower bound, 11.5 Gyr, in Figure 2.6.);

(3) the cosmological constant ($\Omega_\Lambda < 0.75$), based on gravitational lens statistics [41, 42], and the matter density ($\Omega_m > 0.2$), based on observed light density and cluster mass-to-light ratios [43], or by utilization of large-scale structure measurements [44];

(4) the power spectrum [45], which constrains $\Gamma = \Omega_{\text{matter}} h = 0.25 \pm 0.05$ for n near unity (assuming that missing matter is cold dark matter);

(5) gas in clusters [46], $\Omega_{\text{baryon}} h^{3/2} / \Omega_{\text{matter}} = 0.07 \pm 0.03$, combined with the nucleosynthesis constraint [47], $\Omega_{\text{baryon}} h^2 = 0.015 \pm 0.005$, to yield $\Omega_{\text{matter}} h^{1/2} = 0.21 \pm 0.12$.

Others may have chosen values different from those discussed above, but only substantial changes would alter the basic conclusions. One should take note, however, that we have omitted the recent bulk flow measurements of Lauer and Postman [48] from the list of constraints because they remain controversial and may already be contradicted by other related large-scale measurements.

The range of concordance consistent with these constraints is indicated in Figure 2.6. The heavily shaded region, both with and without vertical stripes, are for flat models with $\Omega_\Lambda > 0$. The vertical stripes indicate the somewhat more restricted subregion that applies for open models with $\Omega_\Lambda = 0$, with the reduction being due to a different age constraint.

The cosmic microwave background anisotropy provides important, new constraints that can resolve some of the remaining ambiguities, including open vs. flat CDM $+\Lambda$ models. Large-angular scale measurements can be used to constrain the spectral index n, which fixes the spectral shape, and an overall normalization, conventionally chosen to be σ_8, the root-mean-square mass fluctuations averaged over eight h^{-1} Mpc spheres. The value of σ_8 required to explain the distribution of large-scale structure and velocities [49] over 1-100 Mpc scales is $\sigma_8 = (0.56 \pm 0.06)(\Omega_m)^{-0.56}$. The inverse, $b = 1/\sigma_8$, is known as the *bias*, the ratio of the root-mean-square fluctuations in galaxies compared to those in total mass. The COBE amplitude is another measure of normalization, but based on the greater distances probed by the cosmic background radiation. The extrapolation from COBE to a normalization at shorter distance scales depends on n, the fractional contribution of energy density fluctuations to cosmic background radiation fluctuations, and the values of Ω_m Ω_Λ, H_0, and Ω_B [50]. The latter parameters determine how fluctuations on $8h^{-1}$ Mpc scales evolve relative to those on COBE scales. Any given point in the concordance region of Figure 2.6 corresponds to fixed Ω_m, Ω_Λ

and H_0. Big bang nucleosynthesis then determines Ω_B, as described above. Since inflation sets a relation between n and the energy density fluctuation contribution [6, 13, 14], matching COBE normalization to the astronomical normalization fixes the remaining undetermined parameter, the spectral index, n. The extrapolation differs, however, depending upon whether the universe is flat or open and upon the value of the cosmological constant.

Figure 2.7 illustrates the permitted range in the n-σ_8 plane for flat CDM + Λ models. For flat models, the most notable feature is that the range of n, between 0.9 and 1.2, is well within the COBE constraint on the spectral index and within the range allowed by inflation although concordance with COBE and inflation were not required as *ab initio* constraints. For σ_8, the range varies between biased (less than unity) to anti-biased (greater than unity). It is likely that the addition of further constraints from large-scale simulations will narrow this range, pushing the limits towards the lower left-hand corner of the concordance region in the n-σ_8 plane.

For open models, the COBE fit [51] requires shifting to n greater than unity, $1.05 < n < 1.4$. The larger value of n is required to compensate for the fact that there is less amplitude growth in open models compared to flat models. So, for a given COBE normalization, the primordial power spectrum must have greater amplitude at small angular scales to be as effective in explaining the observed

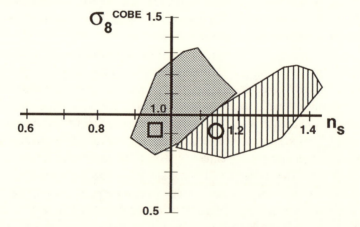

Figure 2.7: The range of concordance in the σ_8-n_s plane for spatially flat CDM +Λ models with non-zero cosmological constant (grey) and open models (striped). The flat models are in a range of n_s that is consistent with inflation, whereas the open models are pushed towards significant positive tilts ($n_s > 1$) that are difficult to achieve in open inflation models. The square and circle correspond to the representative flat and open models shown in Figure 2.6.

large-scale structure of the universe. This corresponds to setting $n > 1$. Requiring $n > 1$ is a significant, new constraint on open models stemming from this analysis. The range is barely compatible with inflation, creating a new challenge for open inflation models.

With the application of both the astrophysical constraints and the constraints from COBE, all three degrees-of-freedom (Ω_Λ, h, n) are fixed to within a narrow range. But, there is additional information available from measurements of intermediate- ($1°$) and small-angular ($10'$-$30'$) scale microwave background anisotropy measurements. These serve as redundant tests of the concordance range that can test competing models, such as cosmic defect or open universe models.

It is worth noting that the cold dark matter plus cosmological constant models are quite different from the standard cold dark matter model that was compared to the data in the previous section. The new models include a significant spectral tilt, a significant gravitational wave contribution, a large value of Ω_Λ, a rather different value of h, and a significantly larger baryon density. Each one of the parameter changes by itself produces a significant change in the predicted microwave background and anisotropy spectrum and here they are all combined.

How different is the spectrum predicted for the concordance CDM + Λ model from standard ($\Lambda = 0$) CDM and how much does it agree or disagree with present data? The answer, shown in Figure 2.8, is rather surprising. The net result of changing all of the parameters is to get a prediction which, over most of the multipoles, is virtually indistinguishable from standard CDM! It is a textbook example of what some call *cosmic confusion* [19]: the fact that it is possible to have different choices of parameters produce nearly identical microwave background anisotropy spectra. Before, the examples were constructed artificially to illustrate the effect; here, confusion has popped out automatically from the concordance analysis.

Given the close agreement of the standard CDM and concordance CDM + Λ models and the fact that the standard CDM model already agreed well with intermediate- and small-angular scale measurements, the CDM +Λ models do also. The open squares in Figure 2.8 represent the predictions based on the concordance model and compare favorably with the experimental results themselves [52]. Amusingly, at the one place where the concordance model predictions diverge somewhat from the standard cold dark matter model, the smaller angular scales, the results are marginally more consistent with the concordance models (e.g., the White Dish experiment [27]). Figure 2.8 also shows the spectra for a representative open model in the concordance region as well as a cosmic texture model. The open model is clearly in poor agreement with the present microwave background limits at the smaller angular scales.

For the moment, the present consistency of observations with CDM + Λ mod-

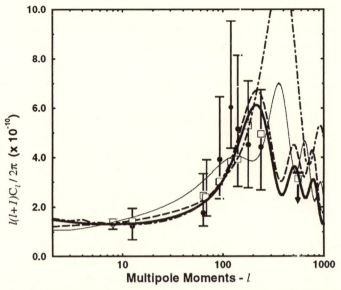

Figure 2.8: Predicted cosmic background radiation power spectrum for the representative models: standard cold dark matter ($n = 1$, $h = 0.5$, $\Omega_\Lambda = 0$; dashed line); cosmic textures (thin solid line); a flat model in the concordance region of Figure 2.6 ($n = 0.96$, $h = 0.65$, $\Omega_m = 0.35$, $\Omega_\Lambda = 0.65$, $\sigma_8 = 0.87$; heavy solid line); and an open model in the concordance region, ($n = 1.15$, $h = 0.625$, $\Omega_m = 0.375$; dot-dashed curve). The latter two are the models indicated by squares and circles, respectively, in Figures 2.6 and 2.7. The boxes represent the theoretical predictions for present experiments for the flat model. Note that the power spectrum for the flat model is remarkably similar to the prediction for standard cold dark matter, except at smaller angular scales ($\ell > 250$), where the flat model is marginally more consistent with present observational limits. Open models predict that the leftmost peak occurs at higher ℓ and, consequently, has more power at small angular scales. Cosmic textures predict a Doppler peak shifted to higher $\ell \approx 350$, also.

els should be regarded only as an intriguing tease. At this point, roughly fifteen observations have been used to constrain the three degrees of freedom (Ω_Λ, h, n), but many of the measurements are imprecise. Dramatic improvements in microwave background anisotropy measurements and astrophysical observations are anticipated over the next decade and then the comparisons can be made with much greater confidence.

2.5   A NEW AGE?

The advent of new observational technologies is opening a new age in our understanding of the Universe. The last section illustrated the power of invoking a combination of astrophysical and microwave background constraints. Over the next decade, an improved and more extensive suite of observations will be used to quantitatively and redundantly test our models. Cosmic concordance will decide the theoretical framework for explaining large-scale structure formation (inflation or alternatives) and will determine the values of cosmological parameters.

If inflationary cosmology survives the concordance tests, the field of cosmology will be transformed. To date, the observationally successful concepts in cosmology are derived from the hot big bang picture, a model born from general relativity, atomic physics and nuclear physics. With the verification of inflationary cosmology, particle cosmology would emerge from its embryonic stage of theoretical speculation and be brought forth into the light of observation to become an established part of our understanding of the Universe.

The verification of inflation would bring renewed focus on remaining open questions: What is the dark matter in the Universe (see the contribution of D. Spergel to these proceedings)? How was the matter-antimatter asymmetry generated? What determines the vacuum density? If inflation is correct, these questions were decided by phenomena that took place after inflation completed. Since the reheat temperature after inflation is much less than the Planck scale, it means that the answers to these questions do not rely on quantum gravity or initial conditions. We could be confident that the answers belong, instead, to particle physics.

The verification of inflation would bring benefits to our understanding of particle physics as well. Measuring the amplitude of the microwave background fluctuations would reveal the characteristic energy scale of inflation likely related to some symmetry-breaking phase transition associated with particle physics interactions near the Planck scale (10^{17} GeV or so). If the measurements were also to verify a non-zero value for the cosmological constant, it would be a profound discovery. The prejudice of most particle physicists is that the true vacuum has zero cosmological constant. One possibility, consistent with this prejudice, is that the present vacuum is a false one and that future symmetry breaking is possible. In this case, the present vacuum is unstable and decay to the true vacuum lies ahead. Whether the present vacuum is true or false, the discovery implies a vacuum energy density today of $(1 \text{ meV})^4$ or so, perhaps related to a new low-energy particle physics threshold. It is intriguing that the characteristic energy is comparable to the neutrino mass suggested in proposed solutions to the solar neutrino problem (see contribution by J. Bahcall, these proceedings). Could the fundamental problems of particle stellar astrophysics and particle cosmology be related?

Whether particle cosmology comes of age as considered here is one possibility stemming from the remarkable range of new observations. The most exciting possibilities are surprises that have not been anticipated. What is absolutely sure is that cosmology is entering a historic period which will define our view of the Universe. The next decades will be regarded by future generations as a unique epoch of grand discovery. Students deciding their future research directions should consider what a rare scientific adventure cosmology offers at this time.

ACKNOWLEDGMENTS

Special thanks are due the many collaborators who contributed to the work discussed herein, Jerry Ostriker, Dick Bond, Rob Crittenden, Rick Davis, George Efstathiou, George Smoot, Mike Turner, and Hardy Hodges.

I would like especially to thank John Bahcall, who has been a friend, advisor, and inspiration to me. Among our discussions, I recall vividly a lunch conversation while I was a visitor to the Institute for Advanced Study in Spring, 1990. Viewing me as an advocate of particle cosmology, John challenged me to specify the observational tests of its stimulating speculations. Up to that point in time, my focus, as well as many in the field, had been on the theoretical underpinnings. I was very dissatisfied with my response, which seemed too vague and reliant on far-future developments. Several years later, while sitting in the audience at the Spring Meeting of the American Physical Society listening to the announcement of the COBE experimental team of the discovery of anisotropy in the cosmic microwave background, my conversation with John sprung to mind. The opportunity to meet his challenge had arisen much sooner than I had anticipated. The encounter with John had sensitized me to the historic opportunities that lay ahead, resulting in refocusing my research on predictions and observational tests of inflation.

This research was supported by the DOE at Penn (DOE-EY-76-C-02-3071); by the John Simon Guggenheim Foundation; and by National Science Foundation Grant NSF PHY 92-45317 and Dyson Visiting Professor Funds at the Institute for Advanced Study, Princeton.

BIBLIOGRAPHIC NOTES

The following list of references provides an introduction for students to the basic principles of theoretical cosmology, especially the inflationary model of the universe, and to the interpretation and analysis of the key observational tests:

- **Guth, A. H., & Steinhardt, P. J. 1984, Sci. American, 250, 90.** An intuitive introduction to the big bang picture and inflationary cosmology which includes useful images.

- **Guth, A. H. 1981, Phys. Rev. D, 23, 347.** The classic paper in which the horizon, flatness, and monopole problems of the standard big bang model are clearly laid out, and the possibility of solving the problems in an inflationary universe is discussed. Guth shows that "old inflation" picture in this paper is not viable because inflation is ceaseless. Nevertheless, the presentation is so powerful that it inspired several attempts to revive the idea, culminating several years later in viable "new inflation" models of Linde and Albrecht & Steinhardt.

- **Bardeen, J., Steinhardt, P. J., & Turner, M. S. 1983, Phys. Rev. D, 28, 679.** One of four papers to emerge from the 1982 Nuffield Meeting (Cambridge) showing that inflation can generate cosmological density perturbations by stretching microscopic quantum fluctuations. The distinctive features of this paper are its use of a gauge-invariant formalism to derive the spectrum, now standardly adopted, and its demonstration that the spectrum is nearly, but not perfectly, scale-invariant in realistic models.

- **A. D. Linde 1990, Particle physics and inflationary cosmology, (Harwood Publishers).** Thorough introduction to the theoretical underpinnings of inflationary cosmology by one of its founders. Although the account is missing some theoretical ideas that were developed shortly after its publication, it is a worthwhile read because of the highly original viewpoint and charm of the author.

- **Peebles, P. J. E. 1994, Physical Cosmology (Princeton: Princeton University Press).** A tour-de-force volume on observational and theoretical cosmology by one of the grandmasters of the field.

- **Smoot, G. F., et al. 1992, ApJ, 396, L1.** The announcement by the COBE (Cosmic Background Explorer) collaboration of the discovery of anisotropy in the cosmic microwave background. The paper galvanized the theoretical and observational cosmological community by demonstrating the possibility of testing ideas about events proposed to take place during the first instants after the big bang.

- **Steinhardt, P. J. 1995, Int. Journal of Mod. Phys., A10, 1091.** A review of the the predictions of inflationary cosmology and the principles for testing inflation using measurements of the cosmic microwave background anisotropy.

- **Hu, W., & Sugiyama, N. 1995, ApJ, 444, 489.** An excellent introduction to the physical principles that underlie the cosmic microwave background anisotropy. The work explains in an intuitive way the features that arise from highly precise numerical codes for computing the anisotropy power spectrum.

- **Bond, J. R 1996, in Cosmology and Large Scale Structure, ed. R. Schaeffer, J. Silk, M. Spiro, & J. Zinn-Justin (Amsterdam: North Holland).** An advanced treatise on the theory and observation of the cosmic microwave background, recommended reading once the introductory lessons in Hu and Sugiyama have been absorbed.

- **Ostriker, J. P., & Steinhardt, P. J. 1995, Nature, 377, 600.** A recent paper which reviews the current observational constraints and shows how they can work in concert to determine the theoretical framework and parameters that describe our universe.

- **Strauss, M. A., & Willick, J. A. 1995, Physics Reports, 261, 271.** A thorough review of the state of theoretical and observational knowledge about large-scale structure and velocities.

BIBLIOGRAPHY

[1] Guth, A. H. 1981, Phys. Rev. D, 23, 347.

[2] Linde, A. D. 1982, Phys. Lett. B, 108, 389; Albrecht, A., & Steinhardt, P. J. 1982, Phys. Rev. Lett., 48, 1220.

[3] Guth, A. H., & Steinhardt, P. J. 1989, "The Inflationary Universe" in The New Physics, ed. P. Davies (Cambridge: Cambridge Univ. Press), 34.

[4] Linde, A. D. 1990, Particle Physics and Inflationary Cosmology (New York: Harwood Academic Publishers).

[5] Bardeen, J., Steinhardt, P. J., & Turner, M. S. 1983, Phys. Rev. D, 28, 679; Guth, A. H., & Pi, S.-Y. 1982, Phys. Rev. Lett., 49, 1110; Starobinskii, A. A. 1982, Phys. Lett. B, 117, 175; Hawking, S. W. 1982, Phys. Lett. B, 115, 295.

[6] Steinhardt, P. J. 1995, in Proceedings of the Snowmass Workshop on the Future of Particle Astrophysics and Cosmology, eds. E. W. Kolb, & R. Peccei (Singapore: World Scientific); 1995, IJMPA, A10, 1091.

[7] Steinhardt, P. J. 1990, Nature, 345, 41.

[8] Crittenden, R., & Steinhardt, P. J. 1992, Phys. Lett. B, 293, 32.

[9] See, for example, title and discussion in Bardeen, J., Steinhardt, P. J., & Turner, M. S. 1983, Phys. Rev. D, 28, 679.

[10] Rubakov, V. A., Sazhin, M. V., & Veryaskin, A. V. 1982, Phys. Lett. B, 115, 189.

[11] Starobinskii, A. I. 1985, Sov. Astron. Lett., 11, 133.

[12] Abbott, L. F., Wise, M. 1984, Nucl. Phys. B, 244, 541.

[13] Davis, R. L., Hodges, H. M., Smoot, G. F., Steinhardt, P. J., & Turner, M. S. 1992, Phys. Rev. Lett., 69, 1856.

[14] Lucchin, F., Matarrese, S., & Mollerach, S. 1992, ApJ, 401, L49; Salopek, D. 1992, Phys. Rev. Lett., 69, 3602; Liddle, A., & Lyth, D. 1992, Phys.Lett. B, 291, 391; Sahni, V., & Souradeep, T. 1992, Mod. Phys. Lett., A7, 3541; Lidsey, J. E., & Coles, P. 1992, MNRAS, 258, 57P; Krauss, L., & White, M. 1992, Phys. Rev. Lett., 69, 869.

[15] Smoot, G. F., et al. 1992, ApJ, 396, L1; Bennett, C. L., et al. 1994, ApJ Lett., submitted; Gorski, K. M., et al. 1994, ApJ, 430, L89.

[16] Crittenden, R., Davis, R. L., & Steinhardt, P. 1993, ApJ, 417, L13.

[17] Crittenden, R., Bond, J. R., Davis, R. L., Efstathiou, G., & Steinhardt, P. J. 1993, Phys. Rev. Lett., 71, 324.

[18] Kamionkowski, M., Spergel, D. N., & Sugiyama, N. 1994, ApJ, 426, L57.

[19] Bond, J. R., Crittenden, R., Davis, R. L., Efstathiou, G., & Steinhardt, P. J. 1994, Phys. Rev. Lett., 72, 13.

[20] Watson, R. A., et al. 1992, Nature, 357, 660; Hancock, S., et al. 1994, Nature, 367, 333.

[21] Ganga, K., et al. 1993, ApJ, 410, L57; Ganga, K., et al. 1994, ApJ, 432, L15.

[22] Gaier, T., et al. 1992, ApJ, 398, L1; Schuster, J., et al. 1993, ApJ, 412, L47; Gundersen, J., et al. 1994, ApJ, 443, L57.

[23] Wollack, E. J., et al. 1993, ApJ, 419, L49; Netterfield, C. B., et al. 1995, ApJ, 445, L69.

[24] Dragovan, M., et al. 1994, ApJ, 355, L67; Ruhl, J. E. et al. 1995, ApJ, 453, L1.

[25] Cheng, E. S., et al. 1994, ApJ, 422, L37; Cheng, E. S., et al. 1996, 456, L71.

[26] Meinhold, P., et al. 1993, ApJ, 409, L1; Gunderson, J., et al. 1993, ApJ, 413, L1; Clapp, A. C., et al. 1994, ApJ, 433, L57; Devlin, M., et al. 1994, ApJ, 430, L1; Tanaka, S. T. et al. 1995, preprint astro-ph/951206.

[27] Tucker, G. S., Griffin, G. S., Nguyen, H., & Peterson, J. B. 1993, ApJ, 419, L45.

[28] Readhead, A. C. S., et al. 1989, ApJ, 346, 556.

[29] Coulson, D., Ferreira, P., Graham, P., & Turok, N. 1994, Nature, 368, 27.

[30] Albrecht, A., Coulson, D., Ferreira, P., Magueijo, J. 1995, IMPERIAL-TP-94-95-30.

[31] Crittenden, R. G., & Turok, N. 1995, PUPT-1545.

[32] Krauss, L., & Turner, M. S. 1994, astro-ph/9404003.

[33] Ostriker, J. P., & Steinhardt, P. J. 1995, Nature, 377, 600.

[34] Turner, M. S., Steigman, G., & Krauss, L. 1984, Phys. Rev. Lett., 52, 2090; Turner, M. S. 1991, Physica Scripta, T36, 167; Peebles, P. J. E. 1984, ApJ, 284, 439; Efstathiou, G., et al. 1990, Nature, 348, 705; Kofman, L., & Starobinskii, A. A. 1985, Sov. Astron. Lett., 11, 271.

[35] Fukugita, M., Hogan, C. J., & Peebles, P. J. E. 1993, Nature, 336, 309.

[36] Freedman, W., et al. 1994, Nature, 371, 757.

[37] Riess, A. G., Press, W. H., & Kirshner, R. P. 1995, ApJ, 438, L17.

[38] Hamuy, R., et al. 1995, ApJ, 109, 1.

[39] Jimenez, R., et al. 1995, preprint.

[40] Bolte, M., & Hogan, C. J. 1995, preprint.

[41] Carroll, S. M., Press, W. H., & Turner, E. L. 1992, ARA&A, 30, 499.

[42] Fukugita, M., & Turner, E. L. 1991, MNRAS, 253, 99; Maoz, D., & Rix, H.-W. 1993, ApJ, 416, 425.

[43] Peebles, P. J. E. 1994, Physical Cosmology (Princeton: Princeton University Press).

[44] Strauss, M. A., & Willick, J. A. 1995, Physics Reports, in press.

[45] Peacock, J. A., & Dodds, S. J. 1994, MNRAS, 267, 1020.

[46] White, S. D. M., Navarro, J. F., Evrard, A. E., & Frenk, C. S. 1993, Nature, 366, 429.

[47] Copi, C., Schramm, D. N., & Turner, M. S. 1995, Science, 267, 192.

[48] Lauer, T. R., & Postman, M. 1994, ApJ, 425, 418.

[49] White, S. D. M., Frenk, C. S., & Efstathiou, G. 1993, MNRAS, 262, 1023.

[50] Bond, J. R. 1996, in Cosmology and Large Scale Structure, ed. R. Schaeffer, J. Silk, M. Spiro, & J. Zinn-Justin (Amsterdam: North Holland).

[51] Gorski, K., Ratra, B., Sugiyama, N., & Banday, A. J. 1995, ApJ, 444, L67.

[52] For access to summaries and computation software for flat band-power estimation, use WWW source:
http://dept.physics.upenn.edu/ www/astro-cosmo/cmbr.

Chapter 3

Understanding Data Better with Bayesian and Global Statistical Methods

William H. Press

Harvard-Smithsonian Center for Astrophysics, Cambridge, MA

Abstract

To understand their data better, astronomers need to use statistical tools that are more advanced than traditional "freshman lab" statistics. As an illustration, the problem of combining apparently incompatible measurements of a quantity (the Hubble constant, e.g.) is presented from both the traditional, and a more sophisticated Bayesian, perspective. Explicit formulas are given for both treatments.

3.1 Introduction

Understanding data better is *always* an unsolved problem in astrophysics, although perhaps not in exactly the sense intended by the conference organizers. While other papers in this volume are more specifically directed at individual sub-areas of astrophysical theory, my contribution is intentionally more longitudinal: I hope that it is applicable to *all* the other areas surveyed.

If the spirit of this volume is to present a menu—a movable feast, indeed—of opportunities for thesis projects of smart second-year graduate students, then the opportunity that I would like to offer is one of voluntary self-choice: *Whatever* your choice of area, make the choice to live your professional life at a high level of statistical sophistication, and not at the level—basically freshman lab level—that is the unfortunate common currency of most astronomers. Thereby will we all move forward together.

What do I mean by "freshman lab level" and what do I mean by "sophisticated methods"? In my conference talk, I illustrated with three examples, in each case

49

contrasting an elementary with a more sophisticated framework: chi-square fitting of parameters to a model; estimating correlation functions (in the simplest diagonal case, "error bars") from a data set; and combining independent, and perhaps incompatible, experimental measurements. Here, I will limit my discussion to the third topic only, both because I want to give a somewhat greater level of detail than was possible in the talk (enough detail to actually be useful in practice), and because my collaborative work on the other two topics is, or will be, written up elsewhere [14, 15, 12, 13].

3.2   COMBINING EXPERIMENTAL MEASUREMENTS

We are given a number of supposedly independent measurements of a quantity, say, $H_1, \ldots, H_N$, with error bars on each one, $\sigma_1, \ldots, \sigma_N$. We are asked for the best estimate of the underlying quantity, call it H_0. (If this notation slyly reminds you of the Hubble constant, you are right!)

The conventional ("freshman lab") answer is to construct an average, weighted by the inverse variance of the individual observations,

$$H_0 = \frac{\sum_{i=1}^{N} H_i/\sigma_i^2}{\sum_{i=1}^{N} 1/\sigma_i^2} \tag{3.1}$$

Equation (3.1) can be derived in any number of ways. For example, it is the maximum likelihood estimator in the case where each measurement has a normal distribution.

If you had a fairly advanced freshman lab, you also learned the formula for the variance of the estimator H_0, namely,

$$\sigma_0^2 = \frac{1}{\sum_{i=1}^{N} 1/\sigma_i^2} \tag{3.2}$$

(If all the σ_i's are equal, this says that the combined standard deviation is $1/\sqrt{N}$ times the individual standard deviations.)

You can derive equation (3.2) from equation (3.1) yourself simply by applying the Var( ) operator to equation (3.1) and using the rules

$$\text{Var}(\alpha x) = \alpha^2 \text{Var}(x) \tag{3.3}$$

and

$$\text{Var}(x + y) = \text{Var}(x) + \text{Var}(y) \tag{3.4}$$

where α is a constant, and x and y are independent random variables. The variance that comes out is, of course, σ_0^2. The variance operator Var() is the usual one, defined by

$$\text{Var}(x) = \langle (x - \langle x \rangle)^2 \rangle \tag{3.5}$$

No matter how good your freshman lab instructor, it is almost a sure bet that he or she didn't show you the even more important formula for testing whether the individual measurements *are in fact compatible*, namely,

$$\chi^2 = \sum_{i=1}^{N} \frac{(H_i - H_0)^2}{\sigma_i^2} \tag{3.6}$$

If this value of χ^2 is *not* compatible with $N-1$ degrees of freedom (i.e., far *outside* the range $(N-1) \pm \sqrt{2(N-1)}$), then the estimate H_0 of equation (3.1) has simply no justification at all. The input values H_i are simply incompatible.

Equation (3.6) derives from the fact that a chi square variable is the sum of squares of (zero-mean, Gaussian) quantities divided by their respective variances. However, you also have to know that you are supposed to treat H_0 as having zero variance, and instead reduce the number of degrees of freedom from N to $N-1$. (Somewhere around this point, rules of thumb give way to the better procedure of actually proving theorems. See [8], §§ 4.1 and 11.4.)

Currently, a generally recognized example of an incompatibility between multiple measurements and their respective claimed error bars is the Hubble constant. This has engendered a sometimes ferocious, and not always fact-based, debate about "which measurements to throw out". Of course it would be best to understand the physical basis for incompatibility between differing measurements. However, it is also true that, even absent such understanding, we do not have to throw all statistical analysis out the window. We can instead, as we will next illustrate, construct a well-posed statistical framework that allows apparently incompatible measurements to be combined in a useful way.

3.3 BAYESIAN COMBINATION OF INCOMPATIBLE MEASUREMENTS

We again have independent measurements $H_1, \ldots, H_N$, with claimed error bars on each one, $\sigma_1, \ldots, \sigma_N$. (We will call this, collectively, the data "D".) But now we want to be sophisticated enough to recognize that *some of the error bars are wrong*, due to (for example) unrecognized systematic effects, or unjustified optimism on the part of the observer. Is there any sensible, yet formal, way to sort all this out?

Here is a method that Chris Kochanek and I have worked out, with an unabashedly Bayesian derivation:

Suppose that p_i is the probability that measurement i is "correct" (in the sense of having accurate error bars), so that $1 - p_i$ is the probability that it is wrong (including the possibility that there are systematic errors, non-negligible in comparison to the quoted error bars). In this section we consider the model that all the p_i's are the same, $p_i = p$. That is, p is the "community-wide probability of doing a correct observation" at a certain epoch and in a certain field of science.

Let $\mathbf{v}$ be a vector of length N whose ith component is either one or zero, signifying that the ith experiment is correct or wrong, respectively. Of course we don't know either the p or $\mathbf{v}$ a priori.

There are three laws of probability that we will need to apply repeatedly. I can never remember their conventional names, so I will instead use descriptive unconventional ones. All three laws can be "proved" by drawing careful Venn diagrams, or by using the "frequentist" definitions for probability

$$P(A) \equiv \frac{\text{Number of Events with Property A}}{\text{Total Number of Events}} \tag{3.7}$$

and for conditional probability

$$P(A|B) \equiv \frac{\text{Number of Events with Properties A and B}}{\text{Number of Events with Property B}} \tag{3.8}$$

The laws are:

Law of Anding

$$P(ABC) = P(A)P(B|A)P(C|AB) \tag{3.9}$$

That is, the probability of A and B and C is the product of three terms: the probability of A alone, the conditional probability of B *given* A, and the conditional probability of C *given* A *and* B.

Law of De-anding

If a set of hypotheses B_i are exhaustive and disjoint, so that $\sum_i P(B_i) = 1$, then

$$P(A) = \sum_i P(AB_i) \tag{3.10}$$

That is, we can recover the total probability of A from the sum of the more restrictive probabilities of A *and* B_i.

Bayes' Rule

If the set $\{H_i\}$ is an exhaustive and disjoint set of hypotheses, while D is some data, then

$$P(H_i|D) \propto P(D|H_i)P(H_i) \tag{3.11}$$

where $P(H_i)$ are the prior probabilities of the hypotheses (that is, their probabilities before the new data D was gleaned). The constant implicit in the proportionality sign is determined by requiring the sum of the left-hand side (over i) to be unity. (One often saves effort in probability calculations by computing proportionalities and leaving the overall normalization to the end.)

In our problem, a complete hypothesis is a particular value H_0 *and* a particularly value for p *and* a particular assignment of ones and zeros in $\mathbf{v}$. If, however, our primary interest is in H_0, then we use the law of de-anding and get

$$P(H_0|D) = \sum_{p,\mathbf{v}} P(H_0 p \mathbf{v}|D) \qquad (3.12)$$

The (formal) sum over p of course becomes an integral if p is (as it is) a continuous variable; the sum over $\mathbf{v}$ denotes a discrete sum over all 2^N possible values of the vector.

Next apply Bayes' Rule,

$$P(H_0|D) \propto \sum_{p,\mathbf{v}} P(D|H_0 p \mathbf{v})P(H_0 p \mathbf{v}) \qquad (3.13)$$

and the law of anding,

$$P(H_0|D) \propto \sum_{p,\mathbf{v}} P(D|H_0 p \mathbf{v})P(H_0)P(p|H_0)P(\mathbf{v}|H_0 p) \qquad (3.14)$$

The four factors in (3.14) are now individually tractable. The first factor, $P(D|H_0 p \mathbf{v})$ is the probability of the data *given* (as $\mathbf{v}$) which experiments are right or wrong. For independent experiments one might model this as something like

$$P(D|H_0 p \mathbf{v}) = \prod_{v_i=1} P_{Gi} \prod_{v_i=0} P_{Bi}$$

$$= \exp\left[\sum_{v_i=1} \frac{-(H_i - H_0)^2}{2\sigma_i^2}\right] \times \exp\left[\sum_{v_i=0} \frac{-(H_i - H_0)^2}{2S^2}\right] \qquad (3.15)$$

Here P_G and P_B are the probability distributions of "good" and "bad" measurements, respectively. S is a large (but finite) number characterizing the standard deviation of "wrong" measurements (e.g., plausible range in which a wrong measurement could have survived the refereeing process). Notice that since we are given $\mathbf{v}$, there is no additional information in p, so it does not appear on the right-hand sides of equation (3.15).

In other contexts, the use of Gaussians in an equation like equation (3.15) might prompt squeals of horror from the illuminati: Don't Gaussians always underestimate the tail probabilities of real measurements? And won't subsequent results

therefore be quite sensitive to the Gaussian assumption? While the answer to the
first question is of course "yes", the answer to the second is in fact "no". As long
as S is chosen to be adequately large, equation (3.14) will never be dominated by
factors far out on the tails of the σ_i's, because the dominating probability soon
comes from the terms in the sum where the component of $\mathbf{v}$ is zero (so that S,
rather than σ_i is controlling). Relative insensitivity to tail probabilities is one of
the appealing features of this formulation.

The second and third factors in equation (3.14) are our priors on H_0 and p. It is
unlikely that our prior on p (the probability of a typical experiment being "correct")
depends on the value of, say, the Hubble constant, so, in fact, $P(p|H_0) = P(p)$.
(The issue is not whether a particular experiment's chance of correctness depends
on H_0—obviously it does—but, rather, whether the single value p that charac-
terizes the current state of experimentation generally is somehow coupled to the
expansion rate of the Universe—which seems unlikely!)

The fourth factor is also, on inspection, trivial: Given p, the probability of
a particular value for the vector $\mathbf{v}$ is just the product of a factor of p for each 1
component times a factor of $(1 - p)$ for each 0 component.

Rearranging the summations, we can now rewrite equation (3.14) as

$$P(H_0|D) \quad \propto \quad P(H_0) \sum_p P(p) \sum_{\mathbf{v}} \left[\prod_{v_i=1} P_{Gi}p \right] \left[\prod_{v_i=0} P_{Bi}(1-p) \right]$$

$$\propto \quad P(H_0) \sum_p P(p) \prod_i \left[pP_{Gi} + (1-p)P_{Bi} \right] \qquad (3.16)$$

The last proportionality actually sums over all 2^N possible values of $\mathbf{v}$ and turns
a 2^N computational problem into an N one! (It is with some shame that I admit
to having done some actual computer calculations, with $N = 15$, before this sim-
plification was forcefully pointed out to me by Kochanek.)

Equation (3.16) is a complete, computationally feasible, prescription for calculating the probability distribution for the desired value H_0 given the mutually
incompatible data. With no other information, one can take the priors $P(H_0)$ and
$P(p)$ to be uniform, and compute just the indicated sum (actually an integral since
p is continuous) over the indicated products. (If you have greater a priori faith
in your experimental colleagues, you can, of course, take a prior distribution for
$P(p)$ that is more skewed towards higher values of p.) There is some judgment
involved in the choice of S in equation (3.15), but over a wide range the answers
are typically insensitive to that choice.

One can see that the method is something like a maximum likelihood method
that attributes to each measurement not a Gaussian, but a weighted sum (with
weights p and $1 - p$) of a "good" Gaussian and a "bad" Gaussian. What makes

the method manifestly Bayesian, however, is that it then integrates over all values of p, weighted by the prior $P(p)$. The traditional equation (3.1) is the mean of the distribution obtained in the limit that $P(p)$ is a delta function at $p = 1$, and with a uniform prior $P(H_0)$.

You can also go back and sum the four inner factors in equation (3.14) over H_0 and $\mathbf{v}$ to get the probability distribution for the variable p, or you can sum over H_0, p, and those $\mathbf{v}$'s with $v_j = 1$ (or $v_j = 0$) to get the probability that experiment j is "good" (or "bad"). It is left as an exercise for the reader to derive the simpler forms analogous to equation (3.16) for these cases. In all cases, you get the constant in front by demanding that probabilities sum to 1.

Another easy exercise is to write down the formula for the probability that *none* of the experiments is good, that is, the total probability in the $\mathbf{v} = 0$ vector component. A value > 0.05 then indicates that *no* result is supported at the 5% confidence level minimally required for the reporting of scientific results.

Equation (3.16) is *not* a "tail-trimming" scheme that simply throws away outlier measurements. In some cases of actual data, the probability $P(H_0|D)$ will be multi-modal, with bumps of probability "protecting" certain outliers. In other cases, where there is a sufficient central core of mutually reinforcing values with mutually compatible error bars, the outliers will not be so protected. I have played around with the method on several data sets, including both published Hubble constant measurements (see § 3.5, below) and measurements of R_0 (the distance to the Galactic center, [11]). In my experience the method is robust, and it gives results that are justifiable in terms of common sense. [1]

One important feature of the method is worth pointing out explicitly: Suppose there is a body of consistent results clustered around one value, but also a maverick outlier at an inconsistent value. In the conventionally weighted average of equation (3.1), the maverick is able to draw the average as close as he wants to his or her value, no matter how good the evidence on the other side, simply by publishing unrealistically small error bars. By contrast, with equation (3.16), smaller error bars (after a certain point) yield *decreasing* weight for the maverick value, because it becomes a "bad" data point with increasingly high probability. There are good statistical, and also good sociological, reasons for adopting a combining procedure with this characteristic.

It is of course true that, for the formulation of the problem given in this paper, a maverick can "stuff the ballot box" by repeated, mutually consistent (but wrong) measurements. Eventually those values would prevail. This is because we have not allowed explicitly for the possibility of correlated systematic errors in different

---

[1] For R_0, I get $7.7 < R_0 < 8.4$ (Kpc) for the 95% confidence interval, and $7.9 < R_0 < 8.2$ as the 50% confidence interval. For H_0, see § 3.5.

experiments. A good problem for a graduate student would be to generalize the method (e.g., given some correlation matrix for the systematic errors of different experiments) so that repeated measurements, if highly correlated in their systematics, get only a single "vote". We have not done this.

3.4   ANOTHER VARIANT OF THE METHOD

Kochanek (private communication) has pointed out a variant method: Instead of assuming a single probability p applicable to the universe of experiments, one assumes an individual (though still unknown) probability p_i for each experiment, that is, a vector $\mathbf{p}$. Then, instead of equation (3.16), we have

$$P(H_0|D) \propto P(H_0) \sum_{\mathbf{p}} \prod_i P(p_i|H_0)[p_i P_{Gi} + (1 - p_i)P_{Bi}] \qquad (3.17)$$

Here the sum over $\mathbf{p}$ is actually a multidimensional integral over each of the p_i's individually, but each applies to only a single term in the product, so

$$P(H_0|D) \propto P(H_0) \prod_i \left\{ \int_0^1 dp_i\, P(p_i)[p_i P_{Gi} + (1 - p_i)P_{Bi}] \right\} \qquad (3.18)$$

If the priors on p_i are a simple function, then the integrals can be done explicitly. Uniform priors, for example, give the exceptionally simple result

$$P(H_0|D) \propto P(H_0) \prod_i \frac{1}{2}(P_{Gi} + P_{Bi}) \qquad (3.19)$$

Comparing this with equation (3.16) one sees that the difference is that the integral over p outside the product has been replaced by the average of p, namely $1/2$, inside the product.

Kochanek and I have debated whether equation (3.16) is superior to equation (3.19) or vice versa. The conceptual difference is that, with all experiments sharing a single value of p, equation (3.16) is able to use their mutual compatibility or incompatibility to estimate p in a non-trivial way, for example concentrating the probability distribution for p strongly near 1 for highly concordant measurements. Equation (3.19), by contrast, has no information on any p_i other than the prior, so it in effect uses $p_i \sim 1/2$. However, in numerical trials on actual data, there is not much difference in the inferred best-estimate values of H_0 for results of the two methods.

3.5   RESULTS FOR THE HUBBLE CONSTANT

Although the intent of this paper is to discuss the statistical method, rather than the details of its application to the problem of the Hubble constant, it is probably a good idea for me to show some concrete results. Figure 3.1 shows the probability distribution for the Hubble constant H_0 that derives from equation (3.15) with $S = 30$ and a set of 13 reputable measurements of H_0 taken from the literature. The included measurements were chosen to include a variety of different techniques (including Type Ia supernovae, Type II supernovae, novae, globular clusters, Sunyayev-Zel'dovich effect, surface brightness fluctuations, planetary nebulae, Virgo cepheids, Tully-Fisher, and $D_N - \sigma$) and—where controversy exists—the values from more than one competing group. While there are surely correlations among the systematics in several of the measurements (notably the cepheid calibration), there are also methods that are entirely uncorrelated (e.g., Type II supernovae). See [3] for some further details.

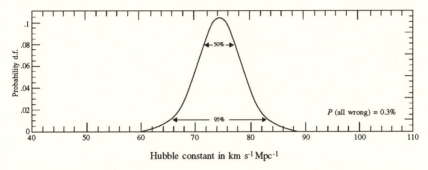

Figure 3.1: Bayesian probability distribution for the Hubble constant, derived from 13 different observations.

One sees that a reasonably narrow distribution is obtained, with a 50% confidence interval of $72 < H_0 < 77$, and 95% confidence interval $66 < H_0 < 82$ (km s^{-1} Mpc^{-1}). The probability that all of the observations are wrong (vector $\mathbf{v} = 0$) is 0.3%, a comfortingly small value. The superficial resemblance of the result to a Gaussian is a result, not an input constraint. (Indeed, if one arbitrarily decreases the quoted error bars of the experiments by a factor of 3, the "Gaussian" breaks up into a tri-modal distribution with "low", "medium", and "high" values for H_0.)

The abscissa of Figure 3.2 shows the input values and error bars assumed for the 13 measurements. (I am intentionally omitting a detailed list of references to avoid, or at least evade, an outburst of controversy. The point is to illustrate a statistical technique, not give a definitive review of H_0.) The ordinates of Figure 3.2 are the posterior estimates for the probability that each measurement has its $v_i = 1$

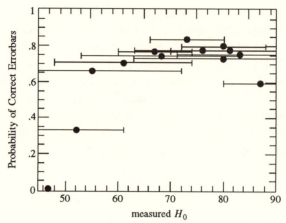

Figure 3.2: Projected onto the horizontal axis, this figure shows the values and error bars that were used to obtain the results of Figure 3.1. The vertical axis gives, for each measurement, the posterior estimate that it is "correct" (that is, compatible with its own error bars given the weight of evidence of all the other measurements).

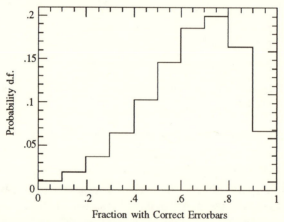

Figure 3.3: Posterior distribution for p, the prior probability that a given measurement of the Hubble constant is correct.

(is "right") rather than $v_i = 0$ (is "wrong"). One sees that most measurements are estimated to have about a 75% chance of being "right".

Finally, Figure 3.3 shows the posterior estimate for $P(p)$, the probability distribution for the prior probability of a random H_0 measurement being correct. It peaks at around 75%, but has a significant tail extending to zero. Note that the Bayesian nature of our method makes it, in effect, an *average* over this distribution of p, rather than assume any single value.

3.6 CONCLUSION

John Bahcall, a friend and mentor to me for more than 20 years, has always spurred those around him to choose important problems (which he often himself suggests), to get involved with real data (which he often himself supplies), and to bring the most powerful analytic tools to bear on the problem at hand. This paper is addressed at the last of these three imperatives. John's own work sets a mighty standard in all three respects, and his work will be studied for effective techniques and important conceptualizations long after the specific data are (as they should be) obsoleted by future observational advances.

BIBLIOGRAPHIC NOTES

Three good starting points for astrophysicists who want to learn some statistics, at a not totally unsophisticated level, are [8], [2], and [9], the latter two (alas) now somewhat out of date.

A good introduction to Bayesian inference, from an astronomical perspective is [7]. Several other papers in the same volume [4] will also be of interest. If you are a doubter who thinks that there is a "subjectivity" in Bayesian statistics that is not present in conventional frequentist approaches, you should look at [1], which shows that the equivalent (or worse) subjectivity is present in conventional approaches and is merely better camouflaged.

There is a series of conferences on "Maximum Entropy and Bayesian Methods" whose proceedings frequently contain papers of interest to astronomers. Recent volumes are [16] and [10].

An astronomical area in which Bayesian analysis has recently shown itself to be powerful is the determination of the mass of the Milky Way from orbital information on satellites. See [6] and, more recently (with some of the same viewpoints as this paper), [5].

BIBLIOGRAPHY

[1] Berger, J. O., & Berry, D. A. 1988, American Scientist, 76, 159.

[2] Eadie, W. T., Drijard, D., James, F. E., Roos, M., & Sadoulet, B. 1971, Statistical Methods in Experimental Physics (Amsterdam: North-Holland).

[3] Ellis, G. F. R., et al (10 authors) 1996, in J. Ehlers and S. Gottloeber, eds., Proceedings Dahlem Workshop on the Structure of the Universe (Chichester: J. Wiley and Sons).

[4] Feigelson, E. D., & Babu, G. J. (eds.) 1992, Statistical Challenges in Modern Astronomy (New York: Springer-Verlag).

[5] Kochanek, C. S. 1996, The Mass of the Milky Way Galaxy, ApJ, 457, 228.

[6] Little, B., & Tremaine, S. 1987, ApJ, 320, 493.

[7] Loredo, T. J. 1992, Statistical Challenges in Modern Astronomy, eds. E. D. Feigelson, & G. J. Babu (New York: Springer-Verlag), p. 275.

[8] Lupton, R. 1993, Statistics in Theory and Practice, (Princeton: Princeton University Press).

[9] Martin, B.R. 1971, Statistics for Physicists, (New York: Academic Press).

[10] Mohammed-Djafari, A., & Demoment, G. (eds.) 1993, Maximum Entropy and Bayesian Methods, proceedings of the XII International Workshop on Maximum Entropy and Bayesian Methods, Paris, 1992 (Dordrecht: Kluwer).

[11] Reid, M. J. 1993, ARA&A, 31, 345.

[12] Riess, A. G., Press, W. H., & Kirshner, R. P. 1995, ApJ, 438, L17.

[13] Riess, A. G., Press, W. H., & Kirshner, R. P. 1995, in preparation.

[14] Rybicki, G. B., & Press, W. H. 1992, ApJ, 398, 169.

[15] Rybicki, G. B., & Press, W. H. 1995, Phys. Rev. Lett., 74, 1060.

[16] Smith, C. R., Erickson, G. J., & Neudorfer, P. O. (eds.) 1992, Maximum Entropy and Bayesian Methods, proceedings of the XI Workshop on Maximum Entropy and Bayesian Methods, Seattle, 1991 (Dordrecht: Kluwer).

CHAPTER 4

LARGE-SCALE STRUCTURE IN THE UNIVERSE

NETA A. BAHCALL

Princeton University Observatory, Princeton, NJ

ABSTRACT

How is the universe organized on large scales? How did this structure evolve from the unknown initial conditions to the present time? The answers to these questions will shed light on the cosmology we live in, the amount, composition and distribution of matter in the universe, the initial spectrum of density fluctuations, and the formation and evolution of galaxies, clusters of galaxies, and larger scale structures.

I review observational studies of large-scale structure, describe the peculiar velocity field on large scales, and present the observational constraints placed on cosmological models and the mass density of the universe. While major progress has been made over the last decade, most observational problems in the field of large-scale structure remain open. I will highlight some of the unsolved problems that seem likely to be solved in the next decade.

4.1 INTRODUCTION

The existence of large-scale structure in the universe has been known for over half a century (e.g., [49, 62]). A spectacular increase in our understanding of this subject has developed over the last decade, led by observations of the distribution of galaxies and of clusters of galaxies. With major surveys underway, the next decade will provide new milestones in the study of large-scale structure. I will highlight what we currently know about large-scale structure, emphasizing some of the unsolved problems and what we can hope to learn in the next ten years.

Why study large-scale structure? We all want to know what the "skeleton" of our universe is, that is, how does the universe look on the largest scales we can investigate? It is natural to ask: on what scales does the universe become homogeneous and isotropic, as expected from the cosmological principle? In addition, we shall see that detailed knowledge of the large-scale structure provides constraints on:

- the formation and evolution of galaxies and larger structures;

- the cosmological model of our universe (including the mass density of the universe, the nature and amount of the dark matter, and the initial spectrum of fluctuations that gave rise to the structures seen today).

What have we learned so far, and what are the main unsolved problems in the field of large-scale structure? I discuss these questions in the sections that follow. I first list some of the unsolved problems.

There are many fundamental problems on which progress is likely to be made in the next decade. In fact, as we go through what is known about large-scale structure, you will see that most of what we want to know is a challenge for the future. Here is a partial list of some of the most interesting unsolved problems.

- Quantify the measures of large-scale structure. How large are the largest coherent structures? How strong is the clustering on large scales (e.g., as quantified by the power spectrum and the correlation functions of galaxies and other systems)?

- What is the topology of large-scale structure? What are the shapes and morphologies of superclusters, voids, filaments, and their networks?

- How does large-scale structure depend on galaxy type, luminosity, surface brightness? How does the large-scale distribution of galaxies differ from that of other systems (e.g., clusters, quasars)?

- What is the amplitude of the peculiar velocity field as a function of scale?

- What is the amount of mass and the distribution of mass on large scales?

- Does mass trace light on large scales? What is in the "voids"?

- What is the mass density, $\Omega_m \equiv \rho_m/\rho_{\text{crit}}$, of the universe?

- What is the baryon fraction of the universe, Ω_b/Ω_m?

- How does the large-scale structure evolve with time?

- What are the implications of the observed large-scale structure for the cosmological model of our universe and for structure formation? (e.g., What is the nature of the dark matter? Does structure form by gravitational instability? What is the initial spectrum of fluctuations that gave rise to the structure we see today? Were the fluctuations Gaussian?)

Two-dimensional surveys of the universe analyzed with correlation function statistics [31, 39] reveal structure to scales of at least $\sim 20h^{-1}$ Mpc. Large redshift surveys of the galaxy distribution reveal a considerably more detailed structure of superclusters, voids, and filament network extending to scales of ~ 50–$100h^{-1}$ Mpc ([29, 30, 16, 28, 22, 19, 27]). The most recent and largest redshift survey, the Las Campanas Redshift Survey ([35]; see also [36]), is presented in Figure 4.1; it reveals the "cellular" nature of the large-scale galaxy distribution. The upcoming Sloan Digital Sky Survey (SDSS), expected to begin operation in 1997 (see § 4.6), will provide a three dimensional map of the entire high-latitude northern sky to $z \sim 0.2$, with redshifts for approximately 10^6 galaxies. Figure 4.2 presents a simulation of one expected redshift slice from the SDSS survey; the slice represents $\sim 6\%$ of the total survey. This survey, and others currently planned, will provide the large increase in the survey volume required to resolve some of the unsolved problems listed above (see § 4.6 for details).

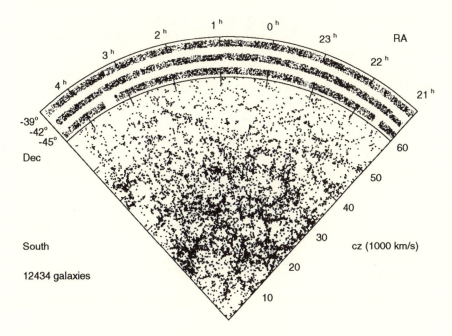

Figure 4.1: Redshift cone diagram for galaxies in the Las Campanas survey [35].

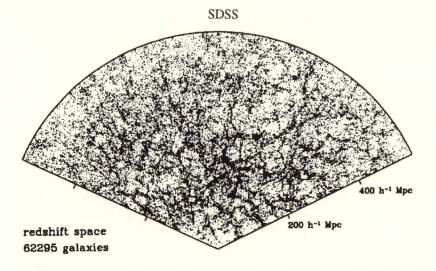

Figure 4.2: The redshift-space distribution of galaxies in a 6° thick slice along the SDSS survey equator from a large N-body cosmological simulation (cold-dark-matter with $\Omega_m h \simeq 0.24$; Gott, et al., in preparation). This slice contains approximately 6% of the 10^6 galaxies in the spectroscopic survey.

One of the exciting new developments in the last decade is the ability to compare the large-scale observations with expectations from different cosmologies using state-of-the-art numerical cosmological simulations [see the contribution by J. P. Ostriker in these proceedings]. Standard cosmological models generally succeed in reproducing the overall large-scale structure that we observe. In Figure 4.3, I present two examples of the observed versus simulated redshift cone diagrams (the latter for a cold-dark-matter, CDM, model). I leave it up to the viewer to decide which figure displays data and which displays a simulation, and how well they match. The "good" news is that a simple cosmological model, starting from a rather uniform distribution of matter after the big-bang, with minute density fluctuations as prescribed by a CDM power spectrum [see the presentations by J. P. Ostriker, P. J. E. Peebles, and P. Steinhardt], can represent the universe we see today reasonably well. This is a real success for cosmology! The "bad" news is that the quantitative match with the simplest standard models is not very good. Model parameters need to be adjusted, some in an ad-hoc manner, in order to fit the data quantitatively.

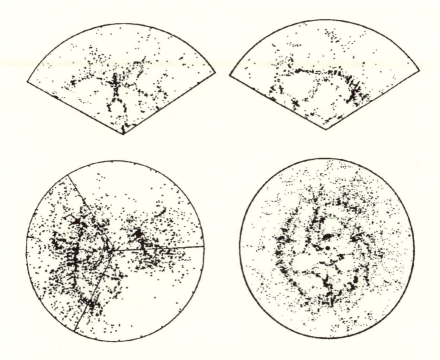

Figure 4.3: Redshift cone diagrams for observed (CfA survey) and simulated (CDM cosmology with $\Omega_m h = 0.3$) galaxy (mass) distributions in the present universe. Which are data and which are simulations? The similarity of the two indicates the underlying success of basic cosmology in reproducing the observed structure.

4.2   CLUSTERING AND LARGE-SCALE STRUCTURE

I summarize in Section 4.2.1 the ways that individual galaxies reveal the large-scale structure of the universe and describe in Section 4.2.2 the study of large-scale structure using clusters of galaxies.

4.2.1   *Galaxies and Large-Scale Structure*

The correlation function (of galaxies, of clusters of galaxies, of quasars) is a powerful statistical measure of the clustering strength on different scales. The two-point spatial correlation function, $\xi(r)$, is the net probability above random of finding a pair of objects, each within a given volume element dV_i, separated by the distance r. The total probability of finding such a pair is given by:

$$dP = n^2(1 + \xi(r))dV_1 dV_2 , \qquad (4.1)$$

where n is the mean density of objects in the sample.

The angular (projected) galaxy correlation function was first determined from the 2D Lick survey and inverted into a spatial correlation function by Groth & Peebles [31]. They find $\xi_{gg}(r) \simeq 20r^{-1.8}$ for $r \lesssim 15h^{-1}$ Mpc, with correlations that drop to the level of the noise for larger scales. This observation implies that galaxies are clustered on at least $\lesssim 15h^{-1}$ Mpc scale, with a correlation scale of $r_o(gg) \simeq 5h^{-1}$ Mpc, where $\xi(r) \equiv (r/r_o)^{-1.8} \equiv Ar^{-1.8}$. More recent results support the above conclusions (with possibly a weak tail to larger scales in the galaxy correlations). One of the recent determinations of the two-point angular galaxy correlation function, using the APM 2D galaxy survey [39, 23], is presented in Figure 4.4. The observed correlation function is compared with expectations from the CDM cosmology (using linear theory estimates) for different values of the parameter $\Gamma = \Omega_m h$. Here Ω_m is the mass density of the universe in terms of the critical density and $h \equiv H_0/100$ km s^{-1} Mpc^{-1} [see the contribution of Peebles]. The different $\Omega_m h$ models differ mainly in the large-scale tail of the galaxy correlations: higher values of $\Omega_m h$ predict less structure on large scales (for a given normalization of the initial mass fluctuation spectrum) since the CDM

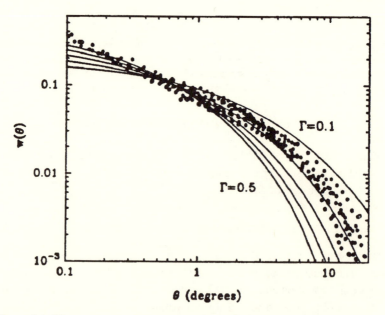

Figure 4.4: The scaled angular correlation function of galaxies measured from the APM survey plotted against linear theory predictions for CDM models (normalized to $\sigma_8 = 1$ on $8h^{-1}$ Mpc scale) with $\Gamma \equiv \Omega_m h = 0.5, 0.4, 0.3, 0.2$ and 0.1 [23].

fluctuation spectrum (see below) peaks on scales that are inversely proportional to $\Omega_m h$. It is clear from Figure 4.4, as was first shown from the analysis of galaxy clusters (see below), that the standard CDM model with $\Omega_m = 1$ and $h = 0.5$ does not produce enough large-scale power to match the observations. As Figure 4.4 shows, the galaxy correlation function requires $\Omega_m h \sim 0.15$–0.2 for a CDM-type spectrum, consistent with other large-scale structure data.

The power spectrum, $P(k)$, which reflects the initial spectrum of fluctuations [see the discussion by P. Steinhardt in these proceedings] that gave rise to galaxies and other structure, is represented by the Fourier transform of the correlation function (e.g., [45]). One of the recent attempts to determine this fundamental statistic using a variety of tracers is presented in Figure 4.5 ([42]; see also [56, 24, 41, 36]). The determination of this composite spectrum assumes different normalizations for the different tracers used (optical galaxies, IR galaxies, clusters of galaxies). The different normalizations imply a different bias parameter b for each of the

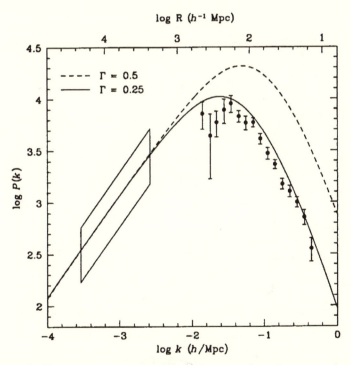

Figure 4.5: The power spectrum as derived from a variety of tracers and redshift surveys, after correction for non-linear effects, redshift distortions, and relative biases; from [42]. The two curves show the Standard CDM power spectrum ($\Gamma = 0.5$), and that of CDM with $\Gamma = 0.25$. Both are normalized to the COBE fluctuations, shown as the box on the left-hand side of the figure.

different tracers [where $b \equiv (\Delta\rho/\rho)_{\text{gal}}/(\Delta\rho/\rho)_m$ represents the overdensity of the galaxy tracer relative to the mass overdensity]. Figure 4.5 also shows the microwave background radiation (MBR) anisotropy as measured by COBE [50] on the largest scales ($\sim 1000h^{-1}$ Mpc) and compares the data with the mass power spectrum expected for two CDM models: a standard CDM model with $\Omega_m h = 0.5$ ($\Omega_m = 1, h = 0.5$), and a low-density CDM model with $\Omega_m h = 0.25$. The latter model appears to provide the best fit to the data, given the normalizations used by the authors for the different galaxy tracers.

The next decade will provide critical advances in the determination of the correlation function and the related power spectrum. The large redshift surveys now underway (§ 4.6) will probe the power spectrum of galaxies to larger scales than currently available and with greater accuracy. These surveys will bridge the gap between the current optical determinations of $P(k)$ of galaxies on scales $\lesssim 100h^{-1}$ Mpc and the MBR anisotropy on scales $\gtrsim 10^3 h^{-1}$ Mpc. This bridge

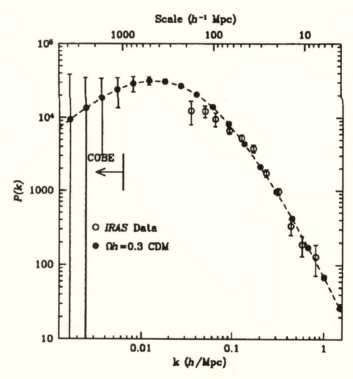

Figure 4.6: Estimate of the galaxy power spectrum that can be obtained with the Sloan Survey. The solid points and dashed line show the linear theory power spectrum of $\Omega_m h = 0.3$ CDM, with error bar estimates appropriate for the Sloan Survey. Open points are from [24]. The scales probed by COBE are shown.

will cover the critical range of the spectrum turnover, which reflects the horizon scale at the time of matter-radiation equality. An example of what may be expected from the Sloan survey is shown in Figure 4.6; a clear determination of $P(k)$ over a wide range of scales is expected in the near future. This will enable the determination of the initial spectrum of fluctuations at recombination that gave rise to the structure we see today and will shed light on the cosmological model parameters that may be responsible for that spectrum (such as $\Omega_m h$ and the nature of the dark matter). In the next decade, $P(k)$ will also be determined from the MBR anisotropy measurements on small scales ($\sim 0.1°$ to $\sim 5°$) [see P. Steinhardt's contribution in these proceedings], allowing a most important overlap in the determination of the galaxy $P(k)$ from redshift surveys and the mass $P(k)$ from the MBR anisotropy. These data will place constraints on cosmological parameters including $\Omega(= \Omega_m + \Omega_\Lambda), \Omega_m, \Omega_b, h$, and the nature of the dark matter itself.

4.2.2 Clusters and Large-Scale Structure

The correlation function of clusters of galaxies efficiently quantifies the large-scale structure of the universe. Clusters are correlated in space more strongly than are individual galaxies, by an order of magnitude, and their correlation extends to considerably larger scales ($\sim 50h^{-1}$ Mpc). The cluster correlation strength increases with richness ($\propto$ luminosity or mass) of the system from single galaxies to the richest clusters [10, 2]. Here "richness" refers to the number of cluster galaxies within a given linear radius of the cluster center and within a given luminosity range (e.g., [1]). The correlation strength also increases with the mean spatial separation of the clusters [53, 4]. This dependence results in a "universal" dimensionless cluster correlation function; the cluster dimensionless correlation scale is constant for all clusters when normalized by the mean cluster separation.

Empirically, two general relations have been found [12] for the correlation function of clusters of galaxies, $\xi_i = A_i r^{-1.8}$:

$$A_i \propto N_i , \tag{4.2}$$

$$A_i \simeq (0.4d_i)^{1.8} , \tag{4.3}$$

where A_i is the amplitude of the cluster correlation function, N_i is the richness of the galaxy clusters of type i, and d_i is the mean separation of the clusters. Here $d_i = n_i^{-1/3}$, where n_i is the mean spatial number density of clusters of richness N_i in a volume-limited, richness-limited complete sample. The first relation, equation (4.2), states that the amplitude of the cluster correlation function increases with cluster richness, i.e., rich clusters are more strongly correlated than poorer clusters. The second relation, equation (4.3), states that the amplitude of the cluster correlation function depends on the mean separation of clusters (or, equivalently, on their number density); the rarer, large mean separation richer clusters are more strongly correlated than the more numerous poorer clusters. Equations (4.2) and (4.3) relate to each other through the richness function of clusters, i.e., the number density of clusters as a function of their richness. Equation (4.3) describes a universal scale-invariant (dimensionless) correlation function with a correlation scale $r_{o,i} = A_i^{1/1.8} \simeq 0.4d_i$ (for $30 \lesssim d_i \lesssim 90h^{-1}$ Mpc).

There are some conflicting statements in the literature about the precise values of the correlation amplitude, A_i. Nearly all these contradictions are caused by not taking account of equation (4.2). When apples are compared to oranges, or the clustering of rich clusters is compared to the clustering of poorer clusters, differences are expected and observed.

Figure 4.7 clarifies the observational situation. The $A_i(d_i)$ relation for groups and clusters of various richnesses is presented in the figure. The recent automated cluster surveys of APM [20] and EDCC [40] are consistent with the predictions of equations (4.2) and (4.3), as is the correlation function of X-ray selected ROSAT clusters of galaxies [47, 7]. Bahcall and Cen [7] show that a flux-limited sample of X-ray selected clusters will exhibit a correlation scale that is smaller than that of a volume-limited, richness-limited sample of comparable apparent spatial

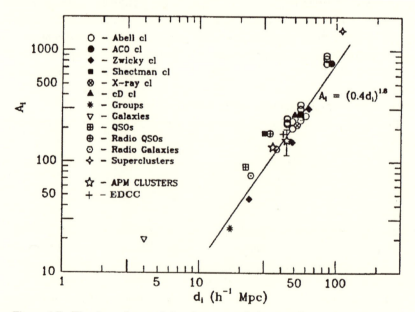

Figure 4.7: The dependence of the cluster correlation amplitude on mean cluster separation [12].

density since the flux-limited sample contains poor groups nearby and only the richest clusters farther away. Using the richness-dependent cluster correlations of equations (4.2) and (4.3), Bahcall and Cen [7] find excellent agreement with the observed flux-limited X-ray cluster correlations of Romer et al. [47]. Comparison of the observed cluster correlation function with cosmological models strongly constrains the model parameters (see below).

The observed mass function (MF), $n(> M)$, of clusters of galaxies, which describes the number density of clusters above a threshold mass M, can be used as a critical test of theories of structure formation in the universe. The richest, most massive clusters are thought to form from rare high peaks in the initial mass-density fluctuations; poorer clusters and groups form from smaller, more common fluctuations. Bahcall and Cen [6] determined the MF of clusters of galaxies using both optical and X-ray observations of clusters. Their MF is presented in Figure 4.8. The function is well fit by the analytic expression

$$n(> M) = 4 \times 10^{-5}(M/M^*)^{-1}\exp(-M/M^*)h^3 \text{ Mpc}^{-3}, \qquad (4.4)$$

with $M^* = (1.8 \pm 0.3) \times 10^{14}h^{-1} M_\odot$, (where the mass M represents the cluster mass within $1.5h^{-1}$ Mpc radius).

Bahcall and Cen [5] compared the observed mass function and correlation function of galaxy clusters with predictions of N-body cosmological simulations

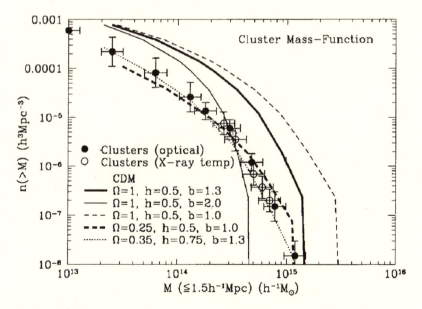

Figure 4.8: The mass function of clusters of galaxies from observations (points) and cosmological simulations of different $\Omega_m h$ CDM models [5, 6].

of standard ($\Omega_m = 1$) and nonstandard ($\Omega_m < 1$) CDM models. They find that none of the standard $\Omega_m = 1$ CDM models, with any normalization, can reproduce both the observed correlation function and the mass function of clusters. A low-density ($\Omega_m \sim 0.2$–0.3) CDM-type model, however, provides a good fit to both sets of observations (see Figs. 4.8–4.11). The constraints on Ω_m are model dependent; a mixed hot + cold dark matter model, for example, with $\Omega_m = 1$, is also consistent with the cluster data.

The observed cluster mass function is presented in Figure 4.8 together with the results of CDM simulations. The standard CDM model ($\Omega_m = 1$) is presented for several bias parameters b [where b represents the overdensity of galaxies relative to the mass overdensity; see § 4.2.1]; only the unbiased case, $b \sim 1$, is consistent with the COBE results. This unbiased model, however, is excluded by the observed mass function: it predicts a much larger number of rich clusters than is observed. The larger bias model of $b \sim 2$, while providing a more appropriate mean abundance of rich clusters, is too steep for the observed mass function; it is also incompatible with the COBE results.

The low-density, low-bias CDM model, with $\Omega_m \approx 0.25$ and $b \approx 1$, (with or without Λ) is consistent with the observed mass function (Fig. 4.8).

How does the CDM model agree with the observations of cluster correlations? Bahcall and Cen [5] used the N-body ($N \sim 10^7$) simulations to determine the

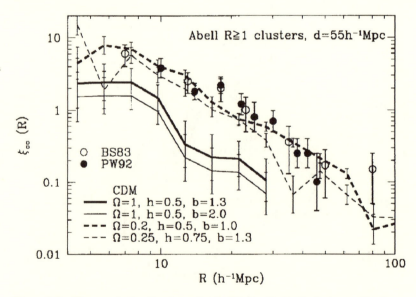

Figure 4.9: The correlation function of rich ($R \geq 1$) Abell clusters, with $d = 55h^{-1}$ Mpc, from observations [10, 43] and simulations [5].

model cluster correlation function as a function of cluster mean separations d. The CDM results for clusters corresponding to the rich Abell clusters (richness class $R \geq 1$) with $d = 55h^{-1}$ Mpc are presented in Figure 4.9 together with the observed correlations [10, 43]. The results indicate that the standard $\Omega_m = 1$ CDM models are inconsistent with the observations; they cannot provide either the strong amplitude or the large scales ($\gtrsim 50h^{-1}$ Mpc) to which the cluster correlations are observed. Similar results are found for the APM and EDCC clusters.

The low-density, low-bias model is consistent with the observed cluster correlation function. It reproduces both the strong amplitude and the large scale to which the cluster correlations are detected. Such a model is the only scale-invariant CDM model that is consistent with both cluster correlations and the cluster mass function.

The dependence of the observed cluster correlation on d was tested in the simulations ([5]). The results are shown in Figure 4.10 for the low-density model. The dependence of correlation amplitude on mean separation is clearly seen in the simulations. To compare this result directly with observations, we plot in Figure 4.11 the dependence of the correlation scale, r_o, on d for both the simulations and the observations. The low-density model agrees well with the observations, yielding $r_o \approx 0.4d$, as observed. The $\Omega_m = 1$ model, while also showing an increase of r_o with d, yields considerably smaller correlation scales and a much slower increase of $r_o(d)$.

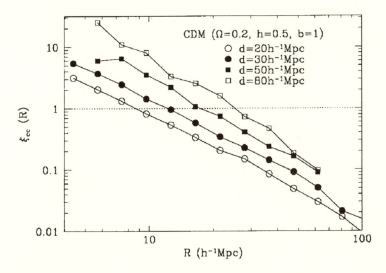

Figure 4.10: Dependence of the model cluster correlation function on mean cluster separation d.

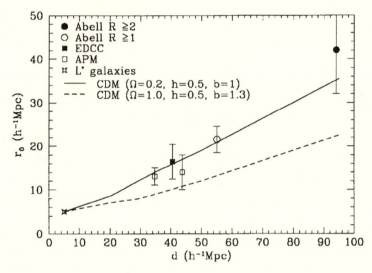

Figure 4.11: The correlation length of the cluster correlation function as a function of mean cluster separation, from observations and cosmological simulations [5].

What causes the observed dependence on cluster richness [Eqs. (4.2–4.3)]? The dependence, seen both in the observations and in the simulations, is most likely caused by the statistics of rare peak events, which Kaiser [33] suggested as an explanation of the observed strong increase of correlation amplitude from galaxies to rich clusters. The correlation function of rare peaks in a Gaussian field increases with their selection threshold. Since more massive clusters correspond to a higher threshold, implying rarer events and thus larger mean separation, equation (4.3) results. A fractal distribution of galaxies and clusters would also produce equation (4.3) (e.g., [53]).

4.3   PECULIAR MOTIONS ON LARGE SCALES

How is the mass distributed in the universe? Does it follow, on the average, the light distribution? To address this important question, peculiar motions on large scales are studied in order to directly trace the mass distribution. It is believed that the peculiar motions (motions relative to a pure Hubble expansion) are caused by the growth of cosmic structures due to gravity [see the presentations by Ostriker, Peebles, and Steinhardt]. A comparison of the mass-density distribution, as reconstructed from peculiar velocity data, with the light distribution (i.e., galaxies) provides information on how well the mass traces light [21, 52]. The basic underlying relation between peculiar velocity and density is given by

$$\vec{\nabla} \cdot \vec{v} = -\Omega_m^{0.6} \delta_m = -\Omega_m^{0.6} \delta_g / b \qquad (4.5)$$

where $\delta_m \equiv (\Delta\rho/\rho)_m$ is the mass overdensity, δ_g is the galaxy overdensity, and $b \equiv \delta_g/\delta_m$ is the bias parameter discussed in § 4.2. A formal analysis yields a measure of the parameter $\beta \equiv \Omega_m^{0.6}/b$. Other methods that place constraints on β include the anisotropy in the galaxy distribution in the redshift direction due to peculiar motions (see [52] for a review).

Measuring peculiar motions is difficult. The motions are usually inferred with the aid of measured distances to galaxies or clusters that are obtained using some (moderately-reliable) distance-indicators (such as the Tully-Fisher or $D_n - \sigma$ relations), and the measured galaxy redshift. The peculiar velocity v_p is then determined from the difference between the measured redshift velocity, cz, and the measured Hubble velocity, v_H, of the system (the latter obtained from the distance-indicator): $v_p = cz - v_H$.

A comparison between the density distribution of IRAS galaxies and the mass-density distribution reconstructed from the observed peculiar motions of galaxies in the supergalactic plane is presented in Figure 4.12 [21, 52]. The distribution of mass and light in this case appear to be similar.

A summary of all measurements of β made so far is presented in Figure 4.13

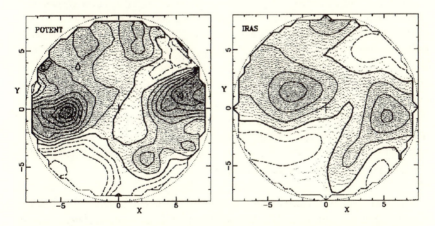

Figure 4.12: A comparison of the mass-density (reconstructed from velocity data) and galaxy-density fields. The left panel is the density field $\nabla \cdot v$ in the supergalactic plane reconstructed from peculiar velocity data. The right panel is the independently determined density field of IRAS galaxies. The smoothing in both panels is 1200 km s^{-1}. The axes are labeled in 1000 km s^{-1}. The Local Group sits at the center of each panel ([21, 52]).

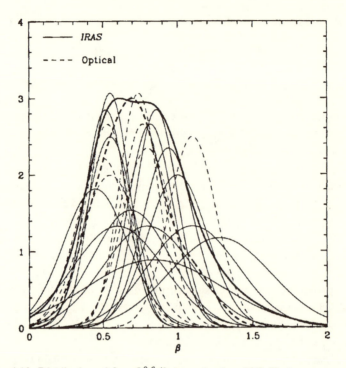

Figure 4.13: Distribution of $\beta = \Omega_m^{0.6}/b$ determinations [52]. Each measurement is shown as a Gaussian of unit integral with mean and standard deviations as given by the different observations. IRAS and optical determinations are plotted with different line types. The (renormalized) sum of all curves (optical and IRAS separately) are shown as the heavy curves.

[52]. The dispersion in the current measurements of β is very large; the various determinations range from $\beta \sim 0.4$ to ~ 1, implying, for $b \simeq 1$, $\Omega_m \sim 0.2$ to ~ 1. No strong conclusion can therefore be reached at present regarding the values of β or Ω_m. The larger and more accurate surveys currently underway, including high precision velocity measurements, may lead to the determination of β and possibly its decomposition into Ω_m and b (e.g., [18]).

Clusters of galaxies can also serve as efficient tracers of the large-scale peculiar velocity field in the universe [8]. Measurements of cluster peculiar velocities are likely to be more accurate than measurements of individual galaxies, since cluster distances can be determined by averaging a large number of cluster members as well as by using different distance indicators. Using large-scale cosmological simulations, Bahcall et al. [8] find that clusters move reasonably fast in all the cosmological models studied, tracing well the underlying matter velocity field on large scales. The clusters exhibit a Maxwellian distribution of peculiar velocities as expected from Gaussian initial density fluctuations. The model cluster 3-D velocity distribution, presented in Figure 4.14, typically peaks at $v \sim 600$ km s^{-1} and extends to high cluster velocities of ~ 2000 km s^{-1}. The low-density CDM model exhibits somewhat lower velocities (Fig. 4.14). Approximately 10% of all model rich clusters (1% for low-density CDM) move with $v \gtrsim 10^3$ km s^{-1}. A comparison of model expectation with the available data of cluster velocities is presented in Figure 4.15 [8]. The cluster velocity data are not sufficiently accurate at present to place constraints on the models; however, improved cluster velocities, expected in the next several years, should help constrain the cosmology.

Cen, Bahcall and Gramann [15] have recently determined the velocity correlation function of clusters in different cosmologies. They find that close cluster pairs, with separations $r \lesssim 10h^{-1}$ Mpc, exhibit strong attractive motions; the pairwise velocities depend sensitively on the model. The mean pairwise attractive cluster velocities on $5h^{-1}$ Mpc scale ranges from ~ 1700 km s^{-1} for $\Omega_m = 1$ CDM to ~ 700 km s^{-1} for $\Omega_m = 0.3$ CDM [15]. The cluster velocity correlation function, presented in Figure 4.16, is negative on small scales—indicating large attractive velocities, and is positive on large scales, to $\sim 200h^{-1}$ Mpc—indicating significant bulk motions in the models. None of the models reproduce the very large bulk flow of clusters on $150h^{-1}$ Mpc scale, $v \simeq 689 \pm 178$ km s^{-1}, recently reported by Lauer and Postman [37]. The bulk flow expected on this large scale

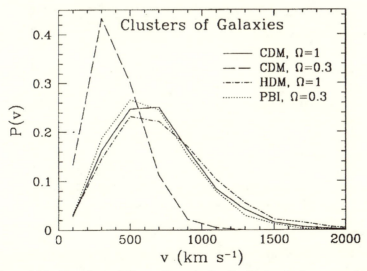

Figure 4.14: The cluster 3-D peculiar velocity distribution in four cosmological models [8].

is generally $\lesssim$ 200 km s^{-1} for all the models studied ($\Omega_m = 1$ and $\Omega_m \simeq 0.3$ CDM, and PBI; [15, 51]).

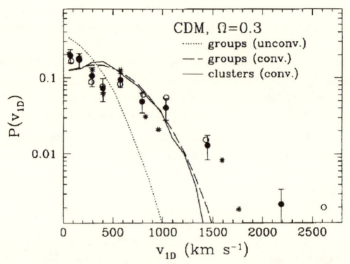

Figure 4.15: Comparison of the currently observed 1-D peculiar velocity distribution of groups and clusters of galaxies with expectation from an $\Omega_m = 0.3$ CDM model. The dotted line is the model expectation; the dashed and solid lines are the model convolved with the observed velocity uncertainties [8].

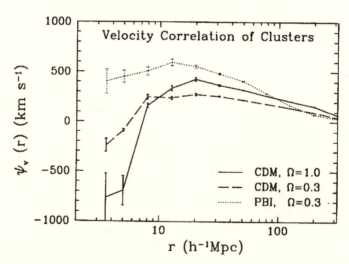

Figure 4.16: The velocity correlation function of rich ($R \geq 1$) clusters for three models [15].

4.4 DARK MATTER AND BARYONS IN CLUSTERS OF GALAXIES

The gravitational potential of clusters of galaxies has traditionally been studied with the aid of the observed velocity dispersion of the cluster galaxies and the virial theorem (e.g. [62, 44]). The cluster potential can also be estimated by the temperature of the hot intracluster gas as determined from X-ray observations (e.g., [32, 48]) and by the weak gravitational lens distortion of background galaxies caused by the intervening cluster mass [55, 34]. Sufficient data are now available from optical observations (galaxy velocity dispersion in clusters), X-ray observations (the temperature of the intracluster gas), and initial estimates from gravitational lensing, so that the results can be intercompared. The cluster mass determinations are, on the average, consistent with each other [3].

The optical and X-ray observations of rich clusters of galaxies yield cluster masses that range from $\sim 10^{14}$ to $\sim 10^{15} h^{-1} M_\odot$ within $1.5 h^{-1}$ Mpc radius of the cluster center. When normalized by the cluster luminosity, a median value of

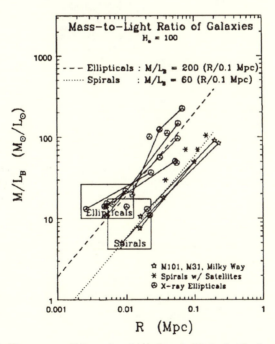

Figure 4.17: The mass-to-light ratio of spiral and elliptical galaxies as a function of scale. The large boxes indicate the typical ($\sim 1\sigma$) range of M/L_B for bright ellipticals and spirals at their luminous (Holmberg) radii. (L_B refers to total corrected blue luminosity.) The best-fit $M/L_B \propto R$ lines are shown [9].

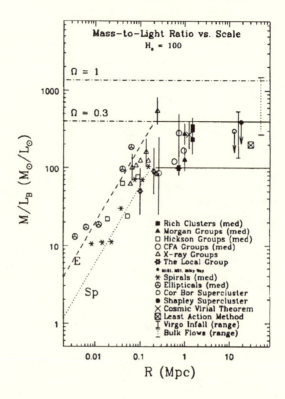

Figure 4.18: A composite mass-to-light ratio of different systems—galaxies, groups, clusters, and superclusters—as a function of scale. The best-fit $M/L_B \propto R$ lines for spirals and ellipticals (from Fig. 4.17) are shown. Typical 1σ scatter around median values are shown. Also presented, for comparison, are the M/L_B (or equivalently Ω_m) determinations from the Cosmic Virial Theorem, the Least Action method, and the range of various reported results from the Virgocentric infall and large-scale bulk flows (assuming mass traces light). The M/L_B expected for $\Omega_m = 1$ and $\Omega_m = 0.3$ are indicated [9].

$M/L_B \simeq 300h$ is observed for rich clusters. This mass-to-light ratio implies a dynamical mass density of $\Omega_{\rm dyn} \sim 0.2$ on $\sim 1.5h^{-1}$ Mpc scale. If, as suggested by theoretical prejudice, the universe has critical density ($\Omega_m = 1$), then most of the mass in the universe *cannot* be concentrated in clusters, groups and galaxies; the mass would have to be distributed more diffusely than the light.

A recent analysis of the mass-to-light ratio of galaxies, groups and clusters [9] suggests that while the M/L ratio of galaxies increases with scale up to radii of $R \sim 0.1$–$0.2h^{-1}$ Mpc, due to the large dark halos around galaxies (see Fig. 4.17), this ratio appears to flatten and remain approximately constant for groups and rich clusters, to scales of ~ 1.5 Mpc, and possibly even to the larger scales of super-clusters (Fig. 4.18). The flattening occurs at $M/L_B \simeq 200\text{--}300h$, corresponding to $\Omega_m \sim 0.2$. This observation may suggest that most of the dark matter is associated with the dark halos of galaxies and that clusters do *not* contain a substantial amount of additional dark matter, other than that associated with (or torn-off from) the galaxy halos, and the hot intracluster medium. Unless the distribution of matter is very different from the distribution of light, with large amounts of dark matter in the "voids" or on very large scales, the cluster observations suggest that the mass density in the universe may be low, $\Omega_m \sim 0.2$ (or $\Omega_m \sim 0.3$ for a small bias of $b \sim 1.5$).

Clusters of galaxies contain many baryons. Within $1.5h^{-1}$ Mpc of a rich cluster, the X-ray emitting gas contributes $\sim 3\text{--}10h^{-1.5}\%$ of the cluster virial mass (or $\sim 10\text{--}30\%$ for $h = 1/2$) [14, 59]. Visible stars contribute only a small additional amount to this value. Standard Big-Bang nucleosynthesis limits the mean baryon density of the universe to $\Omega_b \sim 0.015h^{-2}$ [57]. This suggests that the baryon fraction in some rich clusters exceeds that of an $\Omega_m = 1$ universe by a large factor [60, 38]. Detailed hydrodynamic simulations [60, 38] suggest that baryons are not preferentially segregated into rich clusters. It is therefore suggested that either the mean density of the universe is considerably smaller, by a factor of ~ 3, than the critical density, or that the baryon density of the universe is much larger than predicted by nucleosynthesis. The observed baryonic mass fraction in rich clusters, when combined with the nucleosynthesis limit, suggests $\Omega_m \sim 0.2\text{--}0.3$; this estimate is consistent with $\Omega_{\text{dyn}} \sim 0.2$ determined from clusters.

4.5 Is $\Omega_m < 1$?

Much of the observational evidence from clusters and from large-scale structure suggests that the mass density of the universe is sub-critical: $\Omega_m \sim 0.2\text{--}0.3$ (§§ 4.2–4.4). I summarize these results below:

- The masses and the $M/L(R)$ relations of galaxies, groups, and clusters of galaxies suggest $\Omega_m \sim 0.2\text{--}0.3$ (§ 4.4).

- The high baryon fraction in clusters of galaxies suggests $\Omega_m \sim 0.2\text{--}0.3$ (§ 4.4).

- Various observations of large-scale structure (the mass function, the power spectrum, and the peculiar velocities on Mpc scale) all suggest $\Omega_m h \sim 0.2$ (for a CDM-type spectrum) (§ 4.2).

- The frequent occurrence of evolved systems of galaxies and clusters at high redshifts ($z \sim 1$) also suggest a low-density universe, in which galaxy systems form at earlier times. No quantitative measures are yet available, however; this topic will be further clarified within the next several years using observations with HST, Keck, other deep ground-based observations, and deep X-ray surveys of clusters.

- If $H_o \sim 70$–$80\,\mathrm{km\,s^{-1}\,Mpc^{-1}}$, as indicated by a number of recent observations ([25] and references therein), then the observed age of the oldest stars requires $\Omega_m \ll 1$.

- Peculiar motions on large scales are too uncertain at the present time (suggesting density values that range from $\Omega_m \sim 0.2$ to ~ 1) to shed much light on Ω_m. Future results, based on larger and more accurate surveys, will help constrain this parameter (§ 4.3).

4.6   THE SDSS AND LARGE-SCALE STRUCTURE

I will now describe the general characteristics of the Sloan Digital Sky Survey (SDSS) (in Section 4.6.1) and then tell you (in Section 4.6.2) some of the exciting things we expect to do with this survey using clusters of galaxies, my own favorite. You should know that my choice of topics reflects primarily my own research activity and that the SDSS will do many extraordinary and important things with individual galaxies, quasars, and stars.

4.6.1   *The Sloan Digital Sky Survey*

Large surveys, especially in three dimensions, are vital for the study of clustering and large-scale structure of the universe. The surveys must cover huge volumes in order to investigate the largest-scale structure. They can provide accurate, systematic, and complete data bases of galaxies, clusters, superclusters, voids, and quasars, which can be used to study the universal structure and its cosmological implications. The surveys will allow both broad statistical studies of large samples as well as specific detailed studies of individual systems.

In this section, I summarize some of the fundamental contributions that large-scale surveys will offer for the study of clustering and large-scale structure. I will concentrate on the planned Sloan Digital Sky Survey (SDSS; [61]), which is the largest survey currently planned; however most of the discussion is generally applicable to large-scale imaging and spectroscopic surveys.

The Sloan survey will make a powerful contribution to the investigation of large-scale structure because of the survey's size, uniformity, and the high quality of its photometric and spectroscopic data. These characteristics will allow galaxy clustering in the present-day universe to be measured with unprecedented precision and detail. The survey will be able to resolve some of the unsolved problems listed in § 4.1, as well as address questions we have not yet thought to ask. There will be a lot of problems for students to solve with these data!

The SDSS will produce a complete photometric and spectroscopic survey of half the northern sky (π steradians). The photometric survey will image the sky in five colors (u$'$, g$'$, r$'$, i$'$, z$'$), using a large array of thirty 2048^2 pixel CCD chips; the imaging survey will contain nearly 5×10^7 galaxies to $g' \sim 23^m$, a comparable number of stars, and about 10^6 quasar candidates (selected on the basis of the five colors). The imaging data will then be used to select the brightest $\sim 10^6$ galaxies and $\sim 10^5$ quasars for which high-resolution spectra will be obtained (to $r' \sim 18^m$ and 19^m, respectively) using 640 fibers on two high-resolution ($R = 2000$) double-spectrographs. Both the imaging and spectroscopic surveys will be done on the same 2.5-meter wide-field (3° FOV) special purpose telescope located at Apache Point, NM. The imaging survey will also be used to produce a catalog in five colors of all the detected objects and their main characteristic parameters. In the southern hemisphere, the SDSS will image repeatedly a long and narrow strip of the sky, $\sim 75° \times 3°$, reaching 2^m fainter than the large-scale northern survey. The repeated imaging, in addition to allowing the detection of fainter images, will also be crucial for the detection of variable objects.

Figure 4.2 shows a simulated distribution in redshift space of galaxies expected from the Sloan Survey in a 6° thick slice along the survey equator. This figure, based on a cold-dark-matter cosmological model simulation, corresponds to approximately 6% of the galaxy redshift survey. The redshift histogram of galaxies in the simulated spectroscopic survey peaks at $z \sim 0.1$–0.15, with a long tail to $z \sim 0.4$.

Several thousand rich clusters of galaxies will be identified in the survey, many with a large number of measured galaxy redshifts per cluster (from the complete spectroscopic survey). Several hundred large superclusters will also be identified. This will provide at least a ten-fold increase in the number of clusters and superclusters over presently available samples.

Currently under construction, the test year for the SDSS system will begin observations in 1997. More detailed information on the SDSS project is provided in [61].

4.6.2 Clusters of Galaxies

Large-scale surveys such as the SDSS will provide a much needed advance in the systematic study of clusters of galaxies, which is currently limited by the unavailability of modern, accurate, complete, and objectively-selected catalogs of clusters, and by the limited photometric and redshift information for the catalogs that do exist. The SDSS will select clusters algorithmically. From the 10^6 galaxies in the spectroscopic sample, we can identify clusters by searching for density enhancements in redshift space. We can also identify clusters from the much larger photometric sample (5×10^7 galaxies) by searching for density enhancements in position-magnitude-color space. We expect to find approximately 4000 clusters of galaxies with a redshift tail to $z \sim 0.5$; many of the clusters will have a large number of measured galaxy redshifts per cluster (hundreds of redshifts for nearby clusters), and at least one or two redshifts for the more distant clusters. This complete sample of clusters can be used to investigate numerous topics in the study of clustering and large-scale structure. I briefly illustrate some of these topics below. The solutions of the problems outlined below will involve the participation of many graduate students!

Clusters as Tracers of Large-Scale Structure

Clusters are efficient tracers of the large-scale structure. Studies of the cluster distribution have yielded important results even though the number of cluster redshifts in complete samples has been quite small, ~ 100 to 300. The SDSS will put such studies on a much firmer basis because of the increased sample size (roughly an order of magnitude more redshifts), the objective and automated identification methods, and the use of cluster-finding algorithms that are less subject to projection contamination.

Catalogs of nearby superclusters have been constructed using Abell clusters with known redshifts, revealing structures on scales as large as $\sim 150h^{-1}$ Mpc [11]. There are claims for still larger, $\sim 300h^{-1}$ Mpc structures in the Abell cluster distribution [54]; these are still controversial [46]. The SDSS redshift survey will identify a considerably bigger sample of superclusters and will clarify the possible existence of structure on very large scales. The sizeable cluster sample should also clarify the relation between the cluster distribution and the galaxy distribution. Cluster peculiar velocities, obtained directly from distance indicators such as the $D_n - \sigma$ measurements and indirectly from the anisotropy of the cluster correlation function in redshift space, will provide information about the internal dynamics of superclusters and the mass distribution on large scales.

The high amplitude of the cluster correlation function, $\xi_{cc}(r)$, provided early evidence for strong clustering on large scales (§ 4.2). The complete sample of SDSS clusters, objectively defined from uniform and accurate photometric and spectroscopic data, will allow a measurement of ξ_{cc} free from systematic effects. We can extend the correlation analysis in several ways, examining in detail the richness dependence of ξ_{cc} as well as the dependence of the cluster correlation on other cluster properties (e.g., velocity dispersion, morphology).

The alignment of galaxies with their host clusters, and of clusters with their neighboring clusters, their host superclusters, and other large-scale structures such as sheets and filaments is of great interest. While the alignment of cD galaxies with their clusters is relatively well established [13], alignments at larger scales are still uncertain [17]. The SDSS will provide a very large sample for alignment studies, including accurate position angles and inclinations. The existence of large-scale alignments may place interesting constraints on the origin of structure in various cosmological models [58].

Global Cluster Properties

The complete cluster survey will allow a detailed investigation of intrinsic cluster properties. From the 2-D and 3-D information, we will be able to determine accurately such properties as cluster richness, morphology, density and density profile, core radius, velocity dispersion profile, optical luminosity, and galaxy content, and to look for correlations between these properties. The cluster catalogs will be matched with data in other bands, in particular the X-ray. This will allow a systematic detailed study of global cluster properties and their cross-correlation. Measurements of galaxy density, morphology, velocity dispersion, and X-ray emission will shed light on the nature of the intracluster medium and its impact on the member galaxies, and on the relative distribution of baryonic and dark matter. High surface mass density clusters will be candidates for gravitational lenses; we can search the photometric data for systematically distorted background galaxies, especially in the deeply-observed southern strip, and target these clusters for yet deeper imaging surveys with larger telescopes. These lensing studies will allow mapping out the dark matter distribution within the clusters [55, 34].

The large statistical sample of clusters will produce new insights into a number of issues associated with structure formation and evolution:

- We will be able to calculate velocity dispersions for ~ 1000 clusters with $z \lesssim 0.2$ compared to a few dozen clusters with reliable dispersion measurements currently available. For nearer clusters, we will have well-sampled galaxy density profiles and velocity dispersion profiles; these allow careful cluster mass determinations. The distributions of cluster masses and velocity dispersions, i.e., the cluster mass function and velocity function will be determined accurately.

- With optical luminosities, galaxy profiles, and velocity dispersion profiles, we can measure accurate mass-to-light ratios for a large sample of clusters.

Combined with X-ray observations of clusters, the baryon fraction can be determined more accurately. These provide constraints on Ω_m and the bias parameter of galaxies in clusters.

- The density and dispersion profiles also retain clues about the history of cluster formation. The frequency of subclustering and non-virial structures in clusters [26] tells us whether clusters are dynamically old or young. In a gravitational instability model with $\Omega_m = 1$, structure continues to form today, while in a low-Ω, open universe, clustering freezes out at moderate redshift. Cluster profiles and substructure statistics thus provide a diagnostic for Ω_m, a diagnostic which is independent of the direct measures of the mass density.

- The evolution of the cluster population provides another diagnostic for Ω_m. Current investigations are limited by the small numbers of known high-redshift clusters, and by the fact that high- and low-redshift clusters are selected in different ways. The uniform SDSS cluster sample will greatly improve the current situation. The number of clusters will be large; we will have redshifts for brightest cluster galaxies to $z \approx 0.5$ by extending the main redshift survey limit $\sim 1^m$ fainter for brightest cluster galaxy candidates, and for more distant clusters we can obtain fairly accurate estimated redshifts from the photometry alone (using the colors and apparent luminosity function of member galaxies). The southern photometric survey will probe the cluster population to redshifts above unity.

Large-Scale Motions of Clusters

Peculiar motions of clusters of galaxies can be determined from the SDSS survey using distance indicators such as Tully-Fisher or $D_n - \sigma$ for cluster galaxies, the cluster luminosity function, or the Brightest-Cluster-Member indicator. This will allow a determination of the large-scale peculiar motion of clusters to hundreds of Mpc.

4.7   SUMMARY

Observations of large-scale structure provide a powerful tool for determining the structure of our universe, for understanding galaxy and structure formation, and for constraining cosmological models. While extraordinary progress has been made over the last decade, many important unsolved problems remain. The large observational surveys currently underway provide a way of resolving some of these questions [see also the contribution by P. Steinhardt, these proceedings]. The interested student has a great opportunity to participate in fundamental discoveries.

What problems can we expect to solve in the next decade? I believe that the large redshift surveys of galaxies, clusters of galaxies, and quasars will provide a quantitative description of structure on a wide range of scales, from ~ 1 to $\sim 10^3 h^{-1}$ Mpc, including the determinations of the power spectrum, the correlation function, the topology, and other improved statistics (§§ 4.2,4.6). The dependence of these statistical measures on galaxy type, luminosity, surface brightness, and system type (galaxies, clusters, quasars) will also be determined and used to constrain cosmological models of galaxy formation. Accurate measurements on large scales of the peculiar motions of galaxies and clusters, and the mass distribution on large scales may take longer. Detailed X-ray surveys of galaxy clusters, combined with the large redshift and lensing surveys, will be able to resolve the problem of the baryon fraction in clusters and test suggested explanations (§ 4.4). Deep optical and X-ray surveys, using HST, Keck, ROSAT, ASCA, AXAF, SIRTF, plus new microwave background studies, and other deep ground-based and space-based observations, should allow a considerable increase in our understanding of the time evolution of galaxies, clusters of galaxies, quasars, and large-scale structure. Since the computed evolution of galaxy systems depends strongly on the assumed cosmology and especially on the mass density, with low-density models evolving at much earlier times than high density models, observations of the kind described here will provide some of the most critical clues needed to constrain the cosmological model over the next decade (§§ 4.5,4.6).

The quantitative description of large-scale structure and its evolution, when compared with state-of-the-art cosmological simulations to be available in the next decade (e.g. J. P. Ostriker, these proceedings), will enhance our chances of determining the cosmology of our universe. Will this goal be reached within the next decade? We will have to meet again in a decade and see.

Acknowledgments

It is a pleasure to thank J. N. Bahcall, D. Eisenstein, K. Fisher, J. P. Ostriker, P. J. E. Peebles, H. Rood, D. Spergel, and M. Strauss for stimulating discussions and helpful comments on the manuscript. The work by N. Bahcall and collaborators is supported by NSF grant AST93-15368 and NASA graduate training grant NGT-51295.

BIBLIOGRAPHIC NOTES

- **Bahcall, N. A. 1988, ARA&A, 26, 631.** A detailed review of the large-scale structure of the universe as traced with clusters of galaxies.

- **Bahcall, N. A., & Soneira, R. M. 1983, ApJ, 270, 20.** The first determination of the three-dimensional cluster correlation function and the richness dependence of the cluster correlations.

- **Geller, M., & Huchra, J. 1989, Science, 246, 897.** A clear and exciting summary of the large-scale structure observed in the CFA redshift survey.

- **Maddox, S., Efstathiou, G., Sutherland, W., & Loveday, J. 1990, MNRAS, 242, 43p.** A recent determination of the angular correlation function of galaxies using the APM survey.

- **Peebles, P. J. E. 1980, The Large-Scale Structure of the Universe; 1993, Principles of Physical Cosmology (Princeton: Princeton University Press).** The standard monographs from which students (and professors) have learned the subjects of large-scale structure and cosmology.

- **Strauss, M., & Willick, J. 1995, Physics Reports, 261, 271.** A comprehensive review of the density and peculiar velocity fields of nearby galaxies.

- **York, D., Gunn, J. E., Kron, R., Bahcall, N. A., Bahcall, J. N., Feldman, P. D., Kent, S. M., Knapp, G. R., Schneider, D., & Szalay, A. 1993, A Digital Sky Survey of the Northern Galactic Cap, NSF proposal.** A detailed summary of the planned Sloan Digital Sky Survey.

BIBLIOGRAPHY

[1] Abell, G. O. 1958, ApJS, 3, 211.

[2] Bahcall, N. A. 1988, ARA&A, 26, 631.

[3] Bahcall, N. A. 1995, in AIP Conference Proceedings 336, Dark Matter, eds. S. S. Holt, & C. L. Bennett (New York: AIP), p. 201.

[4] Bahcall, N. A., & Burgett, W. S. 1986, ApJ, 300, L35.

[5] Bahcall, N. A., & Cen, R. Y. 1992, ApJ, 398, L81.

[6] Bahcall, N. A., & Cen, R. Y. 1993, ApJ, 407, L49.

[7] Bahcall, N. A., & Cen, R. Y. 1994, ApJ, 426, L15.

[8] Bahcall, N. A., Gramann, M., & Cen, R. 1994, ApJ, 436, 23.

[9] Bahcall, N. A., Lubin, L., & Dorman, V. 1995, ApJ, 447, L81.

[10] Bahcall, N. A., & Soneira, R. M. 1983, ApJ, 270, 20.

[11] Bahcall, N. A., & Soneira, R. M. 1984, ApJ, 277, 27.

[12] Bahcall, N. A., & West, M. L. 1992, ApJ, 392, 419.

[13] Binggeli, B. 1982, A&A, 107, 338.

[14] Briel, U. G., Henry, J. P., & Boringer, H. 1992, A&A, 259, L31.

[15] Cen, R., Bahcall, N. A., & Gramann, M. 1994, ApJ, 437, L51.

[16] Chincarini, G., Rood, H. J., & Thompson, L. A. 1981, ApJ, 249, L47.

[17] Chincarini, G., Vettolani, G., & de Souza, R. 1988, A&A, 193, 47.

[18] Cole, S., Fisher, K. B., & Weinberg, D. H. 1994, MNRAS, 267, 785.

[19] da Costa, L. N., et al. 1988, ApJ, 327, 544.

[20] Dalton, G., Efstathiou, G., Maddox, S., & Sutherland, W. 1992, ApJ, 390, L1.

[21] Dekel, A. 1994, ARA&A, 32, 371.

[22] de Lapparent, V., Geller, M., & Huchra, J. 1986, ApJ, 302, L1.

[23] Efstathiou, G., Sutherland, W., & Maddox, S. 1990, Nature, 348, 705.

[24] Fisher, K. B., Davis, M., Strauss, M. A., Yahil, A., & Huchra, J. P. 1993, ApJ, 402, 42.

[25] Freedman, W., et al. 1994, Nature, 371, 757.

[26] Geller, M. 1990, in Space Telescope Symposium Series #4, Clusters of Galaxies, eds. W. Oegerle, M. Fitchett, & L. Danly (Cambridge: Cambridge University Press).

[27] Geller, M., & Huchra, J. 1989, Science, 246, 897.

[28] Giovanelli, R., Haynes, M., & Chincarini, G. 1986, ApJ, 300, 77.

[29] Gregory, S. A., & Thompson, L. A. 1978, ApJ, 222, 784.

[30] Gregory, S. A., Thompson, L. A., & Tifft, W. 1981, ApJ, 243, 411.

[31] Groth, E., & Peebles, P. J. E. 1977, ApJ, 217, 385.

[32] Jones, C., & Forman, W. 1984, ApJ, 276, 38.

[33] Kaiser, N. 1984, ApJ, 284, L9.

[34] Kaiser, N., & Squires, G. 1993, ApJ, 404, 441.

[35] Kirshner, R., Oemler, A., Schechter, P., & Shectman, S. 1995, preprint (to be published in ApJ).

[36] Landy, S. D., Shectman, S. A., Lin, H., Kirshner, R. P., Oemler, A. A., & Tucker, D. 1996, ApJ, 456, L1.

[37] Lauer, T., & Postman, M. 1994, ApJ., 425, 418.

[38] Lubin, L., Cen, R., Bahcall, N. A., & Ostriker, J. P. 1996, ApJ, 460, 10.

[39] Maddox, S., Efstathiou, G., Sutherland, W., & Loveday, J. 1990, MNRAS, 242, 43p.

[40] Nichol, R., Collins, C., Guzzo, L., & Lumsden, S. 1992, MNRAS, 255, 21.

[41] Park, C., Vogeley, M. S., Geller, M. J., & Huchra, J. P. 1994, ApJ, 431, 569.

[42] Peacock, J., & Dodds, S. J. 1994, MNRAS, 267, 1020.

[43] Peacock, J., & West, M. 1992, MNRAS, 259, 494.

[44] Peebles, P. J. E. 1980, The Large-Scale Structure of the Universe (Princeton: Princeton University Press).

[45] Peebles, P. J. E. 1993, Principles of Physical Cosmology, (Princeton University Press: Princeton).

[46] Postman, M., Spergel, D. N., Sutin, B., & Juszkiewicz, R. 1989, ApJ, 346, 588.

[47] Romer, A.K., Collins, C., Böhringer, H., Cruddace, R., Ebeling, H., MacGillivray, H., & Voges, W. 1994, Nature, 372, 75.

[48] Sarazin, C. L. 1986, Rev. Mod. Phys., 58, 1.

[49] Shapley, H. 1930, Harvard Obs. Bull. No. 874, 9.

[50] Smoot, G. R., Bennett, C. L., Kogut, A., Wright, E. L., Aymon, J., Boggess, N. W., Cheng, E. S., De Amici, G., Gukis, S., & Hauser, M. G. 1992, ApJ, 396, L1.

[51] Strauss, M., Cen, R., Ostriker, J. P., Postman, M., & Lauer, T. 1995, ApJ, 444, 507.

[52] Strauss, M., & Willick J. 1995, Physics Reports, 261, 271.

[53] Szalay, A., & Schramm, D. 1985, Nature, 314, 718.

[54] Tully, R. B. 1987, ApJ, 323, 1.

[55] Tyson, J. A., Wenk, R. A., & Valdes, F. 1990, ApJ, 349, L1.

[56] Vogeley, M. S., Park, C., Geller, M. J., & Huchra, J. P. 1992, ApJ, 391, L5.

[57] Walker, T. P., Steigman, G., Schramm, D. N., Olive, K. A., & Kang, H. S. 1991, ApJ, 376. 51.

[58] West, M. J., Oemler, A., & Dekel, A. 1989, ApJ, 346, 539.

[59] White, D., & Fabian, A. 1995, MNRAS, 273, 72.

[60] White, S. D. M., Navarro, J. F., Evrard, A., & Frenk, C.S. 1993, Nature, 366, 429.

[61] York, D., Gunn, J. E., Kron, R., Bahcall, N. A., Bahcall, J. N., Feldman, P. D., Kent, S. M., Knapp, G. R., Schneider, D., & Szalay, A. 1993, A Digital Sky Survey of the Northern Galactic Cap, NSF proposal.

[62] Zwicky, F. 1957, Morphological Astronomy (Berlin: Springer-Verlag).

CHAPTER 5

UNSOLVED PROBLEMS IN GRAVITATIONAL LENSING

R. D. BLANDFORD

California Institute of Technology, Pasadena, CA

ABSTRACT

Recent observations of gravitational lenses and their standard theoretical interpretation are briefly reviewed. The prospects for gravitational lenses resolving some of the central questions of cosmology and extragalactic astronomy are assessed. It is suggested that a large enough sample of radio rings could provide a convincing and direct measurement of the Hubble constant, but that it will be very difficult to measure q_0 cosmographically using lenses and that further significant limits on λ are likely to be quite subjective. The opportunity for measuring the strength of large scale density fluctuations in the cosmological mass distribution certainly exists but the task is bedeviled by uncertainties in the evolution of the sources. Studies of the mass distribution in clusters are at a very interesting stage and the prospects look good for extending them so as to witness cluster formation. Likewise, a direct determination of the redshift distribution of the faintest galaxies may be derivable using giant arcs and it may also be possible to trace morphological and stellar evolution in galaxies in this manner. Galaxy-galaxy lensing is proving to be a powerful technique for learning about the sizes of local galaxies and, ultimately, tracing their dynamical histories over a cosmological timescale. Some interesting possible candidates for cosmological dark matter have been or are capable of being ruled out using gravitational lensing arguments. Finally, microlensing arguments have been used to probe AGN emission sites though some uncertainties remain. Gravitational lensing still has many surprises in store for us.

5.1   INTRODUCTION

In recent years, the emphasis of research on gravitational lensing has shifted from discovery and demonstration of the effect to using lenses as powerful tools for addressing some of the most important questions of observational cosmology and extragalactic astronomy. Although gravitational lensing has yet to contribute any unique, generally accepted and important discoveries in this area, it has certainly provided some valuable corroborative evidence for existing cosmological claims and many of us believe that gravitational lensing is on the verge of fulfilling its longstanding promise of providing unique cosmological measurements. In this review, I will emphasize recent and imminent observations and draw attention to a few of the practical problems that will have to be solved to realize the full potential of gravitational lenses. It is to these issues of detail that the "smart, educated, nonspecialists, rather than our usual colleagues", to whom we were instructed to direct our articles, should pay close attention if they wish to assess the significance of a particular claim.

I have structured this article around eight "important, exciting and hopefully soluble problems" to whose solution gravitational lenses may have something to offer. Although I shall try to make this contribution self-contained, it is necessarily telegraphic and the interested reader should consult [44], [2] and [41] for many more details and a more representative bibliography. I begin with a brief overview of the physics of gravitational lensing.

5.2   GRAVITATIONAL LENS OPTICS

Gravitational lensing was predicted theoretically by Eddington, Lodge, Zwicky, Einstein and others, long before the first convincing example of this phenomenon, Q0957+561, was discovered [52]. Since this date, the tally of lenses has increased to (in my subjective opinion) nine secure plus six probable instances of multiple quasar imaging, five secure and two possible cases of radio rings and 25 secure plus at least 10 probable cases of rich clusters exhibiting arcs and arclets (Figs. 5.1, 5.2, 5.3).

In each of these cases elementary geometrical optics suffices to relate the observed angle θ (regarded as a two dimensional vector) to the reduced deflection angle $\alpha = (D_{ds}/D_s)\hat{\alpha}$ (where $\hat{\alpha}$ is the true deflection angle measured at the lens and D_{ds}, D_s are angular diameter distances measured to the source from the lens and the earth respectively. (Generalization to more than one lens plane is straightforward and sometimes necessary.) The general lens equation is then

$$\beta = \theta - \alpha(\theta), \qquad (5.1)$$

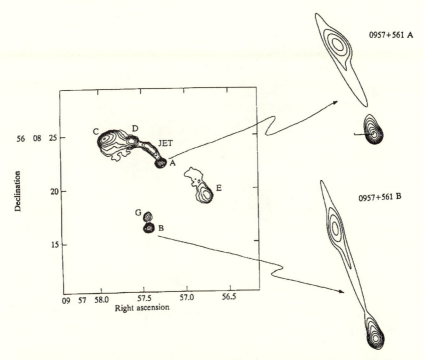

Figure 5.1: a) Radio Image of the first discovered gravitational lens 0957+561 [43]. The radio sources A, B, are coincident with two optical images of a quasar with a redshift $z_s = 1.4$ and separated by $\sim 6''$. The radio source G is identified with the nucleus of a giant elliptical galaxy at a redshift $z_d = 0.36$. This galaxy, in concert with its surrounding cluster is responsible for the lensing. b) VLBI images of the compact radio sources A, B that show that have a similar structure apart from a relative magnification and an inversion as predicted by theory[17]. Montage supplied by Hewitt.

where β measures the true angular position of the source. The actual gravitational deflector itself behaves like a transparent medium with a variable refractive index

$$n = 1 - 2\phi(\mathbf{r})/c^2 \,, \tag{5.2}$$

where ϕ is the Newtonian gravitational potential in Euclidean space. Equivalently, we can form the two dimensional gravitational potential $\psi = \int ds\phi$, where the integral is performed along the line of sight. ψ satisfies Poisson's equation, $\nabla^2\psi = 4\pi G\Sigma$, where Σ is the surface density. The true deflection angle is then

$$\hat{\alpha} = \frac{2\nabla\psi}{c^2} \,. \tag{5.3}$$

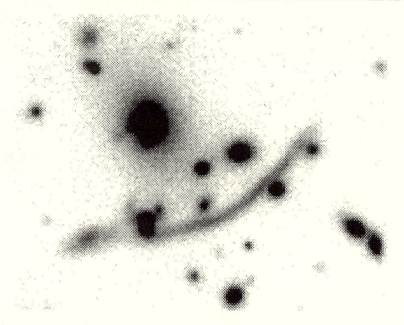

Figure 5.2: Portion of the rich cluster Abell 370, the first example to be discovered exhibiting giant gravitationally lensed arcs [24].

The matrix $\mu_{ij} = \partial\theta_i/\partial\beta_j$ can be computed and describes the local magnification. For a point source, its determinant μ describes the flux magnification as a lens does not change the intensity. This is all that is needed to model a given gravitational lens system.

Some lenses, e.g., stars, can be treated as point masses and $\hat{\alpha} = 4GM/D_d\theta c^2$, of order milliarcseconds for stars in the halo of our Galaxy and microarcseconds in cosmologically distant galaxies. In this case, we can solve a quadratic equation for $\theta(\beta)$ to obtain the locations of the two images formed on opposite sides of the lens. The magnification is given by $|(\theta/\beta)d\theta/d\beta|$ for each image and the combined magnification is given by

$$\mu = \frac{\beta^2 + 2\theta_0^2}{\beta(\beta^2 + 4\theta_0^2)^{1/2}}, \tag{5.4}$$

where $\theta_0 = (4GMD_{ds}/D_dD_sc^2)^{1/2}$. This is the functional form that is observed in Galactic microlensing events [35].

Galaxies and clusters can be most simply modeled using elliptical potentials and have characteristic deflections $\hat{\alpha} \sim 4\pi\sigma^2/c^2$, where σ is the one dimensional velocity dispersion assuming isotropy and virial equilibrium. For distant galaxies, this evaluates to $\sim 1 - 2''$; for clusters, $\hat{\alpha} \sim 10 - 20''$. Given a set of observa-

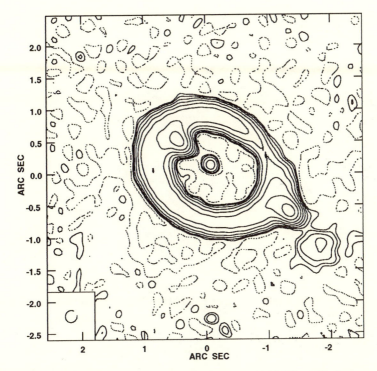

Figure 5.3: The first discovered radio ring MG1131+0456[7]. This image was obtained with the VLA at 8GHz and shows intensity contours as well as polarization.

tions of a multiple-imaged source, we have a typical astronomical inversion problem to recover the true source brightness and the lens potential. Much effort has been devoted to developing techniques to perform this inversion in such a way that the lens model is no more sophisticated than the quality of the observational data can support and the inferences no stronger than the model will allow. (Rigorous attention to these precepts should guard against making premature cosmological claims.) Lens components are incorporated into models including, with increasing sophistication: point masses, power law radial mass density variation, ellipticity, core radii, truncation radii and full model galaxy density distributions. (Note that even when the lens can be seen, the light need not directly trace the mass in either galaxies or clusters.)

A complementary approach to gravitational lensing [2] that contributes some useful intuition is to imagine a null geodesic congruence of rays originating at Earth and propagating backward in time and out into space. Matter inside the beam, both in discrete lenses and in intergalactic space, will contribute isotropic "Ricci" focusing of the rays down to a single focus conjugate to the observer. Mat-

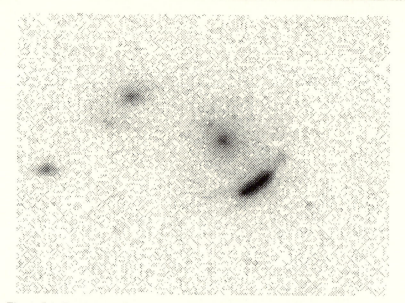

Figure 5.4: Luminous infrared source FSC10214+4724, revealing that much of the flux is almost certainly a gravitationally lensed background galaxy [11].

ter distributed anisotropically outside the beam will contribute a "Weyl" shear which will break the symmetry and cause two foci to form where the rays meet along roughly orthogonal lines. The loci of these foci for different ray congruences lie on caustic surfaces, to which the rays are tangent. Sources that lie close to caustics are strongly magnified and will appear bright and elongated on the sky (Fig. 5.4).

5.3   THE PROBLEMS

5.3.1   *How Old Is the Universe?*

Much effort (and even more publicity) has been devoted to attempts to measure the Hubble constant. It has long been recognized [39] that gravitational lenses provide a direct means for doing this that avoids the cumulative systematic errors associated with the cosmic distance ladder. Given a time delay, Δt, between the variation of the flux in two images separated by θ on the sky, the Hubble constant is given by

$$H_0 = \frac{K\theta^2}{\Delta t} \tag{5.5}$$

where K is a model-dependent constant of order unity. (Roughly half of this delay is contributed by the extra geometrical path length associated with the deflection;

the remainder can be thought of as being due to the reduction of the speed of light in the vicinity of the lens associated with the gravitational refractive index, cf. Eq. [5.2].) In Q0957+561 a strong case has been made for an observed time delay $\Delta t = 1.5$ yr [36]. (A recent analysis [46] has offered an alternative solution $\Delta t = 1.1$ yr. There is little point in debating the matter as Q0957+561A has just undergone a significant variation and we will know who is right in just over a year.) Adopting the former delay and adopting a naive model gives a value $H_0 \sim$ 60 km s^{-1} Mpc^{-1}. Unfortunately attempts to improve upon this have led to large changes in the quoted value. The history is instructive.

Firstly, it was realized that interposing a uniform sheet of matter is equivalent to adding a magnifying lens which will have the effect of overestimating H_0. Measuring the velocity dispersion of the central galaxy can, in principle, break this degeneracy. However, it is necessary to understand the relation between the measured stellar velocity dispersion and that for the dark matter as galaxy surface brightnesses generally fall off with radius faster than the total surface density. For example, an isothermal contribution to the galaxy mass, with an isotropic velocity dispersion, requires a gravitational field at radius r of $-\sigma^2 d \ln \rho / dr$. This depends upon the density gradient and so if the star density falls off more rapidly than the majority of the dark matter density, the dark matter velocity dispersion will be underestimated. In the case of Q0957+561, there were reports that the lensing mass was shared between two quite different clusters and also that an arc had been seen at a location incompatible with simple models. Both reports are now largely discounted. However, in view of the contentious nature of the field, it is doubtful that any determination of H_0 based solely upon 0957+561, however tightly modeled [20], will find general acceptance in the astronomical community at large.

For these reasons many suspect, that the best hope for gravitational lensing to provide a universally accepted value for H_0, lies with the radio rings. Here the major advantage is that instead of dealing with two or four discrete point images, we can use thousands of pixels whose surface brightnesses can be matched to constrain strongly the underlying potential. The lenses associated with rings are generally expected to be formed by isolated spiral galaxies and these are relatively simple to model compared with galaxy clusters. All of this has motivated the development of sophisticated image inversion routines [55]. In addition, there is no fear of microlensing altering the measured relative magnifications, as it can do for optical observations of quasars. Polarization measurements can add a powerful additional constraint on the modeling as both the degree and the direction of the linear polarization like the intensity should be unchanged by propagation through a gravitational lens.

Related large scale radio surveys are underway to try to discover more radio

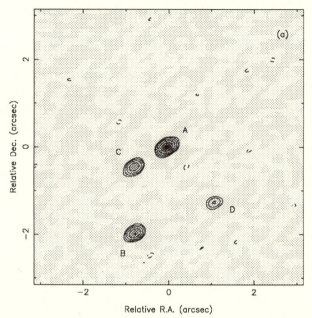

Figure 5.5: Recently discovered quadruply-imaged radio source 1608+656 [32].

rings. The CLASS survey in which $\gtrsim 10^4$ radio sources are mapped at a rate ~ 1 min^{-1} has already found and analyzed one excellent example, 1608+656 [32] and has several, additional strong candidates (Fig. 5.5). If, as is currently projected, this survey finds ~ 30 strong lens candidates, it is hoped that at least five of these will come with lens and source redshifts, strong variability (including "possible hyperlensing" of the central radio components) and simple potentials so that it will be possible to obtain several determinations of H_0. The largest systematic errors are likely to be in the lens modeling rather than the time delay measurements and much work needs to be done in validating inversion techniques for different types of sources. I would guess that it will take of order five lenses modeled to better than ten percent accuracy for this method to be taken seriously. Rings and arcs around individual galaxies are not just confined to radio astronomy. Recently, it has been convincingly demonstrated that the "most luminous object in the universe", FSC 10214+4724, is actually a gravitational lens with an extended arc encircling a galaxy [5] [11].

5.3.2 *What Is the Shape of the Universe?*

It appears to be even harder, in practice, to use gravitational lenses to probe the large scale geometry of the universe. Part of the problem is that it turns out that the distance ratio D_{ds}/D_s that appears in the definition of α is only weakly sensitive to Ω_0. For example with $z_d = 0.5$, $z_s = 1$, the ratio varies by ~ 0.03 as Ω_0 varies between 0.1 and 1. It will be hard to understand the radial variation of a galaxy or cluster velocity dispersion so that a useful determination of Ω_0 would follow.

If there is a large cosmological constant [6], then the fractional variation in the distance ratio can be somewhat larger. With our example, there is a ~ 15 percent variation in the distance ratio as we increase λ in a flat universe from 0 to 0.7. Yet another approach is to analyze multiply-imaged quasars like 2016+112 where two separate lenses are necessary to reproduce the image configuration. Here the relevant quantity to measure is the ratio of distance ratios, $D_{dd'}D_{os}/D_{od'}D_{ds}$, in obvious notation. This quantity varies by typically 0.06 in a model of 2016+112. If the lens model is highly non-linear and mandates a particular value for this ratio, then there is a slight chance that this could be turned into an Ω_0 measurement.

If the program to obtain several time delay determinations of the Hubble constant is successful, then this will also admit a possible Ω_0 determination. For a given model, the constant K in equation (5.5) depends upon Ω_0 and we can use the ratio of the time delays for two different lenses to eliminate H_0 and solve for Ω_0. For example, for one lens with $z_d = 0.5$, $z_s = 1$ and another with $z_d = 1$, $z_s = 3$, this ratio varies by roughly 20 percent as Ω_0 varies between 0.1 and 1.

A rather different method is to use the observed probability distribution of source redshifts, making due allowance for all the subtle observational selection effects and guessing the degree of cosmological evolution. This has mostly been employed to set an upper limit on λ [26]. (As λ increases, the amount of comoving volume per unit redshift for $z_s \sim 1 - 2$ also increases and more sources might be expected. The quoted bounds on λ vary between 0.5 and 0.9 depending upon different subjective judgment about what degree of evolution in the sources is unreasonable [28].

5.3.3 *What Is the Large Scale Distribution of Matter?*

A classical cosmological goal is to relate the cosmography—the geometry of the universe—to the dynamics as characterized by the large scale distribution of matter. One approach to this problem is to compare the observed distributions of multiple-imaged quasars, in both magnitude and separation with the expectations in different standard cosmologies. This probes the mass distribution on scales $\sim 0.01 - 1$ Mpc. On this basis, it has been deduced that too few large separation lenses are observed for a standard, COBE-normalized CDM cosmography (although this model has other shortcomings) [28] [54] [47].

Weak gravitational lensing of high redshift galaxies in "empty" fields provides a useful probe of matter on larger scales $\sim 1 - 100$ Mpc. If mass is distributed in a similar manner to luminous matter, then larger density concentrations will lead to a preferential tangential elongation of the galaxy images, through Weyl focusing (cf. § 5.2). (Giant voids will create a radial pattern.)

We can measure this effect statistically by defining a complex number (or equivalently a two dimensional vector) called the orientation χ whose magnitude is (to first order) the galaxy ellipticity, $1 - b/a$ and whose argument is twice the position angle of the minor axis of the image. We do this for each image and then average over many images to define the mean orientation or "polarization", p

$$p = <\chi> \tag{5.6}$$

$$= \mu_{22} - \mu_{11} - 2i\mu_{12} , \tag{5.7}$$

$$= \frac{2D_{ds}D_d}{D_s c^2} \int ds(\Phi_{,22} - \Phi_{,11} - 2i\Phi_{,12}) \tag{5.8}$$

to linear order and where the integral is along the line of sight and $\Phi_{,ij}$ are the second derivatives of the local Newtonian potential. This is then an estimator of the shear in the gravitational field. For a point mass lens, M at location θ':

$$p(\theta) = \frac{8M D_{ds} \exp[2i\phi]}{c^2 D_d D_s |\theta - \theta'|^2} , \tag{5.9}$$

where ϕ is the position angle of the relative vector $\theta - \theta'$. We can now sum over all masses along the line of sight to relate the rms polarization to an integral over the power spectrum of cosmological density fluctuations.

This effect can be sought statistically by measuring the shapes of tens of thousands of galaxies using giant CCDs under conditions of excellent seeing. Again using CDM as a benchmark and guessing a plausible redshift distribution for the galaxies, we expect to observe a local polarization in the galaxy images of strength ~ 0.04, of which a fraction of order a half, dependent upon the seeing, is likely to survive transmission through the atmosphere and the telescope. What is measured is an upper limit $\sim 0.02 \pm 0.02$ [31]. We appear to be on the brink of an interesting cosmological measurement, analogous to the microwave background fluctuation experiments. Archival HST images are proving to be very useful for this test.

5.3.4   How Are Rich Clusters of Galaxies Formed?

In recent years, the emphasis in rich cluster lensing has been on using the giant arcs (which are examples of strong lensing) and the arclets (where the lensing is weak) as probes of the cluster potential. When a cluster has a central surface density that exceeds the so-called critical density $\Sigma_c = c^2 D_s / 4\pi G D_d D_{ds} \sim 1\,\mathrm{g\,cm^{-2}}$, and has a modestly elliptical shape, a pair of imaginary caustic surfaces will form behind it. Unless the cluster is only a marginal lens, the outer caustic surface will be associated with a radial image merger. If we imagine a galaxy crossing the caustic from the outside, a pair of images will form with fluxes varying inversely with the square root of the distance the galaxy has moved past the caustic surface. As galaxies have finite sizes, their images will be stretched in an approximately radial direction. See [29] for the best example of a radial merger. Far more common, though, are the tangential arcs that will form as the imaginary galaxy crosses the inner caustic surface. Generically, this surface will contain four cusp curves (ribs) extending approximately parallel to the optic axis. When a galaxy crosses the caustic close to a cusp, one bright image will divide into three images. The most spectacular observed arcs are formed close to cusps [44].

One great advantage of studying the giant arcs is that as they are strongly magnified, it is often possible to observe them spectroscopically and measure the redshifts of the source galaxies. By probing arcs formed by galaxies at different redshifts, we have a basis for exploring the densest inner parts of rich clusters and this has been used to make quantitative models of the 2D potential ψ and the associated surface density Σ. In addition it is also possible to measure the cluster core radii, which reflects upon the formation history and the nature of the dark matter. So far, the results suggest that the gravitational core radii are significantly smaller than previously estimated from X-ray measurements and the surface density is correspondingly larger than anticipated. However, overall, the mass distribution in clusters is broadly consistent with the overall distribution of their bright constituent galaxies [30].

The weakly lensed arclets provide a complementary probe of the potential. Their surface density is generally too low and crowding by genuine cluster galaxies is too severe for them to be of much statistical use in the cluster cores. However, the opposite is true in the outer parts and some quite sophisticated inversion procedures have been devised to infer the lens mass distribution [50] [22]. Particularizing equation (5.9) to a single lens plane and generalizing to a continuous distribution of mass of surface density Σ, we obtain

$$p(\boldsymbol{\theta}) = \frac{2}{\pi \Sigma_c} \int \frac{d^2\theta' \Sigma(\boldsymbol{\theta'}) \exp[2i\phi]}{|\boldsymbol{\theta} - \boldsymbol{\theta'}|^2}. \tag{5.10}$$

This integral equation can be inverted to obtain

$$\Sigma(\boldsymbol{\theta}) = \frac{\Sigma_c}{2\pi} \int \frac{d^2\theta' p(\boldsymbol{\theta'}) \exp[-2i\phi]}{|\boldsymbol{\theta} - \boldsymbol{\theta'}|^2} \tag{5.11}$$

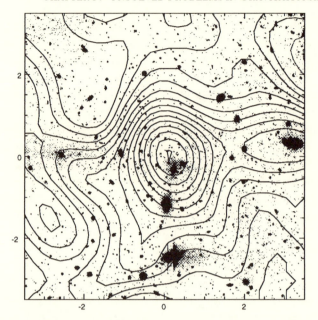

Figure 5.6: Image of MS1224+20 with superposed surface mass density profile deduced from observations of weak lensing [13].

[22]. In order to apply this inversion successfully, the boundaries of the observed area must be treated with care [45]. Overall, the results are broadly consonant with analyses of strong lensing, X-ray emission and microwave background dips and the technique holds considerable promise [13] (Fig. 5.6).

In a truly spectacular illustration of the power of the refurbished Hubble telescope, Kneib et al. (1995), have mapped the four giant arcs and over 100 arclets in the $z = 0.175$ rich cluster Abell 2218 (Fig. 5.7). The advantages over ground-based observation are immediately apparent; several spurious arcs are now seen as close pairs and many faint features can be matched on the basis of their resolved internal structure. There are two principal outputs, a detailed surface density model for the cluster which can be compared with X-ray and Sunyaev-Zel'dovich observations and a redshift distribution for the source galaxies. These are not independent. In addition, the redshift distribution depends upon the assumed cosmological model. The caustic surfaces are more complex than the simple description given above and individual galaxies have a significant effect upon the distortion. This provides a salutary reminder that many clusters contain substructure that does not get erased until they have had time to relax and, conversely, it demonstrates that it may be possible to use the amount of substructure as a measure of Ω_0 [42].

Large concentrations of matter also magnify through Ricci focusing. This has

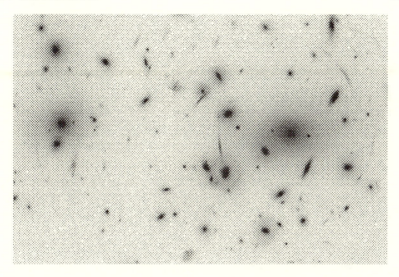

Figure 5.7: HST Observations of the rich cluster Abell 2218, revealing a rich and complex variety of arcs and arclets [25].

a curious effect on galaxy counts, though because the same areal magnification that enables us to observe intrinsically fainter galaxies at a given flux limit also reduces their surface density. The fractional increase in the flux-limited density is $(2.5r - 1)\mu$, where $r = d \log N/dm \sim 0.3$ is the galaxy count slope. The background galaxy density should therefore be slightly lower in cluster centers. Attempts to measure this effect, aided by color differences in galaxy counts, are encouraging [4]. The next step is surely to use all three approaches in concert [34]. Lensing by cosmologically distant compact groups of galaxies is also worth exploring using these techniques [12].

5.3.5   *When Did Galaxies Form and How Did They Evolve?*

Recent counts of galaxies using the Hubble and Keck telescopes [48] [10] have exacerbated a long-standing difficulty. There are too many of them. Put simply, the local density of bright galaxies is roughly 3×10^8 per Hubble volume $((c/H_0)^3)$ and there are ~ 3 comoving Hubble volumes out to a redshift $z \sim 2$ in an Einstein-De Sitter universe. By contrast, galaxies (or at least faint images that are widely supposed to be galaxies) have density $\gtrsim 7 \times 10^5$ sq deg^{-1}, and there are $\sim 40,000$ sq deg on the sky. Therefore there are at least 30 times as many supposed galaxies as might naively be anticipated and there is no sign that the counts are converging. If the galaxy counts continue to increase at the observed rate of 2 per magnitude, then either the proposed ultra-deep exposure with HST or $\sim$ 2 nights of observation under conditions of excellent seeing using the Keck telescope should suffice to attain Eddington's famous estimate of 10^{11} galaxies in the universe.

As explained clearly in Ellis' contribution there is no shortage of rationalizations of this discrepancy. We may live in an open universe, or one with a substantial cosmological constant (cf. § 5.3.2) which has a larger available comoving volume than an Einstein De-Sitter universe. Alternatively there may be a large population of galaxies with sub-L^* luminosity at intermediate redshifts that have disappeared either through merging or fading after the first generation of supernovae expel their gas. Some observers suspect that the local luminosity function will have to be augmented by a large population of low surface brightness galaxies. Whatever the true explanation, a key step towards unraveling this mystery will be to determine the faint galaxy redshift distribution which cannot be measured directly, because we can only obtain redshifts spectroscopically to a limit that is $\gtrsim$ 20 times brighter than we can detect objects. However, we may be able to use gravitational lensing to help us. The giant arcs are often magnified by $\gtrsim$ 20 and yet their sources are often no brighter than the faintest galaxies we can detect. The combination of spectroscopy and lens modeling should furnish a small but fair sample of these galaxies, selected in a calculable manner that depends upon the lensing geometry. In principle, this can be used to understand their evolution empirically. In addition the various weak lensing investigations are also quite sensitive to the adopted galaxy redshift distribution and so an improved understanding of the nature of the signal should also lead to a better understanding of the redshift distribution.

5.3.6   *How Big Are Galaxies?*

Although much effort has been expended in accumulating quantitative evidence for the presence of dark matter in the inner regions of galaxies astronomers have been vague about how much mass galaxies possess *in toto*. This is because it is quite difficult to measure the outer radii of galaxy halos and galaxy masses increase roughly in direct proportion to this scale. One technique for addressing this question is to measure weak galaxy-galaxy lensing. The shapes of individual galaxies should be correlated with their positions in the sense that a background source galaxy will be slightly, tangentially elongated with respect to a closer lens galaxy [49]. Again this effect can be sought statistically and the signal is naturally strongest relative to the noise for galaxy angular separations comparable with their projected dark halo radii $\sim 30''$. (It does not matter that galaxy halos overlap on the sky as the effect is small and the distortions due to different pairs of galaxies add linearly.) This effect has recently been sought in 24^m galaxies on a 10' square CCD image [3] and detected at the 2σ level. The signal is most compatible with a circular velocity for a L^* galaxy $\sim$ 220 km s^{-1} and (with lower significance) a typical halo size $\gtrsim$ 100 kpc. The prospects for improving the signal to noise by a factor 10 using HST to measure the galaxy ellipticities for galaxies up to ten times fainter seem excellent.

Radio rings are turning out to be excellent probes of the inner parts of galaxies. For example, 1654+134 is well modeled by a isothermal galaxy of velocity dispersion $\sim$ 220 km s^{-1} with a small core radius and $M/L \sim$ 20 [27]. As more and larger rings are detected, it should be possible to measure the radial variation of M/L and to set limits on typical galaxy sizes.

5.3.7 Of What Are Galaxies Made?

Gravitational lensing can also probe the granularity of the dark matter that dominates galaxy potential wells through the measurement of microlensing (cf. Alcock, this volume). For a halo of radius R, the effective critical density is $\Sigma_c \sim c^2/4\pi G R$. Now the mass that binds the halo at radius R must have projected surface density $V_c^2/4\pi G$. If a fraction of this mass is in discrete objects of mass M and density N then the optical depth to a strong microlensing event is given by

$$\tau \sim \int ds N \pi D_d^2 \theta_0^2 \sim \frac{\Sigma}{\Sigma_c} \sim \left(\frac{V_c}{c}\right)^2 \sim 5 \times 10^{-7}, \qquad (5.12)$$

(Any doubts that stellar microlensing is occurring should have been laid to rest by the spectroscopic investigation of a candidate microlensing event which showed no spectral variation as the source brightened [1].) It has been argued on the basis of the 50 microlensing events reported by EROS, MACHO and OGLE so far, that the MACHO fraction of our Galactic halo is less than 30 percent [16].

It has also been suggested that halos comprise $\sim$ 10^6 M$_\odot$ black holes. It is possible to argue against this in the case of the 0957+561 lens galaxy on the basis of the similarity of the two VLBI maps of the small scale radio jets [53] [15]. Apparently, less than 10 percent of the mass in this giant elliptical galaxy can be in the form of $\sim$ 10^6 M$_\odot$ black holes.

In fact, the main impact of microlensing is in classical astronomy and it should provide the most reliable estimates of the incidence of binarism in the disk, bulge and halo of our Galaxy. Already one microlensing event has been associated with a binary star. It should also prove to be a powerful method for detecting planets.

5.3.8 How Big Are AGN Ultraviolet Emission Regions?

My last problem concerns the structure of active galactic nuclei. The case for these being powered by massive black holes continues to strengthen (Rees, these proceedings), although the location and nature of the UV continuum source remains a matter of controversy. It is clear that it is possible to set an upper bound on the size of the central continuum source in those lensed quasars that exhibit microlensing variation, like 2237+031 [8]. In this case, a source size $\lesssim 10^{15}$ cm is indicated and adopting a simple model for emission from an accretion disk, it is possible to conclude that the continuum source is probably so compact that a non-thermal emission process is at work [37], but see [21]. It may ultimately be possible to observe color variation in accretion disks, if they are present [56].

Unfortunately, the non-Gaussian character of microlensing variability requires that an individual source must be observed through a very large number of variations to sample it properly. It is therefore worth remembering that most multiple imaged quasars have at least one image that ought to be subject to microlensing variation if it is sufficiently compact [57] and eventually it may be possible to average over lenses rather than time.

5.4   HOW MANY MORE SURPRISES WILL GRAVITATIONAL LENSES PROVIDE?

The history of gravitational lensing has (with a few conspicuous exceptions) been one of surprises. Even those phenomena that were anticipated theoretically, were greeted initially with scepticism. There is therefore a good precedent for taking seriously speculations for further manifestations of gravitational lensing. Among suggestions that deserve both more theoretical analysis and some observational follow up are:

(i) Repeating γ-ray burst sources. This would surely clinch the case for an extragalactic origin, although the converse statement can probably not be made safely (cf. Piran, these proceedings).

(ii) "Femtolensing" of γ-ray bursters, for example by intergalactic objects of mass $\sim 10^{17-20}$ g, where diffractive effects become important [19]. It is possible that a hypothetical population of interstellar comets might lead to diffractive scintillation of white dwarfs in the LMC [51].

(iii) Another effect which seems instrumentally marginal but is certainly worth seeking is to look for microlensing on the pixel scale in more distant galaxies than the LMC such as M31 (cf. Gould, these proceedings).

(iv) At the other end of the cost spectrum are proposals to launch very small spacecraft to determine the distances of microlensing sources parallactically [40].

(v) Extreme examples of microlensing. Perhaps the ultimate possibility is a cosmologically distant γ-ray pulsar crossing a smooth caustic formed by an intergalactic massive black hole. In principal, the magnification could be as large as $\sim 10^8$.

(vi) Exotic cosmologies admit spectacular possibilities like observing antipodes along general lines of sight [18]. There is no evidence that we inhabit such a universe, but it is surely worth keeping a slightly open mind about this possibility.

(vii) As well as limiting the density of $\sim 10^6$ M$_\odot$ black holes in galaxy halos, it is important to impose corresponding limits on the density of intergalactic black holes. A limit $\Omega_{BH} \lesssim 10^{-3}$ in the $\sim 10^6 - 10^9$ M$_\odot$ range is attainable by limiting the occurrence of "phantom" images of compact radio sources. This can be done fairly routinely as maps are made with the VLBA [23, 9].

(viii) Although their contribution to the overall mass density can be limited in other ways, it is still worth seeking optical evidence for cosmological defects like strings and domain walls. The discovery of just one example would have far-reaching implications.

(ix) The most extreme gravitational lens is a black hole. Given a suitably located and sufficiently compact source, gravitational lensing can in principle be used to verify the form of the Kerr metric. This could happen in either an AGN [38] or a compact object binary in our Galaxy [33]. Again the odds are very long, but the benefits of just one example would be dramatic.

One of many valuable lessons that I have learnt from John Bahcall is that one's view of what is worth considering theoretically should not be overly constrained by contemporary observations. If an idea can be explored theoretically, then theorists should be emboldened to do so as they are not usually so perceptive that they can anticipate where it will lead, guess how clever instrumentalists can be, and predict how devious the universe actually is. In this spirit, all of the above ideas are worth considering in more detail and I fully expect that some of them will bear observational fruit.

ACKNOWLEDGMENTS

Support under NSF contract AST93-23375 is gratefully acknowledged.

BIBLIOGRAPHIC NOTES

- **Blandford, R. D., & Narayan, R. 1992, ARA&A, 30, 311.** A review article describing cosmological application of gravitational lensing. Although the optical fundamentals are described *ab initio*, the treatment is compact and the beginning student would be well-advised to begin with Schneider et al. (1992). The discussion of some topics, especially microlensing and cluster lensing is now somewhat out of date.

- **Caroll, S. M., Press, W. H., & Turner, E. L. 1992, ARA&A, 30, 499.** Extremely clearly written and comprehensive discussion of the cosmological constant that gives a deceptively simple physical discussion of the reasons for possibly invoking its presence and a useful and original account of its observational implications. We still argue about whether or not it is needed and yet the observational upper bounds on its numerical value have not come down much recently.

- **Fort, B., & Mellier, Y. 1994, A&A Rev., 5, 239.** More recent review of gravitational lensing by two French observational astronomers who have contributed substantially to our understanding of cluster lensing, a topic on which it is particularly strong.

- **Kaiser, N., & Squires, G. 1993, ApJ, 404, 441.** Elegant derivation of a powerful method for converting observations of the weak polarization of the images of background galaxies induced by a foreground rich cluster into a surface density profile. Although the method has been augmented in recent years, the salient principles are clearly set out in this paper.

- **Paczyński, B. 1986, ApJ, 304, 1.** The paper which stimulated the current intense activity in microlensing. It is worth realizing that at the time this proposal was made, many astronomers were sceptical that microlensing would ever be measured and expected that it would be impossible to see the signal in the presence of the noise caused by intrinsic stellar variability. Not for the first time, the methodology of a branch of physics was brought to bear upon a peculiarly astronomical problem with spectacular results.

- **Refsdal, S. 1964, MNRAS, 128, 307.** First paper to spell out clearly how measurement of a time delay in variations of a background source could be used to measure the Hubble constant.

- **Refsdal, S. 1990, Gravitational Lensing, ed. Y. Mellier, B. Fort, & G. Soucail (Berlin: Springer-Verlag) p. 13.** Another, well-written review of the field of gravitational lensing co-authored by a Norwegian astronomer who championed the effect and anticipated many consequences, now observed, long before Walsh et al. (1979) discovered the first example of this phenomenon.

- **Schneider, P., Ehlers, J., & Falco, E. E. 1992, Gravitational Lenses (Heidelberg: Springer-Verlag).** The only modern, comprehensive textbook on the subject. This is necessary for any serious student of the field. Most of the complementary approaches to understanding the optics of gravitational lensing are discussed in some detail. The observational sections are already starting to look a bit out of date—a back-handed tribute to the book's influence!

- **Walsh, D. Carswell, R. F., & Weymann, R. 1979, Nature, 279, 381.** The first discovery of a *bona fide* gravitational lens. The background to this discovery is in itself an interesting tale recounted by Dennis Walsh elsewhere [Gravitational Lenses, ed. J. M. Moran, J. N. Hewitt, & K.-Y. Lo (Berlin: Springer-Verlag) 1989]. By the time they wrote the paper, the authors were in no doubt about what they had found.

BIBLIOGRAPHY

[1] Bennetti, S., Pasquini, L., & West, R. M. 1995, A&A, in press.

[2] Blandford, R. D., & Narayan, R. 1992, ARA&A, 30, 311.

[3] Brainerd, T. G., Blandford, R. D., & Smail, I. R. 1995, ApJ, submitted.

[4] Broadhurst, T. J., Taylor, A. N., & Peacock, J. A. 1995a, ApJ, 438, 49.

[5] Broadhurst, T. J., & Lehár, J. 1995b, preprint.

[6] Caroll, S. M., Press, W. H., & Turner, E. L. 1992, ARA&A, 30, 499.

[7] Chen, G., & Hewitt, J. N. 1995, preprint.

[8] Corrigan, R. T., et al. 1992, AJ, 102, 34.

[9] Dalcanton, J. J., Canizares, C. R., Granados, A., & Steidel, J. T. 1994, ApJ, 424, 550.

[10] Dickinson, M. 1995, private communication.

[11] Eisenhardt, P., Soifer, B. T., Armus, L., Hogg, D., Neugebauer, G., & Werner, M. 1995, preprint.

[12] De Oliveira, C. M., & Giraud, E. 1995, ApJ, 437, L103.

[13] Fahlman, G., Kaiser, N., Squires, G., & Woods, D. 1994, ApJ, 437, 56.

[14] Fort, B., & Mellier, Y. 1994, A&A Rev., 5, 239.

[15] Garrett, M. A. Calder, R. J., Porcas, R. W., King, L. J., Walsh, D., & Wilkinson, P. N. 1994, MNRAS, 270, 457.

[16] Gates, E. I., Gyuk, G., & Turner, M. S. 1995, Phys. Rev. Lett., 74, 3724.

[17] Gorenstein, M., et al. 1988, ApJ, 334, 42.

[18] Gott, J. R., Park, M-G., & Lee H. M. 1987, ApJ, 338, 1.

[19] Gould, A. 1992, ApJ, 386, L5.

[20] Grogin, N. A., & Narayan, R. 1995, preprint.

[21] Jaroszyński, M., Wambsganss, J., & Paczyński, B. 1992, ApJ, 396, L65.

[22] Kaiser, N., & Squires, G. 1993, ApJ, 404, 441.

[23] Kassiola, A. Kovner, I., & Blandford, R. D. 1991, ApJ, 381, 6.

[24] Kneib, J. P., et al. 1993, A&A, 273, 367.

[25] Kneib, J. P., Ellis, R. S., Smail, I., Couch, W. J., & Sharples, R. M. 1995, ApJ, in press.

[26] Kochanek, C. S. 1993, ApJ, 419, 12.

[27] Kochanek, C. S. 1995, ApJ, 445, 559.

[28] Kochanek, C. S. 1995, ApJ, in press.

[29] Mellier, Y., Fort, B., & Kneib, J.-P. 1993, ApJ, 407, 33.

[30] Miralda-Escudé, J., & Babul, A. 1995, ApJ, 449, 18.

[31] Mould, J. R., et al. 1994, MNRAS, 271, 31.

[32] Myers, S., et al. 1995, ApJ, 447, L5.

[33] Narayan, R., Piran, T., & Shemi, A. 1991, ApJ, 379, L17.

[34] Narayan, R., & Bartelman, M. 1995, preprint.

[35] Paczyński, B. 1986, ApJ, 304, 1.

[36] Press, W. H., Rybicki, G., & Hewitt, J. N. 1992, ApJ, 385, 404.

[37] Rauch, K. P., & Blandford, R. D. 1991, ApJ, 381, L39.

[38] Rauch, K. P., & Blandford, R. D. 1994, ApJ, 421, 46.

[39] Refsdal, S. 1964, MNRAS, 128, 307.

[40] Refsdal, S. 1990, Gravitational Lensing, ed. Y. Mellier, B. Fort, & G. Soucail (Berlin: Springer-Verlag) p. 13.

[41] Refsdal, S., & Surdej, J. 1994, Rep. Prog. Phys., 57, 117. General, up-to-date research report.

[42] Richstone, D., Loeb, A., & Turner, E. L. 1992, ApJ, 393, 477.

[43] Roberts, D. H., Greenfield, P. E., Hewitt, J. N., Burke, B. F., & Dupree, A. K. 1985, ApJ, 293, 356.

[44] Schneider, P., Ehlers, J., & Falco, E. E. 1992, Gravitational Lenses (Heidelberg: Springer-Verlag).

[45] Seitz, C., & Schneider, P. 1995, A&A, 297, 287.

[46] Pet, J., Kayser, R., Refsdal, S., & Schramm, T. 1994, A&A, 286, 775.

[47] Seljak, U. 1995, preprint.

[48] Smail, I., Hogg, D. W., Yan, L., & Cohen, J. G. 1995, ApJ, in press. Deep ground-based galaxy counts.

[49] Tyson, J. A. Valdes, F., Jarvis, J. F., & Mills, A. P. 1984, ApJ, 281, L59.

[50] Tyson, J. A., Valdes, F., & Wenk, R. A. 1990, ApJ, 349, 1.

[51] Ulmer, A., & Goodman, J. 1995, ApJ, 442, 67.

[52] Walsh, D. Carswell, R. F., & Weymann, R. 1979, Nature, 279, 381.

[53] Wambsganss, J., & Paczyński, B. 1994, AJ, 108, 1156.

[54] Wambsganss, J., Cen, R., Ostriker, J. P., & Turner, E. L. 1995, Science, 268, 274.

[55] Wallington, S., Narayan, R., & Kochanek, C. S. 1994, ApJ, 426, 60.

[56] Wambsganss, J., & Paczyński, B. 1991, AJ, 102, 864.

[57] Witt, H. J. Shude, M., & Schechter, P. L. 1995, ApJ, 443, 18.

CHAPTER 6

WHAT CAN BE LEARNED FROM NUMERICAL SIMULATIONS OF COSMOLOGY

JEREMIAH P. OSTRIKER

Department of Astrophysical Sciences, Princeton University, Princeton, NJ

ABSTRACT

Models for the growth of structure are mathematically precise. The initial conditions are unknown to us and are dependent on poorly understood fundamental physics. However, the subsequent developments (given the initial conditions) can be calculated, in principle, to arbitrary accuracy using well verified physical laws and numerical techniques. In fact (as contrasted to "in principle"), a lack of sufficient computing power and of codes which incorporate a sufficiently broad suite of standard physics has, until recently, restricted work to those aspects of the problem which are dominated by a collisionless, hypothetical "dark matter" acting only through Newton's laws of motion and gravity.

Recent technical developments allow one to make more realistic models. The problems amenable to treatment with currently available techniques include prediction of gravitational lensing, the properties and evolution of the Lyman-α clouds, and the properties and evolution of clusters of galaxies (especially X-ray properties). These model-dependent predictions can be compared with a rapidly growing base of observations in order to discriminate amongst proposed models. Even in these limited areas, the uncertainties in the current generation of simulations are considerably larger than the observational uncertainties. But, numerical accuracy is improving rapidly: the standard resolution ($N \equiv$ number of resolution elements per simulation) in published work having gone from $30^3 = 10^{4.4}$ to over $300^3 = 10^{7.4}$ volume elements over the last five years. The present state-of-the-art for hydrodynamical simulations is $512^3 = 10^{8.1}$ volume elements, with $1024^3 = 10^{9.0}$ reachable within two years. Galaxy formation itself, perhaps the most important and challenging problem, appears to be beyond the capabilities of present techniques, in part because it reduces to the currently unsolved problem of star formation. But, for the other areas mentioned above, we are at or near the level of resolution required to address interesting problems.

6.1 INTRODUCTION

Cosmology, as an intellectual discipline, has changed greatly in the last two decades. For nearly half of a century after the discovery of the Hubble law of expansion and the first elucidation of Friedman-Robertson-Walker models in the 1920's and 1930's, cosmology might have been caricatured as the science of two numbers, one representing the age and rate of expansion of the universe, h_0, and one number, q_0, representing the overall geometry and the prospects of eternal expansion or ultimate recollapse.

While inhomogeneities were of course well recognized (after all, we live on a planet, near a star, in a galaxy, which is part of a group of galaxies, etc., and these units are *ipso facto* inhomogeneities), these were thought to be of only local interest, since the universe was known (or believed) to be homogeneous and isotropic on large scales. The inhomogeneities were useful as providers of clocks (e.g., stellar ages) or meter sticks (luminosity distances to first brightest galaxies), which might enable one to determine the two numbers of interest to cosmologists. Aside from this utility, inhomogeneities were a nuisance, to be "averaged over" and not of intrinsic interest. As a measure of this mind set, we can study two classical texts, Peebles' "Physical Cosmology," [1] and Weinberg's "Gravitation and Cosmology" [2] published in 1971 and 1972, respectively. These were by far the most influential texts of their era and are still of great value, but they made little mention of the growth of structure in the universe. But, less than a decade later, by 1980, the first author's text, "The Large Scale Structure of the Universe," [3] was able to present, as a well developed subject, the empirical foundations for the study of irregularities in the observed galaxy distributions along with the beginnings of a theoretical discussion for how they may have developed. And by one decade later still, several texts [4, 5, 6] were beginning to present the, by now, standard pictures for the growth of structure.

These ideas might be summarized very roughly as follows. While we live in a universe that contains a dense photon field, as postulated in hot, big bang models, and also a normal baryonic component, with composition as determined by standard nucleosynthesis, there is also a substantial amount of weakly interacting "dark matter" of unknown composition. Furthermore, the universe is, as was thought for several decades, homogeneous and isotropic on the largest scales, but, somehow, perhaps related to the most basic fundamental ideas of physics, spatial perturbations were imprinted at a very low level ($\sim$ one part in 10^5) at early epochs. These perturbations grew (most efficiently in the weakly interacting, dark component) by gravitational instabilities (lumps contracting and voids expanding) until, at some epoch, the perturbations ($\delta\rho/\rho$) approached unit amplitude on some scales. A variety of highly nonlinear—but completely "classical"—phenomena then began to occur. For example, collapsing gaseous lumps would run into one another, be shock-heated and then radiate some of the resultant thermal radiation in the UV or X-ray bands. The emitted radiation, diffusing through space, could heat and ionize other gas, with innumerable other complex phenomena occurring subsequently. This process of initial perturbations inducing subsequent instabilities leads then to the currently observed universe with its bewildering variety of phenomena, from the 10^5 cm scale of neutron stars to the 10^{27} cm scale of the largest known structures.

The critical new point in this weltanschauung is that the developed complex structure is, in principle, completely predictable—at least in its statistical properties. If we knew the composition of the universe (massless particles, baryons and dark matter), the overall structure (essentially h_0 and q_0) and, most importantly, had a correct statistical description of the perturbations—all at some specified early epoch—then we could imagine putting all of this information into a perfect computer. Next, we could simply apply standard physical theory to compute forward, in order to rediscover the present, observed universe. Correspondingly, any major errors in our original prescription for matter, cosmological model or initial perturbations would be expected to lead to predictions at significant variance with observations.

This new paradigm seems, at first, quite fantastic. In fact, the approach has been amazingly successful! Most of the successes have been based on the realization that predictions on the largest scales should be most secure, since, in this domain, non-linearities will be minimized, and gravity—due to its long range nature—should reign as the supreme arbiter of structure formation.

As an example of how one proceeds, consider the spatial distribution of galaxies. One can Fourier transform the angular positions of galaxies from a large data base, such as was done with the APM survey [7], use information on the brightness of the galaxies to transform from 2-D to 3-D, and then compare with the power spectrum of mass fluctuations predicted by various models after they have been normalized (at early epochs and very large scales) to match the COBE measured microwave background fluctuations. Exercises such as these, although somewhat complicated by the need to make assumptions concerning the relation between galaxy number fluctuations and predicted mass fluctuations (spatial "bias"), are sufficiently definitive to have led to the abandonment of the simplest version of the standard "Cold Dark Matter" (CDM) scenario (cf. [8]). Since mass fluctuations lead to velocity perturbations (from the smooth Hubble flow), measurements of galaxy proper velocities provide an independent test of theories, and one which is slightly less uncertain (but "velocity bias" has been invoked as well as spatial bias). These comparisons have tended to reinforce those conclusions reached by considering only spatial fluctuations, so that now consideration has turned towards models which are variations on the previously standard picture.

But need one be limited to relatively large scale ($\Delta l > 8h^{-1}$ Mpc) measurements and theoretical predictions in the assessment of models? No. Many other phenomena, typically involving matter in gaseous phases, should be amenable to the new paradigm for testing theories. As but two examples, consider X-ray clusters [see N. Bahcall, this volume] and the Lyman-α forest lines [9].

The X-ray clusters typically emit enormous luminosities $\gtrsim 10^{44}$ erg/sec, enabling them to be seen to relatively large redshifts. We know enough now to have confirmed that this emission is primarily bremsstrahlung radiation emitted by hot gas (with line emission enhancements) that has been shock heated as it fell into the deep potential wells of these clusters. In a more sophisticated and accurate version of the story, the gas was heated as it fell, along with galaxies and dark matter, into regions where the initial gravitational potential wells were deepest. Thus, gas dynamical simulations, which allow for both the baryonic and dark matter components of the universe, can be made, with the results compared to observations (cf. [10, 11]). The distribution of luminosities, cluster sizes and temperatures (and the correlations amongst these quantities) are all computable. The results obtained to date are consistent with those determined from the studies of galaxies. Too many luminous clusters with temperatures that are too high are produced in the standard CDM model. As we will discuss subsequently, this powerful set of tests may now be applied to many different models which are variants of the original CDM picture.

Let us turn now to the other end of the mass, length, and time scales. While the clusters test the zero redshift properties of models for wavelengths on the $1h^{-1}$ Mpc $\rightarrow$ $50h^{-1}$ Mpc scales, the Lyman-α forest is typically seen at redshifts $z = 2$–4 and arises from fluctuations at the $0.050h^{-1} \rightarrow 3.0h^{-1}$Mpc scale. The data becoming available from Keck and other large telescopes are providing an enormous amount of information about fluctuations on these scales. Fully nonlinear simulations by several groups [12, 13, 14] have indicated that popular models do, in fact, reproduce much of this observed structure, with the primary uncertain parameter which goes into the theoretical simulations—the UV background radiation—being determinable directly from the observations via the "proximity effect." Thus, the Lyman-α forest provides another stringent set of tests for the models and, like the X-ray clusters, it relies on detailed hydrodynamic modeling, not just the dynamics of the dark matter.

As these examples make clear, the requirements for a physical situation such that cosmological hydro calculations can be made with some security are simply stated:

a) Most of the baryonic component should be in the gaseous, not condensed (e.g., stellar) state, so that uncertainties in the conversion of gas to stars or galaxies are not important.

b) The principal processes for heating and cooling the gas are either computable, in a self-consistent fashion (e.g., shock heating), or constrained by observations (e.g., the UV and soft X-ray backgrounds), so that the physical state of the gas can be computed with confidence.

c) The dynamical range for the computations must be sufficient to more than bridge the gap from a resolution scale of ΔL, less than the Jeans' length, to a box size L, large enough to provide a fair sample for the structures being studied. At the present time, with present hardware and software, the X-ray clusters and the Lyman-α clouds both (marginally) satisfy these requirements for phenomena that can be investigated with profit.

6.2 SIMULATION METHODS

6.2.1 *Specification of Models*

A specific simulation must adopt a set of parameters for the global cosmology [see the discussions by P. J. E. Peebles and P. Steinhardt in these proceedings], current Hubble constant h_0 ($h \equiv h_0/100 \mathrm{km\ s}^{-1}/\mathrm{Mpc}$), and current values for the mass density (in units of the critical density) in three forms: baryons, dark matter, and cosmological constant ($\Omega_b, \Omega_d, \Omega_\Lambda$ with $\Omega_{\mathrm{tot}} \equiv \Omega_b + \Omega_d + \Omega_\Lambda$). Further, the dark matter must be classified as to whether it was hot or cold (relativistic or non-relativistic) at decoupling. If there was a period during which the dark matter was cold, i.e., non relativistic and contained most of the mass density before decoupling (the requirement being that the predominant species had a mass particle above a few eV), then gravitational perturbations can grow in this component, while the baryonic gas is still frozen, mechanically coupled to the much more uniform relativistic photon fluid via the Compton drag process. Since most investigators adopt the value for the baryon density, fixed by conventional light element nucleosynthesis ($\Omega_b h^2 = 0.0125 \pm 0.025$, [15]), Ω_b is not treated as a separate parameter once h has been specified. Typically, the models are broken into two categories—flat or open—dependent on whether Ω_{tot} equals unity or is less than unity (and typically in the range $0.1 \leq \Omega_{\text{tot}} \leq 0.4$).

Then one specifies the nature of the perturbations imposed at early epochs, dividing them into "Gaussian" (in the distribution of amplitudes with phases random) or "non-Gaussian" (either locally non-Gaussian, or globally with phases correlated over a horizon, as with cosmic strings). Then one specifies the nature of the perturbations, which are conventionally divided into "adiabatic" (e.g., a simple, local compression or expansion of all components) or "isocurvature" variants (where matter and radiation are oppositely perturbed to leave a variation in entropy per baryon but maintain uniformity in the total energy density). Finally, the spectrum of perturbations is specified, with power law initial states $P_k \propto k^n$ being popular and the typical choice of n close to the standard, $n = 1$, [16], for which perturbations enter the horizon, at early times, with fixed amplitude.

At this point the reader may be bewildered at the range of choices permitted, and think that, with so many apparently free parameters, the subject is almost like a game without rules, or with rules so loosely specified that anyone can claim to have won at any time. Such an unconstrained game is not worth playing. In fact, the range of models considered is strongly circumscribed by both the commendable taste of scientists for simplicity and the strong constraints provided by observation. Specifically, most current investigation is limited to models that are defined by three parameters which are all variants of the "standard" cold dark matter scenario.

It is useful to first define that standard scenario precisely [see the discussion by Steinhardt in these proceedings], even though it is almost certainly incorrect. Motivated by earlier dimensional analysis [16] and following the more recent, simple notions of an early inflationary period [17], one assumes a spectrum of perturbations with $n = 1$ precisely. Only one (cold) type of non-baryonic dark matter is presumed to exist and the cosmological constant is taken as zero, with the total producing a flat universe: $\Omega_{\text{tot}} = \Omega_b + \Omega_{\text{CDM}} = 1$. Such a model is defined by two parameters, the amplitude, (symbolically) A, and the Hubble constant h. These parameters can be fixed by direct observations: A can be determined from microwave background fluctuations measured by the COBE satellite and h from direct measurements of the Hubble constant. Since most recent estimates of the latter are in the range $h = 0.70 \pm 0.15$, it is difficult to credit very low values of h which are preferred for standard CDM. As a compromise, the "standard" CDM scenario modelers have tended to adopt $h = 0.5$ (but see Bartlett, et al. [18] who presents arguments for a much lower value), which yields an age of less than 13 billion years that is marginally consistent with the age determinations for the oldest galactic stars [19]. The model defined in this fashion fails to match observations on several counts (summarized, with references, in [8]).[1] These include an inability to match the properties of the clusters of galaxies, too high a small scale velocity dispersion and the wrong slope for the large scale galaxy-galaxy correlation function (see also N. Bahcall in these proceedings).

This failure, the disagreement between calculation and observation, has led to detailed examination of several variants of the CDM scenario that are defined by three parameters $(A,\ h,\ p)$. For the "tilted" scenario "TCDM," p represents $n - 1$, the slope of the power spectrum; for the open scenario "OCDM," p represents the value of $\Omega_t\ <\ 1$; for the Λ model "LCDM," p represents Ω_Λ with flatness maintained ($\Omega_{\text{tot}}\ =\ \Omega_b + \Omega_{\text{CDM}} + \Omega_\Lambda\ =\ 1$); and for the mixed dark matter model, p represents the mass in the neutrino component Ω_ν (with $\Omega_{\text{tot}} = \Omega_b + \Omega_{\text{CDM}} + \Omega_\nu\ =\ 1$). One way to specify these three parameter models is to use the prescription advocated in [20], relying on three of the more accurately defined observational constraints:

(1) from CBR fluctuations (primarily the COBE results): $A \times (1 \pm 8\%)$

(2) from the abundance of rich clusters: $\eta \equiv \sigma_8 \Omega^{0.56} = 0.56 \times (1 \pm 11\%)$

(3) from the slope of the galaxy-galaxy correlation function: $\Gamma \equiv \Omega_m h = 0.25 \times (1 \pm 20\%)$.

The second constraint has been obtained by Bahcall, Cen, and Ostriker [21] and by White et al. (1993) [22], the third by Peacock and Dodds [23]. Using these three constraints, the variant models denoted above are fairly well-tied down to give power spectra which are observationally fixed at both short and long wavelengths. All properties of the model then are specified to an accuracy of approximately 15% [the accuracy of the original specification of the model—see (1), (2), (3) above], so an accurate calculation of their properties followed by comparison to observations (at least as accurate as $\pm 15\%$) might lead to a statistically significant conflict, i.e., falsification of the model.

Now a few words on technical details. The power spectrum in the linear domain must be computed using a "transfer function" which tracks the growth of the amplitude in waves of each component (Ω_b, Ω_ν, Ω_{CDM}, Ω_γ) through decoupling to the point when the numerical simulation is to begin. A clear and well motivated discussion of transfer functions is presented by [5]. A commonly used source for the CDM transfer function is [24]. At the current time, there are available, by anonymous FTP, codes for computing transfer functions for a broad variety of models from [25].

---

[1] No attempt to provide a comprehensive review of the literature will be made in this brief paper, which will emphasize the Princeton contribution to the subject (as best evaluated by the reviewer) without prejudice to the preponderance of excellent work done elsewhere.

Finally, one must specify the initial mass distribution at some starting epoch. This can be done most simply by displacing particles from a regular grid [26] with velocities determined by the Zeldovich approximation. Recently, more sophisticated "glass" initial conditions have come into vogue (cf. [27] for a recent example).

6.2.2 Physical Processes and Numerical Methods

Hydro

One must make an early choice between a particle based, Lagrangian approach (e.g., SPH [28]) or a mesh based Eulerian model (cf. for example, [29]). This is not the place for a detailed discussion of the merits and drawbacks of the two approaches; they are addressed (with numerous references to the technical literature) in the context of comparisons between methods addressing a realistic simulation, by [30]. In any case, the continual drive to ever greater dynamic range has led to the development of hybrid schemes which combine the useful attributes of the two approaches.

Gravity

If the mass density is defined at N points in space, then a direct determination of the gravitational potential via Poisson's integral equation requires $O(N^2)$ operations. As this places an intolerable (and unnecessary) computational burden on the whole calculation [other parts scale as $O(N)$], many algorithms have been developed which reduce the problem to $O[N \ln(N)]$ operations. For example, if the mass is defined on a regular grid, then one can use efficient Fourier transform methods (e.g., FFT) to transform to k space, solve Poisson's differential equation there (using periodic boundary conditions) and then transform back. While this is the fastest method used at present (for given N), the regular grid it requires is clearly not the best way of specifying the mass distribution. Other less rapid techniques are available for unstructured grids that can be designed to put a finer mesh in regions of higher density. Alternatively, to reach higher resolutions, one can compute forces of particles within a cell directly (P^3M [31]); here the cost becomes very high when the number of particles in any cell becomes large. Finally, there are completely unstructured methods, the prototype of which is the Tree algorithm [32] which takes advantage of the fact that, from a distance, the potential at any finite group of particles can be accurately represented by a multipole expansion. An interesting recent hybrid of the particle mesh and tree algorithms has been developed by Xu [33], which can be implemented efficiently on a parallel machine architecture. In typical codes, the time spent on gravity is comparable to the time spent on all other aspects of the problem in a given timestep. The reason for this is that, for each mass element, gravity requires $O[\ln(N)]$ operations, which is comparable to the number of other variables.

Atomic Physics

There exist numerous good atomic physics texts to guide the student [34], as a closely related set of problems has been studied for decades in the context of the interstellar medium. The only word of warning offered is that one must be careful about employing short cuts. For example, one can write, under some circumstances, the cooling rate per unit volume of gas having a specified chemical composition as $n^2 \Lambda(T)$ with the "cooling function," $\Lambda(T)$ tabulated by several groups. But the hidden assumption, that ionization is purely determined by collisional processes, is typically very poor in the low-density intergalactic medium, where background radiation fields can ionize gas efficiently (at a rate $\propto n$, not n^2). Furthermore, the normal assumption of ionization equilibrium can be badly off in low density environments. As a consequence of these complications, the Princeton group has typically taken the brute force approach of computing for each timestep and each mass element the change in the numbers of each relevant ionized state. Thus, symbolically $dQ_n/dt = -Q_n \int \sigma_\nu J_\nu \, dV - Q_n N_e I(T) + Q_{n+1} N_e \alpha(T)$, where the first term represents photoionization, the second collisional ionization, and the third recombination from the next higher state of ionization. Since the number of important species is limited, if one is working with gas of a primeval composition, the cost here (aside from the cost of the programmer's time) is less than that of the gravitational computation.

Photoionization is a crucial process, so it is very important to calculate the mean radiation field properly. One must allow for emission of photons, absorption processes, and the cosmological dilution and red shifting processes. If one computes only the mean (frequency dependent) radiation field, $J_\nu(t)$, then the computational burden is not large. As far as I know, no one yet has attempted to compute the radiation field as a function of both space and direction, $I_\nu(\vec{\theta}, \vec{r}, t)$, but some simpler treatments which allow for shielding are now implemented [35, 36].

A detailed presentation of all of the relevant equations for treating the heating and cooling of a hydrogen-helium plasma (with $T > 10^{3.5} K$) is contained in [37].

Magnetic Fields

Magnetic fields, even if not primordially existent, will be generated naturally by the "Bierman battery" whenever shocks cause a significant angle between pressure and density gradients:

$$\frac{\partial \vec{B}}{\partial t} - \nabla \times \left(\vec{v} \times \vec{B} \right) = \frac{\vec{\nabla} p \times \vec{\nabla} \rho}{\rho^2} \left(\frac{cMe}{e} \right) . \tag{6.1}$$

Recently, Kulsrud et al. [38] rediscovered a remarkable result implied by Biermann's original work [39]. If the vector fields for both cyclotron frequency $\vec{\omega}_{cl} \equiv \left(e\vec{B}/M_e c \right)$ and the vorticity $\vec{\omega} \equiv \nabla \times \vec{v}$ are initially zero, and dissipation can be neglected, and, moreover, the magnetic forces are unimportant dynamically, then $\vec{\omega}_{cl} = -\vec{\omega}$, at all times and places. Both quantities satisfy the same partial differential equation and initial conditions. Thus, an accurate calculation of the velocity field—neglecting magnetic terms altogether—allows one to compute the magnetic fields, provided the latter are small. A first computation by Kulsrud et al. [38] shows that average fields of order 10^{-21} gauss are generated, but that the fields in clusters of galaxies will be orders of magnitude larger, as turbulence leads to a rapid amplification of field strength until equipartition is reached.

Galaxy/Star Formation and Secondary Radiation Fields

Galaxy/Star Formation. As the growth of structure proceeds, we know, empirically, that condensation to ever higher densities leads to the formation of stars and quasars. Given the current state of our knowledge, we cannot model this evolutionary process accurately. But the problem cannot be avoided; some scheme *must* be invented. The reason is that, since any numerical code has finite spatial and temporal resolution, situations will occur when the code simply cannot cope with the physics of collapsing, fragmenting matter. The computations would necessarily halt if no prescription were included in the code to treat regions which were

$$\text{a) contracting; } \vec{\nabla} \cdot \vec{v} < 0 \tag{6.2}$$

and

$$\text{b) gravitationally unstable; } M_{\text{Jeans}} < M_{\text{gas}} \tag{6.3}$$

and

$$\text{c) cooling rapidly; } t_{\text{cool}} < t_{\text{dynamical}} . \tag{6.4}$$

Such mass elements cannot be followed by the code. In the Princeton approach to the problem, we have allowed these mass elements to condense out as subgalactic units treated as particles. Thus, if all of conditions (1)–(3) are satisfied (and some minimal overdensity criterion, as well, is adopted for convenience), then we remove from the cell a mass in gas

$$\Delta M_{\text{gas}} = -\frac{M_{\text{gas}}}{\rho} \frac{\Delta t}{t_{\text{dyn}}} , \tag{6.5}$$

where Δt is the timestep t_{dyn}, the free fall time within the cell $[\propto (G\rho_{cell})^{-1/2}]$. The mass removed from the gas phase is used to create a new particle with mass $= \Delta M_{gas}$ and a velocity equal to the cells' hydrodynamic velocity. Thus, the operation conserves mass and momentum, but, of course, it does not conserve energy. The energy radiated during the successive collapse and fragmentation phases, which we cannot capture, has been included in the computation as the hidden binding energy of the created particles. We can track these subgalactic units and group them, at any time into "galaxies," with designated "age" (time since creation), mass, and angular momentum.

One would hope that this prescription for forming the condensed component would be robust, i.e., that it would not depend on the spatial or temporal resolution of the code nor on the numerical coefficients that might be included on the right hand sides of equations (6.3), (6.4) and (6.5). The notion behind this hope is the expectation that once gas starts to collapse, it will not stop, but will continue at an accelerated pace. If so, it should not matter when, in the collapsing phase, one steps in and labels the material as designated for the condensed state. Further, the process should be self-regulatory to some extent; as gas is removed from the contracting component via equation (6.5), the remaining gas has lower density and will not as easily satisfy criteria (6.3) and (6.4). Actual numerical tests are only mildly reassuring. As one changes the prescription within reasonable bounds, the results for the amount of condensed (galactic) material formed do change, but by moderate amounts, so that we feel it possible to compute the ρ_{gal} to within approximately a factor of two.

Secondary Radiation Fields. Since the condensed matter presumably is now divided into the familiar categories of stars, interstellar gas, quasars, etc., it will necessarily emit radiation which will affect the remainder of the computation.

Here we have two choices. First, we could simply rely on our observational estimates of the background radiation in fields $J_\nu(z)$, and we could say that, after the onset of galaxy formation, we will put in, by hand, a background with those properties. This option has been chosen by the Weinberg, Hernquist, & Katz collaboration [41]. It has the virtue of simplicity, but ambiguities exist in specifying the time dependence (should $J_\nu(z)$ be taken to be proportional to the rate of galaxy formation, the lagged rate, or the integrated rate...?), and radiation transfer (we do not know observationally the background in just those spectral regions near atomic edges, where absorption is most important but knowledge of J_ν is most critical). The Princeton group has chosen a second alternative: to allow the condensed component to emit radiation with some appropriate source function S_ν and then to compute $dJ_\nu = (S_\nu - \rho\kappa_\nu J_\nu + \text{cosmological terms})\, dt$.

An efficiency of turning condensed barionic matter into radiation must then be specified $\epsilon_{rad} \equiv \Delta E_{rad}/\Delta M_{condensed}c^2$, which, while picked to reasonably match star forming regions like Orion (ϵ is approximately equal to $10^{-4.5}$), is essentially arbitrary and must be adjusted to produce approximately the observed background radiation field, J_ν. The spectral slope of the source function is designed to have a soft component appropriate to hot stars and a comparable hard, power law component designed to mimic AGN spectra.

But, the point cannot be made too strongly that it is essential to utilize *some* method for adding an ionizing background radiation field since (a) it exists in nature, and (b) it strongly affects the thermal state of gas at temperatures $< 10^5$ degrees K.

For some purposes, knowledge of $J_\nu(z)$ is vital but can be constrained by observations, as in the computation of the gaseous component that produces the Lyman-α clouds. For some other purposes $J_\nu(z)$ is essentially irrelevant, as in the computation of the properties of the X-ray clusters ($T > 10^7$ K). But, for the all important purpose of computing galaxy formation itself, the issue is extremely complex. Recent work by Steinmetz and by Navarro & Steinmetz [40] and WHK [41] indicates that the computed formation of low mass systems $M_{gal} < 10^9 \ M_\odot$ is strongly dependent on the background radiation field, but ordinary ($M_{gal} \approx 10^{11} - 10^{12} \ M_\odot$) systems are virtually unaffected. The situation is further complicated by the fact that a smooth radiation field cannot be assumed; the forming galaxies are themselves local sources of radiation. Much more work will need to be done before the picture becomes clear.

6.3   RESULTS: COMPARISON WITH OBSERVATIONS

6.3.1   *Hot Components*

X-ray Clusters

Several groups have by now computed the X-ray luminosity emitted by the hot, dense regions observationally identified with clusters of galaxies. These are regions, vertices where Zeldovich-like pancakes or ribbons intersect, where the potential wells are deepest and where gas, dark matter, and galaxies collect into massive assemblages. The velocity dispersion is typically comparable to or greater than 1000 km s^{-1} and the temperatures are greater than 10^8 K. The fact that such distinct regions, with the appropriate properties, form in all plausible cosmological scenarios is a first triumph of cosmological hydrodynamics. Had these giant X-ray emitting regions not been observed first [42], they would have been an early prediction of the simulations. The number of such regions per unit volume as a function of luminosity, $N(L_x)$, predicted by a specific scenario, can now be used to test the model. Also available are the distribution of temperatures, $N(T_x)$, the correlation between these observables, $T_x(L_x)$, the core radii of the X-ray emitting gas, R_x, and numerous other quantities such as, for example, the ratio of gas-to-total mass (as a function of radius) within the X-ray clusters.

It is numerically difficult to compute a large enough volume so that it contains a fair sample of clusters. The requirement is that the volume be greater than $(200h^{-1} \text{ Mpc})^3$ with a resolution length small enough to compute the luminosity of specific clusters accurately (Δ l is less than $20h^{-1}$ kpc). Thus the required dynamic range necessary for a secure calculation is 10^4. All errors, if this dynamic range is not reached, conspire to underestimate the number of computed high luminosity clusters. Thus, a result, such as that by Cen and Ostriker [10] indicating that the COBE normalized standard CDM model produces too many high temperature, luminous X-ray clusters, is useful in ruling out some cosmological models. Other work by Bryan et al. [11] for the CHDM model, and by Cen and Ostriker for the LCDM model [10] indicates that successful models may exist, but numerically better work is needed before either result is confirmed.

Background Hot Gas

There is also a much larger volume and mass of gas in the temperature range $10^{6.0}\text{K}-10^{7.5}\text{K}$, which would be associated observationally with more moderate density enhancements (groups and weak clusters of galaxies). For this gas, line emission and cooling are important and resolution is still more important than it is for the hotter gas. Preliminary results by Cen, Kang, Ostriker, & Ryu [43] indicate that the emission from this gas corresponds well with one component of the observed soft X-ray background [44].

6.3.2 *Warm Components*

Lyman-α Forest

The neutral hydrogen detected in gas at roughly 10^4 K, as it produces absorption lines between us and distant quasars, at the rest frame Lyman-α wavelength, has been identified observationally as the "Lyman-α Forest" and treated, because of its clumpy nature, in heuristic models, as distinct "clouds" of uncertain geometry. The confinement mechanism has remained unknown, with candidates being gravity in mini-halos, external gas pressure or no-confinement (i.e., free expansion). Recent detailed simulation work by at least three groups [45] has indicated that the "clouds" arise naturally in popular cosmogonic scenarios, that there is some truth in all of the earlier conceptual models, but that the dominant mechanism is ram pressure confinement of infalling gas. Filament-like caustics form from small wavelength perturbations going non-linear. These caustics become weakly shocked, Zeldovich pancake-like ribbons of gas, which are overdense by a factor or order 10^1–10^2, with the temperature being maintained by photoheating processes at about $10^{4.5}$K. Not only are the individual and group statistical properties of the Lyman-α clouds reproduced well by the simulations, but it appears that the spatial correlations [46] are well modeled also. Study of the Lyman-α clouds should be a fruitful field for further work, because of the vast body of cosmological information accumulated in the detailed observational absorption line studies now becoming amenable to theoretical analysis.

Gas Near Galaxies

The gaseous components discussed so far are either so hot as to be essentially unaffected by stellar sources (X-ray clusters), or so far from them as to be unreachable except by the dilute metagalactic radiation field (Lyman-α clouds). In addition, there will be the all important over-dense components in temperature range $10^{4.0}$–$10^{5.5}$ within the caustic surfaces from which galaxies will form and which may contain a significant fraction of the total baryonic component of the universe (since galaxy formation is, like star formation, notably inefficient). Here the high resolution SPH work by Evrard, Summers, & Davis [47], Katz, Weinberg, & Hernquist [48] or Steimetz [49] is best. The problem is, however, very difficult, since the detailed output from massive stars must be known (UV, X-rays, SN ejecta, etc.) and spatially dependent radiative transfer is required. This area is wide open to further investigation, but the technical problems are formidable.

6.3.3 *Cold Condensed Components*

Perhaps easier to treat is the truly cold component which cools to the opaque neutral atomic or the molecular state or proceeds further into a condensed stellar component. As noted earlier, once gas has started to cool, collapse, and fragment in an accelerated fashion, it must end in one of these phases (neutral atomic, molecular or stellar), although which phase cannot be computed absent a good theory of star formation. Codes should be able to compute the masses, angular momenta, and spatial distributions in these cold components—the principle components of galaxies—even when their information concerning the internal states of the objects is poor. Thus, it is encouraging that even the first computations (cf., Cen and Ostriker [50]) reproduced a plausible galaxy mass function and the density-morphology relations found in the real universe. Specifically, if one identifies the most gas-poor systems as "ellipticals", they tend, statistically, to have stars with the oldest average age, to live in the regions of highest baryon density, and to be (on average) most massive. The computed gas-poor systems are strikingly like normal elliptical galaxies and gas-rich systems like spirals or irregulars. To be believed, this work must be confirmed by higher resolution studies and by other investigators using different approximations with regard to the physical modeling.

6.4   Conclusions, Prospects, and More Questions

First, the most important conclusion. *The theories currently being discussed are true scientific theories, capable of verification or falsification. Some* of the consequences of these theories, even in the extreme nonlinear regime (e.g., X-ray clusters), are computable to sufficient accuracy that valid and successful comparisons with observations are possible. Now to some more detailed and highly tentative conclusions:

1) Open Models Work Best ($\Lambda \neq 0$?)

The arguments presented here are summarized in two papers (with references therein to work of other groups). Ostriker and Steinhardt [20] examine the general astrophysical constraints and Ostriker and Cen [51] explore the implications of the numerical simulations. While the arguments are far from conclusive, the preponderance of evidence favors an open universe, with or without a significant cosmological constant (see also N. Bahcall and P. J. E. Peebles in these proceedings). The primary criteria are two-fold: (1) open models tend to form structure earlier, in accord with results from HST gravitational lensing from clusters, Gunn-Peterson measurements, etc., and (2) open models, with lower mass density (but given $\delta\rho/\rho$), have less gravitational lensing per unit length [52] and lower small scale velocity dispersions [51], in better accord with observations.

2) Pop III Should Exist (early Z contamination).

The universe was reionized and reheated somewhere in the time interval $20 \gtrsim z \gtrsim 5$, and the best candidate for the energy and ionizing photon sources is massive stars, which also tend to explode and contaminate the surroundings with heavy elements. These arguments have been made, in semi-analytical forms, by numerous investigators over the last two decades. Recent numerical work [36], shows that there is a distinct population (PopIII?), which is responsible for the ionization, reheating, and contamination of the universe. Even those parts of the IGM far from the caustics (where galaxies form) contain "Lyman α clouds" with metal abundances of $Z = 10^{-2} Z_\odot$.

3) Lyman-α Forest Clouds.

What are the "Lyman-α clouds" observed in absorption between us and distant quasars? The most likely possibility is that the clouds are gas collecting, due to gravitational forces, in weakly shocked, nearly isothermal regions with ribbon-like topology. As time proceeds, new and larger ribbons form and the earlier ones fragment and dissipate, so the smaller pressure-dominated "clouds" tend to be transient structures. The larger clouds are dominated by gravity, fragment, and collapse to galaxies. Several groups performing numerical simulations find that the models naturally produce approximately the correct number density, column density, redshift and line width. It is premature to say which models fare best when compared to observations.

4) Galaxy Formation ("bias" small but real).

Galaxies form as the larger ribbon-like density enhancements fragment. They then collect, with dark matter and preponderance of the baryons (still in gaseous form), into "clusters" at the vertices where the ribbons intersect and the potential wells are deepest. Those systems which start earliest (and have, at $z = 0$, the oldest mean "age") are ultimately among the most massive galaxies, live in the regions of highest density, are associated with the smallest amount of cold gas, and have the highest ratio of stars to dark matter. These are—quantitatively—properties of elliptical galaxies and of spiral bulges. In general, the observed correlations of galaxy properties with mean stellar age and with environment are reproduced in the simulations.

5) Feedback (UV, SN).

Supernovae within galaxies can affect at most the neighboring few $(h^{-1}\,\mathrm{Mpc})^3$ and, within this region, may be able to amplify (or de-amplify) galaxy formation. The UV output is more potent. The global UV background heats gas to $10^{4.5}$ K and thus, by increasing the Jeans' mass, tends to produce a cutoff of galaxy formation below $10^{8.5}\,M_\odot$. The local background (both UV and SM)—during galaxy formation—is probably important in regulating star formation. It has not been included in any detailed work to date.

6) X-rays from Hot Gas.

Very hot, shocked gas collects at the vertices where we see the X-ray "clusters." The computed models reproduce observations fairly well, and, as they improve, may be expected to discriminate among cosmological scenarios.

In addition, the slightly less hot but far more widespread gas in the caustics and weaker clusters contributes significantly to the soft X-ray ($h\nu <$ 1 KeV) diffuse cosmic background.

Finally, let us turn to some of the even more open questions. What are the best numerical techniques for modeling the various components? Numerical methods are the counterpart to instrumentation in observational astronomy. The activity is absolutely vital; progress depends entirely on continuing technical work, but that work is unglamorous (except to the workers) and is often unrewarded. For example, the jury is still out on the question of whether mesh based or particle based codes will be more successful in the larger domain, and hybrid schemes are being developed which combine virtues of both methods.

Second, one must ask the always central question of which questions should be addressed? This arena is where scientific genius expresses itself most clearly. Galileo's contribution was not to the accurate measurement of acceleration, but was the realization that acceleration, not velocity or position, was the quantity to study. The crucial issue is to find quantities that

1) can be computed with some accuracy,
2) have been or can be observed with sufficient ease, so that statistical measures can be determined (since only statistical predictions can be made—remember the random phase assumption),
3) have some scientific significance.

On the astronomical side we can ask what can "explain" the observed properties of the universe, the regularities observed in the properties of galaxies, clusters and the intergalactic absorption clouds? We are interested, in this astronomical context, primarily in the most robust conclusions of the models—the results that do *not* depend on the specific scenario, because this is required for the explanations to be reliable.

But, if we are interested in cosmology per se, we should look in the opposite direction, as we want to know which output properties are most sensitive to input variations. Which calculated properties are *least* robust with respect to the choice of models? These quantities will enable us best to distinguish among competing scenarios, to separate among the models the losers from the potential winners.

There is a plethora of questions in both categories now being addressed, and many more will be treatable in the near future, given the rapid and inevitable march of technology. I end with a note of envy. Given the inevitable turning of the pages of the calendar, I will need to leave much of the most exciting work to the next generation, many of the best of which have attended this conference or are reading this book!

BIBLIOGRAPHIC NOTES

- **Peebles, P. J. E. 1971, Physical Cosmology (Princeton: Princeton University Press); Weinberg, S. 1972, Gravitation and Cosmology (New York: Wiley & Sons).** These are the two classic texts on homogeneous cosmology. The student needs to master the basic elements of at least one of these texts in order to understand the language of cosmologists.

- **Peebles, P. J. E. 1980, The Large Scale Structure of the Universe (Princeton: Princeton University Press); Peebles, P. J. E. 1993, Principles of Physical Cosmology (Princeton: Princeton University Press).** The observational and theoretical basis of modern cosmological studies is clearly, simply, and presciently presented in these two indispensable monographs.

- **Padmanahban, T. 1993, Structure Formation in The Universe (Cambridge: Cambridge University Press).** Excellent and relatively up-to-date with a mathematical bias.

- **Ostriker, J. P. 1991, Development of Large Scale Structure in The Universe (Cambridge: Cambridge University Press).** This provides a brief and clear, if somewhat eccentric, review of the subject for first year graduate students.

BIBLIOGRAPHY

[1] Peebles, P. J. E. 1971, Physical Cosmology (Princeton: Princeton University Press).

[2] Weinberg, S. 1972, Gravitation and Cosmology (New York: Wiley & Sons).

[3] Peebles, P. J. E. 1980, The Large Scale Structure of the Universe (Princeton: Princeton University Press).

[4] Peebles, P. J. E. 1993, Principles of Physical Cosmology (Princeton: Princeton University Press).

[5] Efstathiou, G. 1990, Cosmological Perturbations, in Physics of the Early Universe, ed. J. A. Peacock, A. F. Heavens, & A. T. Davies (Edinburgh: Edinburgh University Press), 361.

[6] Padmanahban, T. 1993, Structure Formation in the Universe (Cambridge: Cambridge University Press).

[7] Bauch, C. M., & Efstathiou, G. 1994, MNRAS, 267, 323.

[8] Ostriker, J. P. 1993, ARA&A, 31, 689.

[9] Press, W. H., & Rybicki, G. B. 1993, ApJ, 418, 585.

[10] Cen, R., & Ostriker, J. P. 1994, ApJ, 429, 4 and 430, 83.

[11] Bryan, G. L., Klypin, A., Loken, C., Norman, M. L., & Burns, J. O. 1994, ApJ, 437, L5.

[12] Cen, R., Miralda-Escudé, J., Ostriker, J. P., & Rauch, M. 1994, ApJ, 437, L9.

[13] Hernquist, L., Katz, N., Weinberg, D. H., & Miralda, J. 1996, ApJ, 457, L51.

[14] Zhang, Y., Anninos, P., & Norman, M. L. 1995, ApJ, 453, L57.

[15] Walker, T., Steigman, G., Schramm, D. N., Olive, K. A., & Kang, H. 1991, ApJ, 376, 51.

[16] Derived separately via dimensional analysis by Peebles, P. J. E., & Yu, J. T. 1970, ApJ, 162, 815; Harrison, E. R. 1970, Phys. Rev. D, 1, 2726; Zeldovich, Ya. B. 1972, MNRAS, 160, 1p.

[17] Guth, A. H., & Pi, S.-Y. 1982, Phys.Rev. Lett., 49, 1110; Starobinskii, A. A. 1982, Phys. Lett. B, 117, 175; Bardeen, J., Steinhardt, P., & Turner, M. S. 1983, Phys. Rev. D, 28, 679.

[18] Bartlett, J. G., Blanchard, A., Silk, J., & Turner, M. S. 1995, Science, 267, 980.

[19] Bolte, M., & Hogan, C. J. 1995, Nature, 376, 399.

[20] Ostriker, J. P., & Steinhardt, P. 1995, Nature, 377, 600.

[21] Bahcall, N. A., Cen, R. Y., & Ostriker, J. P. 1996, ApJ, 462, L49.

[22] White, S. D. M., Efstathiou, G., & Frenk, C. S. 1993, MNRAS, 262, 1023.

[23] Peacock, J., & Dodds, S. J. 1994, MNRAS, 267, 1020.

[24] Bardeen, J. M., Bond, J. R., Kaiser, N., Szalay, A. S. 1986, ApJ, 304, 15.

[25] Bertschinger, E. 1995, astro-ph/9506070.

[26] Efstathiou, G., Davis, M., Frenk, C. S., & White, S. D. M. 1985, ApJS, 57, 241.

[27] White, S. 1994, in Clusters of Galaxies, proceedings of the XXIX Rencontres de Moriond, Meribel, Savoie, France, March 12–19, 1994 (Gif Sur Yvette: France).

[28] Monaghan, J. J., & Gingold, R. A. 1983, J. Comp. Phys., 52, 374; Hernquist, L., & Katz, N. 1989, ApJS, 70, 419.

[29] Cen, R. Y., Jameson, A., Liu, F., & Ostriker, J. P. 1990, ApJ, 362, L41.

[30] Kang, H., Ostriker, J. P., Cen, R. Y., Ryu, D., Hernquist, L., Evrard, A. E., Bryan, G. L., & Norman, M. L. 1994, ApJ, 430, 83.

[31] Hockney, R. W., & Eastwood, J. W. 1988, in Computer Simulations Using Particles (New York: Wiley & Sons); Efstathiou, G., Davis, M., Frenk, C. S., & White, S. D. M. 1985, ApJS, 57, 241.

[32] Barnes, J. E., & Hut, P. 1986, Nature, 324, 446.

[33] Xu, G. 1995, ApJS, 98, 355.

[34] See, for example, Osterbrock, D. E. 1989, Astrophysics of Gaseous Nebulae and Active Galactic Nuclei (Mill Valley, CA: University Science Books).

[35] Katz, N., Weinberg, D. H., Hernquist, L., & Miralda-Escudé, J. 1996, ApJ, 457, L57.

[36] Gnedin, N., & Ostriker, J. P. 1996, in preparation.

[37] Cen, R. 1992, ApJS, 78, 341.

[38] Kulsrud, R. M., Ryu, D., Cen R., & Ostriker, J. P. 1996, ApJ, submitted.

[39] Biermann, L. 1950, Z. Natureforsch, 5a, 65.

[40] Steinmetz, M. 1995, Proc. 17th Texas Symposium on Relativistic Astrophysics, Annals of the New York Academy of Sciences, 759, 628; Navarro, J. F., & Steinmetz, M. 1996, astro-ph/9605043.

[41] Weinberg, D., Hernquist, L., & Katz, N. 1996, astro-ph/9604175 and references therein.

[42] Gursky, H., Kellogg, E., Murray, S., Leong, C., Tananbaum, H., & Giacconi, R. 1971, ApJ, 167, L81.

[43] Cen, R., Kang, H., Ostriker, J. P., & Ryu, D. 1995, ApJ, 451, 436.

[44] Gendreau, K. C., et al. 1995, PASJ, 47, L5.

[45] Cen, R., Miralda-Escudé, J., Ostriker, J. P., & Rauch, M. 1994, ApJ, 437, L9; Hernquist, L., Katz, N., Weinberg, D. H., & Miralda-Escudé, J. 1996, ApJ, 457, L51; and Zhang, Y., Anninos, P., & Norman, M. 1995, ApJ, 453, L57.

[46] Miralda-Escudé, J., Cen, R., Ostriker, J. P., & Rauch, M. 1996, ApJ, in press.

[47] Evrard, A. E., Summers, F. J., & Davis, M. 1994, ApJ, 422, 11.

[48] Katz, N., Weinberg, D., & Hernquist, L. 1996, ApJS, in press.

[49] Steinmetz, M. 1996, MNRAS, 278, 1005.

[50] Cen, R., & Ostriker, J. P. 1993, ApJ, 417, 415.

[51] Ostriker, J. P., & Cen, R. 1996, ApJ, 464, 270.

[52] Wambsganss, J., Cen, R., Ostriker, J. P., & Turner, E. L. 1995, Science, 268, 274.

CHAPTER 7

THE CENTERS OF ELLIPTICAL GALAXIES

SCOTT TREMAINE

Canadian Institute for Theoretical Astrophysics, University of Toronto,
Toronto, Canada, and Institute of Astronomy, Cambridge, England

ABSTRACT

The properties of distant quasars and the stellar kinematics of nearby galaxies independently suggest that many, perhaps most, galaxies contain central black holes of
mass 10^6–$10^9 M_\odot$. The distribution of black-hole masses as a function of galaxy
luminosity and type, and the influence of the black holes on the structure of the
central regions of galaxies are important unsolved problems.

7.1 INTRODUCTION

Most astronomers believe that quasars are active galactic nuclei (AGNS), and that
the power source for AGNS is accretion onto a massive black hole (BH). The supporting arguments (Rees 1984, Blandford et al. 1991) include the high efficiency
of gravitational energy release through disk accretion onto a BH compared to other
power sources; the rapid variability of some AGNS, which implies a compact source;
and the apparent superluminal expansion in some radio sources, implying relativistic outflow which is most naturally produced in a relativistic potential well.
Moreover most other plausible power sources eventually evolve into BHS so these
objects are likely to be present even if they were not the power source.

The comoving density of quasars is a strong function of redshift, declining
by a factor of 10^2–10^3 from $z = 2$ to the present (Hartwick and Schade 1990).
Thus many local galaxies should contain "dead quasars"—massive central BHS that
show no sign of activity because they are starved of fuel.

These simple arguments suggest several unsolved problems: *Are massive black*

137

holes present in the centers of nearby galaxies? What is the distribution of black-hole masses as a function of galaxy luminosity and type? How are the structure and dynamics of galaxies in their central regions related to the central black hole?

7.1.1   Black Holes and Quasars

The local energy density in quasar light is (Chokshi and Turner 1992)

$$u = 1.3 \times 10^{-15} \text{ erg cm}^{-3}. \tag{7.1}$$

If this energy is produced by burning fuel with an assumed efficiency $\epsilon \equiv \Delta E / (\Delta M c^2)$, then the mean mass density of dead quasars must be at least (Sołtan 1982, Chokshi and Turner 1992)

$$\rho_\bullet = \frac{u}{\epsilon c^2} = 2.2 \times 10^5 \left(\frac{0.1}{\epsilon}\right) \text{M}_\odot \text{ Mpc}^{-3}, \tag{7.2}$$

assuming that most of the fuel is accreted onto the BH, and that the universe is homogeneous and transparent.

The mass of a dead quasar may be written

$$M_\bullet = \frac{L_Q \tau}{\epsilon c^2} = 7 \times 10^8 \text{M}_\odot \left(\frac{L_Q}{10^{12} \text{L}_\odot}\right) \left(\frac{\tau}{10^9 \text{ y}}\right) \left(\frac{0.1}{\epsilon}\right), \tag{7.3}$$

where L_Q is the quasar luminosity and τ is its lifetime. An upper limit to the lifetime is the evolution timescale for the quasar population as a whole, $\sim 10^9$ y; however, upper limits to BH masses in nearby galaxies and direct estimates of the BH masses in AGNs both suggest that the typical masses of dead quasars are $M_\bullet = 10^7$–$10^8 \text{M}_\odot$ (Haehnelt and Rees 1993), so that equation (7.3) suggests that the lifetime of an individual quasar is only 10^7–10^8 y.

To focus the discussion, let us adopt a "strawman" model in which a fraction f of all galaxies contain a central BH and the BH mass is proportional to the galaxy luminosity. Thus $M_\bullet = \Upsilon L$ where Υ is the (black hole) mass to (galaxy) light ratio. The luminosity density of galaxies is $j = 1.5 \times 10^8 \text{L}_\odot \text{ Mpc}^{-3}$ in the blue band (Efstathiou et al. 1988; I assume a Hubble constant $H_0 = 80 \text{ km s}^{-1} \text{ Mpc}^{-1}$); thus

$$\Upsilon = \frac{\rho_\bullet}{fj} = \frac{0.0015}{f} \left(\frac{0.1}{\epsilon}\right) \frac{\text{M}_\odot}{\text{L}_\odot}. \tag{7.4}$$

A second estimate of Υ comes from dividing the typical dead quasar mass derived above, $M_\bullet \approx 10^{7.5} \text{M}_\odot$, by the typical luminosity of a bright galaxy, $L \approx 10^{10} \text{L}_\odot$, to get $\Upsilon \approx 10^{-2.5}$. If this estimate is to be consistent with equation (7.4) then f cannot be far from unity; in other words most or all galaxies must contain massive central BHs (Haehnelt and Rees 1993). A possible concern with ubiquitous central BHS is the absence of significant non-stellar radiation from most nearby galaxies with claimed BHS (Fabian and Canizares 1988, Rees 1990, Kormendy and Richstone 1995); however, Narayan et al. (1995) have argued persuasively that the required low accretion efficiency is a natural consequence of advection-dominated accretion flows.

Detection of a significant sample of these exotic objects—or proof that they are not present—would enhance our understanding of both AGNs and the central regions of all galaxies.

7.1.2 The Sphere of Influence

In the remainder of this article I discuss the dynamical interactions between a massive central BH and the surrounding galaxy. Most of these interactions only require a massive dark object, which need not be a black hole. Thus the term "BH" henceforth refers to any such object although strictly we should use a different acronym such as MDO (Kormendy and Richstone 1995).

The radius r_h of the dynamical sphere of influence of a central BH is found by equating the potential energy from the BH, $GM_\bullet/r$, to the kinetic energy of the stars, $\frac{3}{2}\sigma^2$, where σ is the line-of-sight velocity dispersion. Neglecting factors of order unity,

$$r_h \equiv \frac{GM_\bullet}{\sigma^2}, \qquad \theta_h \equiv \frac{r_h}{d} = 0\rlap{.}''9 \left(\frac{M_\bullet}{10^8 M_\odot}\right) \left(\frac{100\ \mathrm{km\ s}^{-1}}{\sigma}\right)^2 \left(\frac{10\ \mathrm{Mpc}}{d}\right),$$
$$(7.5)$$

where d is the distance to the galaxy. It is natural to expect that the presence of a BH should be reflected in the photometric and kinematic behavior of the galaxy near r_h (Peebles 1972).

The number of galaxies in which the BH sphere of influence exceeds some limiting resolution θ can be estimated using the Faber-Jackson law, $\sigma(L) \simeq \sigma^\star(L/L^\star)^{0.25}$, and the Schechter (1976) luminosity function, which states that the number of galaxies per unit volume with luminosity in the range $[L, L + dL]$ is

$$\phi(L)dL = \phi^\star(L/L^\star)^\alpha \exp(-L/L^\star)dL/L^\star. \qquad (7.6)$$

Taking $\sigma^\star = 220\ \mathrm{km\ s}^{-1}$, $\phi^\star = 0.008\ \mathrm{Mpc}^{-3}$, $L^\star = 1.8 \times 10^{10} L_\odot$, $\alpha = -1.07$ (Efstathiou et al. 1988), and assuming once again that a fraction f of all galaxies host BHS with mass $M_\bullet = \Upsilon L$ yields the estimate

$$N(\theta_h > \theta) = \frac{4\pi}{3} f \int_0^\infty dL\phi(L) \left[\frac{G\Upsilon L}{\sigma^2(L)\theta}\right]^3 \simeq 0.03f \left(\frac{\Upsilon}{0.003}\right)^3 \left(\frac{1''}{\theta}\right)^3.$$
$$(7.7)$$

Thus the number of galaxies in which the sphere of influence is resolved is a strong function of the resolution. Ground-based observations (FWHM $\gtrsim 0\overset{''}{.}3$) are expected to resolve θ_h in at best a handful of galaxies, and in most of these θ_h will be close to the resolution limit. The Hubble Space Telescope (HST; FWHM $\lesssim 0\overset{''}{.}1$) should resolve θ_h in a much larger sample, of order 10^2 galaxies—which of course is one reason why it was built.

7.1.3   Cores and Cusps

Understanding the central structure of a galaxy without a central BH is a prerequisite for investigating the effects of central BHS. A modest initial assumption is that all physical variables vary smoothly near the center and hence can be expanded in Taylor series in a Cartesian coordinate system with origin at the center (as in the Sun, planets, globular clusters, etc.). Then in a spherical galaxy the luminosity density may be written $j = j_0 + j_1 r^2 + O(r^4)$—terms that are odd powers of r vanish because $r = (x^2 + y^2 + z^2)^{1/2}$ is not a smooth function of the Cartesian coordinates near the center—and the surface brightness may be written

$$I(R) = I_0 + I_1 R^2 + O(R^4), \tag{7.8}$$

where r is the radius and R is the projected radius. A galaxy satisfying (7.8) can be said to have an "analytic core" and its "core radius" R_c is defined by the relation $I(R_c) = \frac{1}{2}I(0)$ (e.g., Richstone and Tremaine 1986). Note that $d \log I/d \log R \to 0$ as $R \to 0$ is not sufficient to ensure an analytic core; both the Hubble-Reynolds law $I(R) = I_0 a^2/(R + a)^2$ and de Vaucouleurs' law $I(R) = I_0 \exp(-kR^{1/4})$ satisfy this constraint, but have singular luminosity density as $r \to 0$ because they do not satisfy (7.8).

We shall see below that few if any galaxies have analytic cores; thus the term "core" must have a broader meaning to be useful. Following Lauer et al. (1995), I use "core" to mean a region around the center in which the surface-brightness profile slope $|d \log I/d \log R|$ is markedly smaller than at larger radii, usually less than 0.3. The transition from steep outer slope to shallow inner slope occurs at the "break radius", which is a generalization of the core radius: the radius of maximum curvature in a $\log I$–$\log R$ plot, that is, the radius at which $|d^2 \log I(R)/d(\log R)^2|$ is maximized. The term "cusp" denotes a region in which the logarithmic slope of the surface-brightness profile is constant and non-zero at all radii exceeding the resolution limit. Cores can (and generally do) have shallow cusps.

Much of our intuition about the structure of the centers of stellar systems is based on models with analytic cores. For contrast, let us assume that the stellar density near the center varies as a power-law in radius,

$$\rho(r) = \rho_0(r_0/r)^k, \tag{7.9}$$

where $0 \leq k < 3$ (the second constraint ensures that the enclosed mass is finite). This density distribution produces a surface-brightness cusp, $I(R) \propto R^{1-k}$, for $k > 1$, while for $0 < k < 1$ there is no surface-brightness cusp but the core is not analytic. The mass within radius r is

$$M(r) = 4\pi \int_0^r r^2 \rho(r) dr = \frac{4\pi\rho_0 r_0^k r^{3-k}}{3-k}. \tag{7.10}$$

For simplicity I assume that the velocity-dispersion tensor is isotropic (although similar anisotropic models exist, and exhibit even richer behavior). The velocity dispersion $\sigma(r)$ is found by integrating the equation of hydrostatic equilibrium,

$$\frac{d}{dr}\left[\rho(r)\sigma^2(r)\right] = -\frac{GM(r)\rho(r)}{r^2}, \tag{7.11}$$

to yield

$$\sigma^2(r) = \frac{1}{\rho(r)}\int_r^{r_{max}} \frac{GM(r)\rho(r)}{r^2}dr = \frac{4\pi G\rho_0 r_0^k r^k}{3-k}\int_r^{r_{max}} r^{1-2k} dr, \tag{7.12}$$

where $r_{max} \gg r$ is a measure of the "edge" of the system.

For $k < 1$ the integral is dominated by radii near r_{max} and

$$\sigma^2(r) = \frac{2\pi G\rho_0 r_0^k r_{max}^{2-2k}}{(3-k)(1-k)} r^k. \tag{7.13}$$

In the limit $k \to 0$ the velocity dispersion is constant, and proportional to the square of the size of the constant-density core; this is a crude version of King's celebrated formula (Richstone and Tremaine 1986)

$$\sigma^2 = \frac{4}{9}\pi G\rho_0 R_c^2. \tag{7.14}$$

For $0 < k < 1$ the velocity dispersion decreases as $r \to 0$ but still is determined by r_{max} (because the pressure $\rho\sigma^2$ is dominated by stars with apocenters near r_{max}, while the density is dominated by local stars).

For $k > 1$ the integral in equation (7.12) is dominated by radii near r so that

$$\sigma^2(r) = \frac{2\pi G\rho_0 r_0^k r^{2-k}}{(3-k)(k-1)} = \frac{2\pi G\rho(r)r^2}{(3-k)(k-1)}. \tag{7.15}$$

In this case the velocity dispersion is determined locally (i.e. there is no dependence on r_{max}); the dispersion decreases as $r \to 0$ for $k < 2$ but grows for $2 < k < 3$.

This interesting behavior is described in more detail by Dehnen (1993) and Tremaine et al. (1994), who construct finite spherical systems in which the density near the center obeys equation (7.9).

7.2 PHOTOMETRY

There are now over 60 elliptical galaxies and spiral bulges with HST photometry (Crane et al. 1993, Jaffe et al. 1994, Lauer et al. 1995, Faber et al. 1996). Their surface-brightness profiles can be divided into two classes:

1. "Core" galaxies exhibit a well-resolved core. The slope of the surface-brightness profile within the core is $|d\log I/d\log R| < 0.3$ but most observed slopes are significantly different from zero; in other words few if any of the galaxies contain an analytic core and the luminosity density is growing with decreasing radius at the innermost measured point. Core galaxies are bright, $M_V \lesssim -20$. Examples include M87 and several cD galaxies.

2. "Power-law" galaxies have no detectable core. Their surface-brightness profiles are approximate power laws, with $d\log I/d\log R \simeq -0.8 \pm 0.3$, to the smallest resolvable radius. Power-law galaxies are generally fainter than core galaxies ($M_V \gtrsim -22$) but their luminosity density near the center is higher (Fig. 7.1). Examples of power-law galaxies are M32 and the Galaxy.

This morphology suggests several unsolved problems:

Do the power-law galaxies contain unresolved cores? There is at least one (weak) argument that power-law galaxies have negligible cores. The Galactic bulge is a relatively bright ($M_V \simeq -18.3$; Kent et al. 1991) power-law galaxy, but near-infrared maps of the Galactic center show that its core radius is only $0.15 \pm 0.05\,\mathrm{pc}$ (Eckart et al. 1993). This is much smaller than one would expect from extrapolating the core radius-luminosity correlation observed for bright galaxies. If power-law galaxies have negligible cores, *why do some galaxies have cores, while others do not?*

Dissipationless collapse or merging cannot increase the maximum phase-space density (Carlberg 1986, Tremaine et al. 1986); thus galaxies produced by either process must have analytic cores if the initial phase-space distribution is itself analytic. Then *why are there no analytic cores?* There are several possible explanations: (i) The initial distribution may contain dense, cold regions with very high phase-space density—perhaps compact bulges—that collect at the center of the galaxy during collapse or merger (Hernquist et al. 1993); (ii) The central density may be enhanced by gas infall and subsequent star formation, or other dissipative processes such as viscous evolution of a gaseous disk (Kormendy and Sanders 1992, Mihos and Hernquist 1994). Unfortunately, numerical simulations cannot confirm this explanation because they do not have either sufficiently high spatial resolution or reliable models of gas dynamics and star formation. (iii) There may be a central BH, in which case the velocity dispersion near the center diverges so that high spatial density need not imply high phase-space density (see § 7.2.1 for a specific model).

Galaxy mergers are common in most models of galaxy formation and offer a natural way to form many ellipticals from disk galaxies (Toomre 1977, Wielen 1990, Barnes and Hernquist 1992). Suppose that a faint, power-law galaxy merges with a bright core galaxy. The density near the center in the faint galaxy may be 100 times higher than in the bright galaxy (Fig. 7.1). Thus tidal forces during the merger should not disrupt the central part of the fainter galaxy, which should spiral intact to the center of the bright galaxy, and remain as a dense lump in the middle of the core. *Why are no such structures seen?* One possible answer is that the merger rate is so low that a remnant of this kind is not expected in our sample. The merger rate probably is not strongly dependent on the mass of the smaller galaxy—there are more small galaxies but their orbital decay from dynamical friction is slower— but nevertheless is quite uncertain: (i) Toomre (1977) estimated that roughly 10%

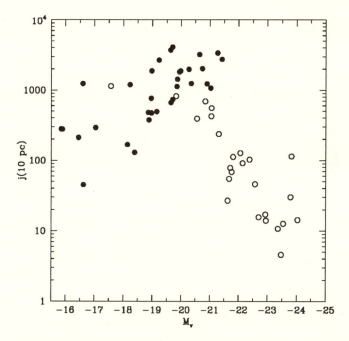

Figure 7.1: Luminosity density as a function of absolute magnitude for a sample of elliptical galaxies and spiral bulges (Faber et al. 1996). The luminosity density is measured in solar luminosities per cubic parsec at radius 10 pc ($H_0 = 80 \text{ km s}^{-1} \text{ Mpc}^{-1}$); in some cases this requires an extrapolation of the observed surface-brightness profile. Open circles denote core galaxies and filled circles denote power-law galaxies.

of giant galaxies have undergone major mergers, based on the frequency of tidal tails and the assumption that most ellipticals are made by mergers; (ii) Tóth and Ostriker (1992) argue that most spiral galaxies cannot accrete more than a few per cent of their disk mass without excessive thickening of the disk (this limit may be too stringent, since it neglects the excitation of bending waves in the disk which are subsequently damped by the halo); (iii) Lacey and Cole (1993) estimate that the fraction of giant galaxies that have consumed a companion is between 4% and 60% depending on the eccentricity of the companion orbit. Given these estimates, the fraction of core galaxies that have merged with a power-law galaxy could be small enough that no such systems are present in our sample, although such low rates are difficult to reconcile with the extensive observational evidence for recent mergers in ellipticals, such as kinematically decoupled cores and shells. A second possible answer to this unsolved problem is that the smaller galaxy may indeed be disrupted by time-varying tidal forces. Weinberg (1996) has argued that tides from the larger galaxy can disrupt a satellite galaxy before it merges if the secondary/primary mass

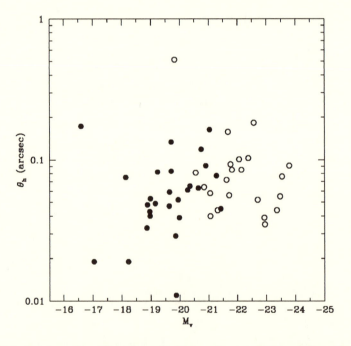

Figure 7.2: Angular radius of the sphere of influence of a central BH (eq. 7.5) as a function of absolute magnitude for a sample of elliptical galaxies and spiral bulges (Faber et al. 1996). The BH mass is assumed to be $M_\bullet = \Upsilon L$ where L is the galaxy luminosity and $\Upsilon = 0.003$. Open circles denote core galaxies and filled circles denote power-law galaxies.

ratio exceeds ~ 0.01. A third possibility, perhaps the most appealing, is that the smaller galaxy is disrupted by a central BH in the large galaxy.

We may estimate the angular radius θ_h of the BH sphere of influence (eq. 7.5) in any galaxy if, following § 7.1.1, we assume that every galaxy contains a central BH of mass $M_\bullet = \Upsilon L$, where $\Upsilon \simeq 0.003$. A dozen galaxies in the sample shown in Figure 7.2 have $\theta_h > 0.1''$ and yet in most of these galaxies there is no feature in the photometric profile that might be identified with this transition radius (an exception is M31, where the edge of the nucleus coincides with $\theta_h \simeq 0''.5$). *Why is there no evidence of central BHS in the photometry?* Perhaps (i) central BHS are only found in a small fraction of galaxies, or (ii) their masses are smaller than we have assumed; or, what is more likely, (iii) the response of the galaxy to a central BH does not generate a clear feature at θ_h, a possibility which is discussed further in the next subsection.

7.2.1   *The Peebles-Young Model*

The Peebles-Young or adiabatic model describes the effect of a central BH on the surrounding galaxy, based on the plausible (but quite possibly wrong) assumptions that the galaxy initially has a spherical analytic core, and that the BH grows slowly compared to the characteristic orbital time of stars in the core, $\lesssim 10^6$ y in (say) the central 100 pc. Thus it might apply, for example, if the BH is formed by accretion of material from a viscous disk formed at kpc scales during the initial collapse of the galaxy (e.g., Haehnelt and Rees 1993).

Given these assumptions, the predicted density distribution close to the BH is easy to derive. The stars that end up here are initially on low-energy orbits. This region of phase space has approximately constant phase-space density f_0 in the initial analytic core. Since phase-space density is conserved as the BH grows adiabatically, the final density of stars bound to the hole is

$$\rho(r) = \int_{E<0} f_0 d\mathbf{v} = \frac{4\pi f_0}{3} v_m^3, \tag{7.16}$$

where $v_m = (2GM_\bullet/r)^{1/2}$ is the escape speed from the BH. If the initial phase-space density is Maxwellian, then $f_0 = \rho_0/(2\pi\sigma^2)^{3/2}$ where ρ_0 is the initial central density and σ is the one-dimensional velocity dispersion; thus (Peebles 1972)

$$\rho(r) = \frac{4\rho_0}{3\sqrt{\pi}} \left(\frac{r_h}{r}\right)^{3/2} \qquad \text{when } r \ll r_h, \tag{7.17}$$

which implies a cusp in surface brightness,

$$I(r) \propto r^{-\gamma} \qquad \text{when } r \ll r_h, \tag{7.18}$$

where $\gamma = \frac{1}{2}$. This result is more robust than its derivation, since other mechanisms of BH formation might also preserve the phase-space density in the core.

Numerical solutions for the Peebles-Young model (Young 1980, Quinlan et al. 1995) confirm that the surface brightness is accurately described by (7.18) with $\gamma = \frac{1}{2}$ at sufficiently small radii. However, the transition to this asymptotic slope is very slow: at radii $\gtrsim 0.01R_c$, where R_c is the core radius of the initial analytic core, the surface-brightness profile shows a smooth transition from the unperturbed profile at $r \gg R_c$ to an approximate power law with slope $\overline{\gamma}$ at $0.01R_c \lesssim r \lesssim R_c$. The slope $\overline{\gamma}$ varies from 0 to $\frac{1}{2}$ depending on the ratio of the BH mass to the core mass; the asymptotic slope of $\frac{1}{2}$ is only approached at much smaller radii, $r \ll 0.01R_c$. There is no clear break in the surface-brightness profile near r_h unless $r_h \ll 0.01R_c$. Since even HST resolves only a limited range inside the break or core radius (the largest break radii of nearby galaxies are $\lesssim 4''$) no sharp feature at r_h should be expected, and the power-law profiles revealed by HST inside the break radius of core galaxies are therefore consistent with Peebles-Young models.

The power-law galaxies have $\gamma \simeq 0.8 \pm 0.3$, significantly greater than the asymptotic slope $\gamma = \frac{1}{2}$ predicted by the arguments above. Nevertheless even these profiles could be explained by the Peebles-Young model, in at least two ways: (i) Quinlan et al. (1995) have shown that the adiabatic growth of a central BH in galaxies with non-analytic cores can produce extended power-law surface-brightness profiles with slopes as large as $\gamma = \frac{4}{3}$; even initial cores that are almost indistinguishable in their initial surface-brightness profile from analytic cores (such as $I(R) = I_0 + I_1 R^2 + I_1' R^2 \log R + \cdots$) can have γ as large as 1. (ii) Even if the core is analytic, power-law profiles with $\gamma \simeq 1$ can also result if the BH mass exceeds the core mass, a plausible supposition since low-luminosity galaxies should have small core masses.

Thus, if all galaxies contain central BHs, the Peebles-Young model can reproduce the main features of the photometric profiles of both core and power-law galaxies; the difference between the two types might simply reflect the relative mass of the initial core and the BH. The converse, however, is not correct: the match between the observed profiles and Peebles-Young models does not imply that massive central BHs are present, since gas dynamics and star formation may generate similar structures. To distinguish these alternatives we need high-resolution kinematic data: if the velocity-dispersion tensor is isotropic, the dispersion should rise as we approach the center if a BH is present, and fall if it is not.

The effects of slow growth of a central BH in an axisymmetric core have not been investigated; triaxial cores are discussed briefly in § 7.4.

7.3   KINEMATIC EVIDENCE FOR CENTRAL BLACK HOLES

The strongest evidence for a nearby massive BH comes from the Sbc galaxy NGC 4258 (Miyoshi et al. 1995, Moran et al. 1995). Water masers have been detected in an edge-on disk $\sim 0.2\,\mathrm{pc}$ from the center of the galaxy. The disk is perpendicular to the radio jet seen at much larger distances, as is expected if the jet emerges from the axis of the disk. The rotation curve of the disk is symmetric and Keplerian; the velocities and centripetal accelerations of the masers imply a central mass of $3.6 \times 10^7 M_\odot$ and incidentally provide the best available distance estimate for the galaxy, $6.4 \pm 0.9\,\mathrm{Mpc}$. Unfortunately, bright maser disks with the favorable geometry found in NGC 4258 are rare.

M87 is a bright, nearby AGN galaxy and has long been regarded as the best site to prospect for a massive BH. The evidence from measurements of the spatial and velocity distribution of the stars remains inconclusive (Kormendy and Richstone 1995). However, HST observations reveal a disk of ionized gas which is approximately perpendicular to the well-known optical jet and which appears to be in circular Keplerian rotation at $\sim 20\,\mathrm{pc}$ from the center; the inferred central mass is $M_\bullet = (2.4 \pm 0.7) \times 10^9 M_\odot$ (Harms et al. 1994). This finding supports (weakly) the Peebles-Young model, which fits the photometric profile (Lauer et al. 1992) accurately if the central BH mass $M_\bullet = (2.6 \pm 0.5) \times 10^9 M_\odot$ (Young et al. 1978), in surprisingly good agreement with the measurement from the gas disk. However, it would be rash to invoke this agreement as an argument for the existence of massive BHs without similar comparisons for other galaxies.

Kormendy and Richstone (1995) review the strong stellar-dynamical evidence for central BHs in six other nearby galaxies: M31, M32, NGC 3115, NGC 3377, NGC 4594, and the Galaxy. They argue that there is evidence for BHs in a fraction $f \simeq 0.2$ of the galaxies surveyed, and that the fraction of galaxies that actually contain BHs is substantially higher because detection is difficult. They suggest that BH mass correlates with luminosity of the elliptical galaxy or the spiral bulge (not the spiral disk), although the correlation is partly a selection effect—small BHs can only be detected in small galaxies. The median ratio of BH mass to the elliptical or bulge luminosity for the eight galaxies is 0.013. Thus the kinematic evidence from nearby galaxies is roughly consistent with the strawman model of § 7.1.1, which assumes that $f \approx 1$ and that the ratio of BH mass to *total* luminosity is $\Upsilon \approx 0.003$.

This research area should advance rapidly in the next few years. Higher resolution kinematic data on the central regions of nearby galaxies will be provided by ground-based telescopes and by the next-generation STIS on HST. High signal-to-noise spectra can yield the complete line-of-sight velocity profile rather than just the mean velocity and dispersion. Several groups are developing improved modeling techniques that generate (i) axisymmetric solutions of the collisionless

Boltzmann and Poisson equations, with (ii) distribution functions depending on all three integrals of motion, that (iii) predict the complete line-of-sight velocity profile.

A challenging unsolved problem is *what is the demography of BHs in galaxies*, that is, the probability $p(M_\bullet, L, T)$ that a galaxy of luminosity L and type T contains a central BH of mass $M_\bullet$. The problem is difficult because a large sample of reliable BH detections is needed, and because of strong selection effects: maser disks or emission-line disks are rare, and the detection of a BH from stellar kinematics is a strong function of the galaxy's photometric profile, velocity dispersion, etc. For example, the existing data suggest that the BH mass is proportional to the bulge luminosity. This correlation probably reflects in part the influence of selection effects on a much broader black-hole mass distribution. A tight correlation between BH mass and bulge luminosity would be difficult to reconcile with the belief that many elliptical galaxies are formed by the merger of disk galaxies (Toomre 1977, Wielen 1990, Barnes and Hernquist 1992), since mergers can convert galactic disks to bulges without any corresponding change in the BH mass.

Finally, we stress once again the distinction between massive dark objects (MDOS) and BHS. MDOS are systems with mass-to-light ratios much higher than normal stellar populations, which may be either relativistic (i.e. BHS) or non-relativistic (e.g., clusters of neutron stars or brown dwarfs). The observations described in this section provide evidence for MDOS; we believe the MDOS are BHS only because of indirect arguments such as those given in § 7.1.1, Kormendy and Richstone (1995), and Maoz (1995).

The kinematic evidence that BHS are guilty of lurking in the centers of nearby galaxies is strong by astronomical standards but perhaps weak by legal ones (or perhaps not; Gastwirth 1992 reproduces a poll of judges showing that the confidence level they assigned to the legal standard "beyond a reasonable doubt" spanned the surprisingly low range 75–95%). Undoubtedly a sharp lawyer could persuade a jury to acquit many individual suspects, and efforts to do so are worthwhile. However, the consequences of erroneous conviction are less severe in astronomy than in law; therefore we should explore the implications of massive BHS for galactic structure whether or not the debate over their existence is fully resolved.

7.4   PHYSICAL PROCESSES

In this section I review some of the physical processes that operate in the region $r \gtrsim 1$ pc that is now accessible to HST in nearby galaxies.

Relaxation. The high stellar densities in Figure 7.1 imply that the relaxation time from star-star encounters is relatively short near the centers of ellipticals. The relaxation times at 10 pc for the galaxies in Figure 7.1 range from 10^{14} y to $10^{11.5}$ y (Faber et al. 1996), too long to be interesting. However, as the density is generally still rising at 10 pc, relaxation and stellar collisions are likely to be important at smaller radii—they are certainly important in several nearby systems such as our own Galaxy (Phinney 1989) and M33 (Kormendy and McClure 1993).

In globular clusters, energy is transported outwards by relaxation, since the velocity dispersion σ decreases outwards (i.e. energy flows from "hot" to "cold" regions), leading eventually to core collapse. In contrast, dynamical models of ellipticals that match the photometry (assuming spherical symmetry, isotropic velocity-dispersion tensor, and constant mass-to-light ratio) show that outside the sphere of influence of the central BH, the velocity dispersion generally *increases* outwards in the central region. Within the BH sphere of influence, the dispersion *decreases* outwards. Thus, if relaxation is important, energy is expected to flow towards the transition radius r_h (eq. 7.5) from both larger and smaller radii. The consequences of this flow have not been explored, although Quinlan (1995) has examined the relaxation-driven evolution of elliptical galaxy models without central BHS.

Our understanding of relaxation in stellar systems is not necessarily complete. The usual estimate of the relaxation time, based on binary encounters between stars, has only been confirmed by N-body experiments for $N < 4 \times 10^3$ over times $< 10^3 t_{\mathrm{dyn}}$, where $t_{\mathrm{dyn}} = r/\sigma$ is the dynamical time (Farouki and Salpeter 1994). The dynamical time at 10 pc is $10^5(100\,\mathrm{km\,s}^{-1}/\sigma)$ y so the age of the galaxy is of order $10^5 t_{\mathrm{dyn}}$ at 10 pc; the luminosity inside 10 pc ranges from 3×10^4 to $4 \times 10^7 L_\odot$ for the galaxies in Figure 7.1, so the effective value of N is 10^5 to 10^8; these values are far outside the range in which our concepts of relaxation have been tested. Other relaxation mechanisms may operate more quickly than binary encounters in some cases: (i) Weinberg (1993) has demonstrated that collective interactions can enhance the relaxation rate from binary encounters by a substantial factor, although he was not able to estimate the enhancement factor for realistic stellar systems. (ii) Angular-momentum relaxation may be enhanced in potentials where two or more of the fundamental frequencies are nearly degenerate, such as the near-Kepler potential close to a central BH (Ostriker 1974, Rauch and Tremaine 1996).

Should the relaxation time be shorter than the age of the galaxy, it can be shown that the stellar surface brightness near the central BH should have a power-law cusp, described by equation (7.18) with $\gamma = \frac{3}{4}$ (Bahcall and Wolf 1976).

Globular Clusters.   The relaxation rate of stars of mass $m_\star$ due to a population of objects with number density n and mass $m \gtrsim m_\star$ is proportional to nm^2. In an elliptical galaxy, the number density of globular clusters relative to stars is $n_{gc}/n_\star \sim 10^{-7.5}$ (Harris 1991), and the typical cluster mass is $m_{gc} \sim 10^5 m_\star$. Thus relaxation from globular cluster encounters is more important than relaxation from stellar encounters, by a factor of $\sim 10^{2.5}$. Over most of a typical galaxy, neither process is significant, but near the center star-globular encounters could be important. The encounters heat up and isotropize the stellar population, while draining energy from the globular cluster orbits (dynamical friction). The clusters spiral in to the center of the galaxy where they are disrupted by tidal forces (from the stellar distribution, a central BH, and other clusters). Tremaine et al.'s (1975) suggestion that the nucleus of M31 was formed by cluster inspiral fails to explain its rapid rotation (van den Bergh 1991) and high metallicity.

A major uncertainty is the number of globulars near the center. In most galaxies, the surface number density of globulars is flat at radii $\lesssim 2$ kpc, while the surface density of stars continues to rise to much smaller radii; thus at radii $\lesssim 1$ kpc the spatial density of globulars is probably flat or even decreasing—perhaps to zero near the center. This deficit may be primordial or may arise through a dynamical process that preferentially destroys clusters at small radii; in the latter case the clusters may have an important influence on the galactic center before they disappear. The most promising dynamical process is tidal disruption of clusters on stochastic orbits that pass too close to a central BH (Ostriker et al. 1989).

Triaxiality.   Dissipationless collapse usually produces triaxial stellar systems and observations such as minor-axis rotation suggest that elliptical galaxies may be triaxial (de Zeeuw and Franx 1991)—although the central regions are probably less so than the outer parts because of gas infall and dissipation. A central BH or density cusp has important consequences for triaxial models (Norman et al. 1985, Gerhard and Binney 1985). Regular box orbits are the "backbone" that supports triaxiality; these orbits pass arbitrarily close to the center and if they are scattered by a central feature they become chaotic so the backbone dissolves. More precisely, the family of regular box orbits is replaced by centrophilic stochastic orbits and centrophobic regular "boxlets" (Miralda-Escudé and Schwarzschild 1989), both of which are less effective supports for triaxiality. Merritt and Fridman (1995) have extended Gerhard and Binney's two-dimensional orbit calculations to three dimensions and find that even weak cusps can destroy strongly triaxial models over the lifetime of a galaxy ($\sim 10^4$ dynamical times at 100 pc). Thus power-law galaxies, and perhaps many core galaxies, are likely to be axisymmetric near their centers.

Off-Center Structures.   The nucleus of the nearby spiral galaxy M31 is offset from
the kinematic and photometric center of the bulge. The offset was revealed by the
Stratoscope II balloon-borne telescope (Light et al. 1974), but a detailed picture of
the nucleus was only obtained two decades later by HST, which showed that the
nucleus contains two separate components, separated by $0.''49$ (Lauer et al. 1993).
The component with the lower surface brightness coincides with the center of the
bulge, while the brighter, off-center component is the nuclear core measured by
Stratoscope. The offset is unlikely to be an artifact of irregular dust obscuration
as there are no color gradients or far-infrared emission; the nucleus is unlikely to be
binary since the binary orbit would decay by dynamical friction in $\lesssim 10^8$ y. A more
promising possibility is that the nucleus contains an eccentric stellar disk orbiting
a central BH (Tremaine 1995); the dynamics and evolution of such disks is another
largely unsolved problem.

Lauer et al. (1995) find that roughly 15% of elliptical galaxies and spiral bulges
observed with HST show lopsided structure (i.e. the bright isophotes do not share a
common center). Most of these are core galaxies, which is perhaps not surprising
since the restoring force $GM(r)/r^2$ approaches zero near the center in galaxies
with relatively flat cores [$k < 1$ in the notation of eq. (7.9)], but diverges near the
center when the density profile is steep.

Possible explanations for lopsided structure include (i) irregular dust obscuration; (ii) an eccentric stellar disk surrounding a central BH; (iii) a binary BH, formed
by the merger of two galaxies containing central BHS (Begelman et al. 1980) or a
steady trickle of BHS from the galactic halo (Xu and Ostriker 1994); (iv) a collective
oscillation of the central part of the galaxy. A result that bears on (iv) is Weinberg's
(1991) discovery that lopsided modes in spherical stellar systems with flat cores
are very weakly damped; thus if an arbitrary spectrum of modes is excited the
lopsided modes will persist much longer than the others.

An unsolved problem is *how far does a "central" BH wander from the center
of the galaxy?* Dynamical friction drags orbiting BHS towards the center but when
does the inspiral stop? This issue was addressed by Bahcall and Wolf (1976), who
estimated the equilibrium separation of a BH from the center of a globular cluster
with an analytic core. The BH is in thermal equilibrium with the core stars, which
have "temperature" $kT = m\sigma^2$, where m is the stellar mass and σ is the one-
dimensional velocity dispersion (which is constant within the core). In the center
of a spherical core with constant density ρ_0, the potential is $U(r) = \frac{2}{3}\pi G\rho_0 r^2$;
equipartition implies that $M_\bullet \langle U(r) \rangle = \frac{3}{2}kT$, so that the mean-square separation
of the BH from the center is

$$\langle r^2 \rangle = \frac{9m\sigma^2}{4\pi G\rho_0 M_\bullet} = \frac{m}{M_\bullet}R_c^2, \qquad (7.19)$$

where R_c is the core radius and we have used King's formula (7.14). This simple argument does not apply to galaxies since they do not have flat cores with constant velocity dispersion. The Brownian motion of BHS near the centers of more realistic galaxy models has been investigated by Quinlan (1995), who finds quite different scaling laws.

Progress in resolving many of these unsolved problems will be driven by large-scale N-body simulations, which are rapidly advancing in power and sophistication. Sigurdsson et al. (1995) now run simulations with $N = 10^6$–10^8 on massively parallel machines, using a self-consistent field method—which unfortunately cannot follow relaxation-driven evolution because it does not compute forces between individual stars. Simulations that actually calculate the force between every star pair are slower, but now benefit from the special-purpose GRAPE computers (Ebisuzaki et al. 1993), which use hard-wired parallel pipelines to speed up the force calculations.

7.5   SUMMARY

Over the past twenty years, most of the discussion of BHS in the centers of elliptical galaxies has focused on whether or not they exist. This issue has become less controversial as the evidence for BHS accumulates. Some of the important remaining unsolved problems include:

- What is the distribution of BH masses as a function of galaxy luminosity and type? In particular, why does the BH mass appear to be roughly proportional to the bulge luminosity?

- What determines the distribution of stars near the center of an elliptical galaxy? In particular, why are there no analytic cores? Do the power-law galaxies contain unresolved cores? If not, why do some galaxies have cores, while others do not? How are these questions related to the presence of a central BH?

- Is there any reliable signature of a central BH in the stellar photometry?

- Under what conditions do eccentric stellar disks form and persist around central BHS? Are such disks the explanation for the double nuclei and lopsided structures seen at the centers of some galaxies?

- How do the low-density cores in giant galaxies survive mergers with small dense companion galaxies?

My research on this subject has been supported in part by grants from NSERC, from the Killam Program of the Canada Council, and from the Raymond and Beverly Sackler Foundation. I thank my HST collaborators Ed Ajhar, Yong-Ik Byun, Alan Dressler, Sandra Faber, Karl Gebhardt, Carl Grillmair, John Kormendy, Tod Lauer, and Doug Richstone for many insights. I have discussed many of the theoretical issues described here with Christophe Pichon, Gerry Quinlan, Kevin Rauch and David Syer, and the manuscript has been improved by suggestions from John Bahcall, Matt Holman, Ramesh Narayan, Jerry Ostriker, Martin Schwarzschild, and especially Roeland van der Marel. Finally, I am grateful to John Bahcall for his constant and enthusiastic support and encouragement over the past two decades.

BIBLIOGRAPHIC NOTES

The following references provide an introduction to many of the issues discussed in this article.

- **Binney, J., & Tremaine, S. 1987, Galactic Dynamics (Princeton: Princeton University Press)**. A graduate-level textbook on galaxy dynamics.

- **Blandford, R. D., Netzer, H., & Woltjer, L. 1991, Active Galactic Nuclei (Berlin: Springer-Verlag)**. The three articles in this book review why massive BHs are believed to be the power source for active galactic nuclei.

- **Kormendy, J., & Richstone, D. O. 1995, ARA&A, 33, 581**. An authoritative, conservative and up-to-date review of the dynamical evidence for massive BHs in the centers of galaxies.

- **Lauer, T. R., et al. 1993, AJ, 106, 1436**. HST observations that first revealed the apparent double structure of the nucleus of M31 and discussed its interpretation.

- **Miyoshi, M., et al. 1995, Nature, 373, 127**. This brief paper, amplified by Moran et al. (1995), presents compelling evidence that the nucleus of NGC 4258 contains a disk of gas in circular Keplerian orbits around a massive central object. At present this is the strongest dynamical evidence for a massive BH in any galaxy.

- **Peebles, P. J. E. 1972, Gen. Rel. Grav. 3, 63**. A remarkably prescient first discussion of the best locations and methods to prospect for BHs.

- **Sołtan, A. 1982, MNRAS, 200, 115**. An elegant derivation of the expected density of massive BHs based on the local density of quasar photons. Updated by Chokshi & Turner (1992).

- **Toomre, A. 1977, in The Evolution of Galaxies and Stellar Populations, eds. B. M. Tinsley & R. B. Larson (New Haven: Yale University Observatory), 401**. An influential and far-sighted paper which provided the first plausible estimate of the frequency of galaxy mergers, and introduced the important hypothesis that most or all elliptical galaxies are created through mergers of disk galaxies.

- **Young, P. J. 1980, ApJ, 242, 1232**. A beautiful paper, which determines the expected shape of the stellar distribution around a BH that grows adiabatically in the center of an isothermal sphere (the Peebles-Young model). A more general version of this calculation is given by Quinlan et al. (1995).

BIBLIOGRAPHY

[1] Bahcall, J. N., & Wolf, R. A. 1976, ApJ, 209, 214.

[2] Barnes, J. E., & Hernquist, L. 1992, ARA&A, 30, 705.

[3] Begelman, M. C., Blandford, R. D., & Rees, M. J. 1980, Nature, 287, 307.

[4] van den Bergh, S. 1991, PASP, 103, 1053.

[5] Blandford, R. D., Netzer, H., & Woltjer, L. 1991, Active Galactic Nuclei (Berlin: Springer-Verlag).

[6] Carlberg, R. G. 1986, ApJ, 310, 593.

[7] Chokshi, A., & Turner, E. L. 1992, MNRAS, 259, 421.

[8] Crane, P., et al. 1993, AJ, 106, 1371.

[9] Dehnen, W. 1993, MNRAS, 265, 250.

[10] Ebisuzaki, T., et al. 1993, PASJ, 45, 269.

[11] Eckart, A., et al. 1993, ApJ, 407, L77.

[12] Efstathiou, G., Ellis, R. S., & Peterson, B. A. 1988, MNRAS, 232, 431.

[13] Faber, S., et al. 1996, in preparation.

[14] Fabian, A. C., & Canizares, C. R. 1988, Nature, 333, 829.

[15] Farouki, R. T., & Salpeter, E. E. 1994, ApJ, 427, 676.

[16] Gastwirth, J. L. 1992, American Statistician, 46, 55.

[17] Gerhard, O. E., & Binney, J. 1985, MNRAS, 216, 467.

[18] Haehnelt, M. G., & Rees, M. J. 1993, MNRAS, 263, 168.

[19] Harms, R. J., et al., 1994, ApJ, 435, L35.

[20] Harris, W. E. 1991, ARA&A, 29, 543.

[21] Hartwick, F.D.A., & Schade, D. 1990, ARA&A, 28, 437.

[22] Hernquist, L., Spergel. D. N., & Heyl, J. S. 1993, ApJ, 416, 415.

[23] Jaffe, W. et al., 1994, AJ, 108, 1567.

[24] Kent, S. M., Dame, T. M., & Fazio, G. 1991, ApJ, 378, 131.

[25] Kormendy, J., & McClure, R. D. 1993, AJ, 105, 1793.

[26] Kormendy, J., & Richstone, D. O. 1995, ARA&A, 33, 581.

[27] Kormendy, J., & Sanders, D. B. 1992, ApJ, 390, L53.

[28] Lacey, C., & Cole, S. 1993, MNRAS, 262, 627.

[29] Lauer, T. R., et al. 1992, AJ, 103, 703.

[30] Lauer, T. R., et al. 1993, AJ, 106, 1436.

[31] Lauer, T. R., et al. 1995, AJ, 110, 2622.

[32] Light, E. S., Danielson, R. E., & Schwarzschild, M. 1974, ApJ, 194, 257.

[33] Maoz, E. 1995, ApJ, 447, L91.

[34] Merritt, D., & Fridman, T. 1996, to be published in ApJ.

[35] Mihos, J. C., & Hernquist, L. 1994, ApJ, 437, L47.

[36] Miralda-Escudé, J., & Schwarzschild, M. 1989, ApJ 339, 752.

[37] Miyoshi, M., et al. 1995, Nature, 373, 127.

[38] Moran, J., et al. 1995, in Quasars and AGN: High Resolution Imaging, to be published in Proc. Nat. Acad. Sci.

[39] Narayan, R., Yi, I., & Mahadevan, R. 1995, Nature, 374, 623,

[40] Norman, C. A., May, A., & van Albada, T. S. 1985, ApJ, 296, 20.

[41] Ostriker, J. P. 1974, unpublished lecture notes.

[42] Ostriker, J. P., Binney, J., & Saha, P. 1989, MNRAS, 241, 849.

[43] Peebles, P. J. E. 1972, Gen. Rel. Grav. 3, 63.

[44] Phinney, E. S. 1989, in The Center of the Galaxy, ed. M. Morris (Dordrecht: Kluwer), 543.

[45] Quinlan, G. D. 1995, presentation at Aspen Center for Physics workshop, "Dynamics of Dense Stellar Systems."

[46] Quinlan, G. D., Hernquist, L., & Sigurdsson, S. 1995, ApJ, 440, 554.

[47] Rauch, K. P., & Tremaine, S. 1996, in preparation.

[48] Rees, M. J. 1984, ARA&A, 22, 471.

[49] Rees, M. J. 1990, Science, 247, 817.

[50] Richstone, D. O., & Tremaine, S. 1986, AJ, 92, 72.

[51] Schechter, P. 1976, ApJ, 203, 297.

[52] Sigurdsson, S., Hernquist, L., & Quinlan, G. D. 1995, ApJ 446, 75.

[53] Sołtan, A. 1982, MNRAS, 200, 115.

[54] Toomre, A. 1977, in The Evolution of Galaxies and Stellar Populations, eds. B. M. Tinsley & R. B. Larson (New Haven: Yale University Observatory), 401.

[55] Tóth, G., & Ostriker, J. P. 1992, ApJ, 389, 5.

[56] Tremaine, S. 1995, AJ, 110, 628.

[57] Tremaine, S., Hénon, M., & Lynden-Bell, D. 1986, MNRAS, 219, 285.

[58] Tremaine, S., Ostriker, J. P., & Spitzer, L. 1975, ApJ, 196, 407.

[59] Tremaine, S., et al. 1994, AJ, 107, 634.

[60] Weinberg, M. D. 1991, ApJ, 368, 66.

[61] Weinberg, M. D. 1993, ApJ, 410, 543.

[62] Weinberg, M. D. 1996, in preparation.

[63] Wielen, R., ed. 1990, Dynamics and Interactions of Galaxies (Berlin: Springer).

[64] Xu, G., & Ostriker, J. P. 1994, ApJ, 437, 184.

[65] Young, P. J. 1980, ApJ, 242, 1232.

[66] Young, P. J., et al. 1978, ApJ, 221, 721.

[67] de Zeeuw, T., & Franx, M. 1991, ARA&A, 29, 239.

CHAPTER 8

THE MORPHOLOGICAL EVOLUTION OF GALAXIES

RICHARD S. ELLIS

Institute of Astronomy, Madingley Road, Cambridge, England

ABSTRACT

I review the general progress made in the study of galaxy evolution concentrating on the impact of systematic ground-based spectroscopic surveys of faint galaxies and high resolution imaging with Hubble Space Telescope. The picture emerging is one where massive regular galaxies have changed little since redshift $z \simeq 1$, whereas there has been a marked decline in the abundance of less massive star forming dwarf galaxies. Reconciling these trends with the conventional hierarchical growth of structure on galactic scales and determining the fate of the star forming dwarfs remains a major unsolved problem in contemporary cosmology.

8.1 INTRODUCTION

Considerable progress has been made in recent years in constraining the recent evolution of normal galaxies through detailed studies of stellar populations of nearby systems and comprehensive photometric and spectroscopic surveys of field and cluster galaxies viewed at large look-back times. An outstanding restriction in the analysis of high redshift data has been the 'comparing apples and oranges' problem. Given the diverse nature of the galaxy population at any redshift, how can similar types be determined unambiguously over a range in look-back time and the evolution of a subset of the population be reliably determined? Let me start by discussing two recent examples of observational progress in the high redshift area which, as a 'Devil's advocate', I will expose to illustrate some weaknesses we seek to eliminate.

In the first example, let us consider the work of Dressler & Gunn (1990) and

159

Aragón-Salamanca et al. (1993) on the respective D_{4000} spectral discontinuities and $V - K$ colours for the reddest galaxies in a number of $z > 0.5$ clusters. By comparing the mean properties of such galaxies with the spectral energy distributions of early-type galaxies observed locally, both teams claim there has been only modest evolution in luminous early-type cluster members since $z \simeq 1$. Whilst it might seem reasonable to suppose that the *reddest* galaxies in each high z cluster are early-type systems, without morphological imaging one cannot exclude the possibility that some of the *bluer* galaxies are also early-type. The simple conclusion derived by Dressler, Aragón-Salamanca and co-workers only truly applies if the early-type population behaves uniformly so that observations of a few examples, selected by colour or spectral class, are representative of the entire population. As there is no *a priori* reason to suppose the local uniformity in the colour-spectral class-morphology plane extends to high redshift, the absence of any one of these observables could seriously undermine any evolutionary tests. Without morphologies, the claim of only modest evolution cannot be rigorously made.

In the second example, we consider spectroscopic surveys of field galaxies (Broadhurst et al. 1988, Colless et al. 1990, Cowie et al. 1992, Glazebrook et al. 1995a) which have collectively reinforced the view that the excess galaxies first noted in the deep photographic and CCD galaxy counts (Tyson 1988) are caused by an increase in the absolute normalisation of the field galaxy luminosity function (LF) with redshift. Some subset of the galaxy population is either disappearing or fading with time. How can examples of this population be located so that their evolution can be tracked independently and studied? Notwithstanding the large investment in redshift surveys, as LFs are statistical by nature, some 'marker' is required to identify the evolving subset.

In this lecture I review the progress in these and related areas that is now possible via the recovery of the originally-anticipated imaging capabilities of Hubble Space Telescope. Morphological classifications are now feasible for remote galaxies at cosmologically interesting redshifts. The morphology of a galaxy is, of course, only some form of visual label with as yet no agreed physical basis. However, I suggest it does provide a good marker for resolving evolutionary impasses such as those discussed above. Conventional photometric and spectroscopic indicators of galaxy class are unstable to sudden changes in the star formation rate and thus it is difficult to use these criteria to reliably connect similar classes across ranges in look-back time. Is a faint blue emission line source the predecessor of a local red quiescent galaxy or new class of object? Although morphologies have disadvantages too (they may also evolve and are difficult to measure at faint limits), the additional dimension is a tremendous bonus in understanding the origin of the Hubble sequence.

8.2 Early Formation of Massive Ellipticals

The repair of HST permits the direct identification of elliptical and S0 galaxies at significant cosmic depths. Figure 8.1 shows a comparison of two images of the distant cluster 0016+16 (z=0.54) taken with similar exposure times and filters using the 4.2m William Herschel telescope in 0.9 arcsec seeing (Smail et al. 1994) and WFPC-2 (as part of the 'Morphs' collaboration of Butcher, Couch, Dressler, Ellis, Oemler, Sharples & Smail). The improvement in the ability to classify faint galaxies is evident.

To illustrate the possibilities further, Figure 8.2 shows a selection of early-type members in 0016+16 drawn from the comprehensive spectroscopic work of Dressler & Gunn (1992); the separation between S0s and ellipticals is apparent. Although inevitably there are some cases where precise classifications remain uncertain, the data offers a number of important possibilities which the Morphs team is exploring.

A particularly interesting, although preliminary, result concerns the photometric homogeneity of the giant elliptical population at these large look-back times. Bower et al. (1992) demonstrated, using precision UBV photometry of ellipticals in the Coma and Virgo clusters, that the intrinsic scatter around the $U - V$ colour-luminosity relation is barely detectable ($\delta(U - V)_o$ <0.033). As the U-band light is a strong constraint on past star-formation rates (SFR), either ellipticals across both clusters shared the same synchronous decline in SFR (a 'divine intervention'

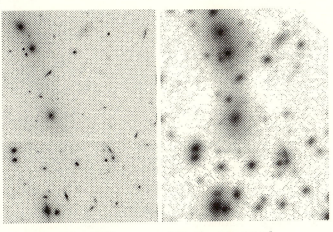

WFPC2 (7 orbits) F814W   WHT (7 hours) I band S–0.9″

Figure 8.1: 0016+16 z=0.54 as viewed by WFPC-2 (left panel, courtesy of the 'Morphs') and the 4.2m WHT in 0.9 arcsec seeing (right panel, from Smail et al. 1994).

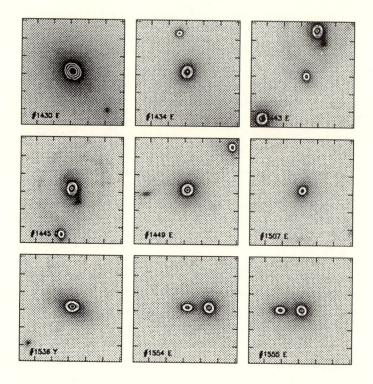

Irs 22–Feb–1995 15:34

Figure 8.2:   A montage of spectroscopically-confirmed early-type members of 0016+16 (z=0.54) from the 'Morphs' WFPC-2 programme, including one spiral (center left); each box is 5 arcsec on a side. The classification of early-types, and even the distinction between Es and S0s, becomes a practical proposition.

hypothesis), or the homogeneity indicates the systems are sufficiently old for individual variations in the time of formation to be negligible. Depending on the degree of synchronicity, Bower et al. claimed ages of between 8 and 12 Gyr. This simple test is remarkably powerful because the constraint depends only on the rate at which the rest-frame U-band luminosity declines with time which, in turn, is governed by main sequence lifetimes which are well-understood, at least on a relative scale.

At $z \simeq 0.55$, by good fortune, the $V - I$ colours from HST map neatly onto rest-frame $U - V$ enabling the same test to be carried out for morphologically-confirmed elliptical members at a significant look-back time. The test has been carried out for several of the clusters in the appropriate redshift range in the Morphs' sample and the scatter is somewhat larger, $\delta(U - v)_o <0.07$, the precise upper limit being determined by the increased photometric uncertainties for such faint sources. Even so, this places a tight constraint on the ages of stars in giant galaxies which is then added to the considerable look-back time.

[A preliminary version of this test was described by Ellis (1992) on the basis of ground-based data alone. Quite apart from the morphological confirmation provided by the improved angular resolution of HST, the WFPC-2 photometry is actually *better* than that of the 4.2m Herschel primarily because the sky background is significantly fainter. It is a misconception that HST is a 'small aperture telescope'. For background-limited imaging the reduced background in the near-infrared renders HST equivalent to a 6.5m telescope independent of its superlative angular resolution.]

Depending on the cosmological model, $\delta(U - V)_o$ <0.07 at z=0.55, when added to the look-back time, implies a *present-day* age of >12 Gyr. Cluster ellipticals thus join globular clusters in presenting a problem for certain cosmological models [cf. the contributions by Peebles and by Steinhardt in this volume]. Indeed, there is evidence that regular red ellipticals are seen in abundance in HST images of clusters beyond $z \simeq 1$ (Dickinson 1996), although spectroscopic work is needed in these fields to be sure of the implications.

These tests make strong statements about the ages of galaxies without directly observing their epoch of formation. A good indication of the likely impact of 8-10 m class telescopes in this area follows from the remarkable absorption line spectrum of the distant radio galaxy 3C65 (z=1.2) obtained with the Keck 10m by Stockton et al. (1995). The high signal/noise in this spectrum details the Balmer lines and Ca II doublet sufficiently clearly to establish an age of >2-3 Gyr at z=1.2. An exhaustive survey of this kind for luminous z=0.5-1 ellipticals without radio emission is possible with the new generation of telescopes and would provide tighter constraints on their likely ages than results discussed here based on broadband colours alone.

The conclusion emerging thus far is that the *stars* that make up massive cluster ellipticals formed in a narrow time interval prior to $z \simeq 3$–5. It is, of course, conceivable that there was some subsequent merging of sub-units to create these galaxies but the associated star formation cannot have been significant at recent times as this would conflict with Bower et al.'s $U - V$ scatter. Kauffmann (1996) addresses this point directly and finds the likely merger rate in hierarchical CDM models can be reconciled with Bower et al.'s data if most of the merging occurred between 0.5< z <1. However, the small scatter in colours now observed in more distant clusters must presumably push any merging to yet higher redshifts.

A final caveat is that the results quoted above might only apply to the rather atypical giant ellipticals in dense cluster regions. Certainly, there is growing evidence that lower luminosity ellipticals and those in lower density environs have suffered more recent star formation (Rose et al. 1994). However, as we will see in § 8.5, there is no evidence for significant evolution in the numbers or luminosities of distant field ellipticals either.

8.3   SLOW EVOLUTION OF MASSIVE DISK GALAXIES

Bergeron & Boissé (1991) and, more recently, Steidel & Dickinson (1995) have demonstrated the value of using the galaxies responsible for the metallic absorption line systems in the spectra of distant QSOs as probes of field galaxy evolution to $z \simeq 1$. Steidel & Dickinson's comprehensive survey of >50 Mg II absorbing galaxies with $\overline{z} \simeq 0.7$ indicates that the typical absorber is a massive disk galaxy whose rest-frame B and K luminosities exceeds $\simeq 0.06L^*$ drawn from a luminosity function (LF) which is remarkably similar to the bright end of that observed locally (although their normalisation is somewhat higher, see § 8.5). The colours of the absorbers likewise span the range seen locally (after allowing for redshift effects). Overall, the survey is consistent with little or no evolution for the absorbing population.

Although these results have received much attention, the sample remains small, particularly in each colour class. The absence of strong luminosity evolution from $z=0$ to 0.7 can only be claimed with confidence for the entire population. Additionally, one might worry that, by selecting galaxies in absorption, only some subset of the population that has 'settled down' is found, with the consequence that a no evolution result is guaranteed. The QSO surveys have certainly found some galaxies which do *not* give rise to absorption lines (so-called 'interlopers'), but many of these are blue star-forming dwarfs (Steidel et al. 1993). That the absorbers and interlopers are non-overlapping photometric samples is perhaps the most interesting result.

The Mg II surveys discussed above demand high resolution QSO absorption line spectroscopy, deep imaging in the vicinity of the QSO and subsequent spectroscopy of candidate galaxies. Without the final stage to confirm a galaxy has precisely the same redshift as that of the Mg II absorption line, no firm association can be made. At higher redshift, this becomes harder both because the galaxies are fainter, and because there is a dearth of suitable features for estimating redshifts with optical spectrographs for $z > 1.3$.

For the purposes of deriving an absorber LF, Aragón-Salamanca et al. (1994) showed that a statistical association will suffice and may be obtained from imaging alone provided (i) the field counts are known in the chosen imaging passband to the limits explored and (ii) the only excess galaxies close to the QSO sightline represent the absorbing population. The latter is a moot point given the possible physical association of galaxies with QSOs but this point could be checked by imaging a control sample of QSOs selected without the appropriate absorption lines. Aragón-Salamanca et al. examined the nature of sources producing clustered CIV absorption lines via deep K-band imaging of $\simeq 11$ QSOs whose absorbers have $1.5 < z < 2.5$. Their K-band LF is similar to that observed locally implying no dramatic luminosity changes to quite high redshifts. However, there is considerable uncertainty in matching the absolute LF normalisations since the excess expected from their choice of clustered CIV lines is difficult to estimate. Steidel & Dickinson have used similar arguments to extend their original Mg II sample beyond $z \simeq 1$ and find equivalent results.

A related development is the search for galaxies producing damped Lyman alpha absorption. These absorbers have a high HI column density and a frequency suggesting they occur in the disks of high redshift galaxies (Wolfe et al. 1987). Can these sources be imaged and, if so, what is their luminosity and colour? As the near-IR k-correction remains modest to quite high z, deep K band images might detect starlight from these disks, particularly if they are luminous systems like their counterpart metallic absorbers at lower redshift. Their detection would also provide some useful constraints on the amount of dust in the high z Universe. A small impact parameter from the QSO sight line is expected if they originate in galaxy disks of high HI column density and so exquisite conditions are important if the galaxy is to be seen next to or 'underneath' the very bright QSO.

Figure 8.3 shows a deep K image taken from the damped Lyα identification programme of Aragón-Salamanca et al. (1996). For 10 suitable QSOs with $z_{abs} \simeq 2$-2.5, several faint candidates have been identified close to the QSO but none implies luminosities much brighter than L^*. HST imaging of such candidates will be particularly useful in clarifying the scale lengths of the star forming component and in constraining the possibility of dust extinction via imaging at shorter rest-wavelengths. When NICMOS becomes available on HST, routine surveys for such galaxies can be contemplated and reliable LFs constructed to quite high z.

Suggestive though these results are, the high redshift absorber programmes need to overcome several problems. Firstly, only for the $z < 1.3$ Mg II samples do we have the vital spectroscopic confirmation. Spectroscopic work is difficult although some progress is possible with broad-band colours and the presence of the Lyman continuum break. However, with high-throughput optical and infrared spectrographs on 8-10m class telescopes it may ultimately be possible to secure the redshifts of the distant Mg II, CIV and Lyα absorbers. It may be that this will only be feasible if the absorbers are star-forming systems with strong emission lines. Another problem is that we may only be seeing a subset of the population at any redshift via these techniques. Part of the galaxy population does not participate in

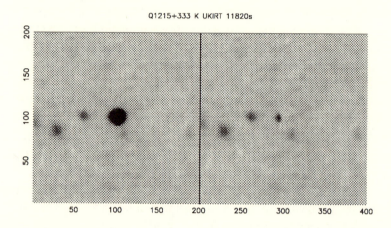

Figure 8.3: The subtraction of the $K \simeq 16$ QSO Q1215+333 from a deep UKIRT image reveals a promising absorber candidate with K=20.2 which may be responsible for the damped Lyα absorption line at z_{abs}=2.001 (Aragón-Salamanca et al. 1995).

the absorption progress and determining that 'interloper' fraction becomes harder as we proceed to higher redshifts. We appear to have been rather lucky in this respect with the MgII absorbers, but this may not be the case for the Lyα systems (Lanzetta et al. 1995). The foregoing is likely to be a major growth area for graduate students fortunate enough to have access to the next generation of telescopes and infrared spectrographs.

8.4   REDSHIFT SURVEYS AND THE DWARF-DOMINATED UNIVERSE

The absence of any significant changes in the bright end of the normal galaxy LF discussed in §§ 8.2–8.3 is difficult to reconcile with the puzzling excess in the field galaxy number counts (Ellis 1993). Elementary considerations suggest the large surface density seen in the deepest work (Metcalfe et al. 1995) implies a dwarf-dominated Universe (Cowie & Lilly 1990). How can this result be consistent with the observed LF today which apparently does not contain a huge population of dwarfs? Either we are seriously undercounting our local Universe (Phillipps & Driver 1995) or there were many more dwarf galaxies in the recent past. The clue to separating these two possibilities lies in securing redshifts for large numbers of faint galaxies. If the local LF has been underestimated, we expect to uncover large numbers of nearby dwarfs.

It might seem something of an embarrassment to admit that the local LF remains so uncertain. However, the evidence for a serious error is only circumstantial. Efstathiou et al. (1988) analysed 326 galaxies in 5 Schmidt-sized fields to $b_J \simeq 17$ and found the Schechter (1976) form appropriate. They determined uncorrected Schechter parameters $< M_B^*, \alpha, \Phi^* >$ of $<$-19.7, -1.07, 0.0156 Mpc^{-3} $>$ (H_o=100 kms sec^{-1} Mpc^{-1}). The more extensive panoramic sparse-sampled APM-Stromlo southern survey of 1769 galaxies (Loveday et al. 1992) found similar parameters $<$-19.7, -1.11, 0.0140 Mpc^{-3} $>$. However, both surveys constrain the faint end slope only to $M_{bJ} \simeq$-16. An upturn at fainter luminosities such as that claimed for the Virgo cluster (Binggeli et al. 1988) cannot be ruled out.

Importantly, a local population of feeble sources would have an Euclidean count slope dominating the faint counts and diminishing any evolution that would otherwise be inferred (Kron 1982). Furthermore, surveys limited at relatively bright apparent magnitudes adopt high surface brightness detection thresholds and may be poorly-suited for finding intrinsically faint galaxies (McGaugh 1994). Clearly, the most reliable constraints on the faint end of the local LF comes from spectroscopy at those faint limits where the contribution can be directly measured (Glazebrook et al. 1995a).

A related uncertainty which has plagued the subject concerns the question of the *absolute* normalisation of the LF. Although the Efstathiou et al. and APM-Stromlo surveys have consistent values of Φ^*, the number counts steepen beyond their apparent magnitude limits, $17 < b_J < 20$, more than can be accounted for by non-evolving models, suggesting southern volumes with $b_J < 17$ may be unrepresentative (c.f. Maddox et al. 1990).

Combining 'benchmark' $b_J < 17$ redshift surveys and deep equivalents within narrow faint apparent magnitude slices is not ideal. Broadhurst et al., Colless et al. & Glazebrook et al. were only able to compare the faint redshift distribution $N(z)$ with empirical predictions based on the bright survey. At no redshift was there a sufficient range in luminosity to examine the form of the LF directly. Moreover, the number of pencil beams was small ($\simeq$5 each) raising the worry that clustering may affect the conclusions.

My colleagues and I have therefore compiled a new 'Autofib' survey (using the robotic fibre positioner built for the Anglo-Australian Telescope—Parry and Sharples 1988) spanning a wide apparent magnitude range in many directions and enabling a direct reconstruction of the LF at various redshifts. It includes $\simeq$700 redshifts from earlier magnitude-limited surveys and $\simeq$1000 new redshifts within $17 < b_J < 22$ (see Table 8.1). Further details of this new survey are contained in Heyl (1994) and preliminary results have been presented in Colless (1995); the full analysis will appear in Ellis et al. (1996).

In estimating absolute magnitudes of faint galaxies, account must be taken of

Table 8.1: The Autofib Redshift Survey

Survey	b_j limits	Area deg^2	Fields	Redshifts
DARS (Peterson et al. 1986)	11.5–16.8	70.80	5	326
BES (Broadhurst et al. 1988)	20.0–21.5	0.50	5	188
LDSS-1 (Colless et al. 1990,1993)	21.0–22.5	0.12	6	100
Autofib bright	17.0–20.0	5.50	16	480
Autofib faint	19.5–22.0	4.70	32	546
LDSS-2 (Glazebrook et al. 1995a)	22.5–24.0	0.07	5	73
TOTAL				1713

the k-correction which depends on the galaxy's (unknown) spectral energy distribution (SED). At a given redshift, $k_{b_J}(z)$ changes by $\simeq 1$ magnitude across the Hubble sequence (c.f. King & Ellis 1985). Heyl (1994) devised a classifier based on cross-correlation of the faint spectra against the wide aperture local spectral catalogue of Kennicutt (1992). Knowing the Kennicutt class which best matches the faint galaxy, the k-correction is determined with reference to King & Ellis' compilation. Realistic simulations suggest the correct spectral class is returned to within ± 1 class for 90% of the cases; 6 classes span the entire sequence.

The large number of faint pencil beams in the Autofib survey leads to new constraints on the *local* LF as well as on its form at high z. So far, 560 galaxies in the survey have $0 < z < 0.1$ but few are less luminous than $M_{b_J} \simeq -16.0$; most galaxies with $b_J > 22$ lie beyond $z \simeq 0.1$. The paucity of low luminosity galaxies severely constrains any possible upturn in the local LF to occur dimmer than $M_{b_J} \simeq -14$. As the photometric data used to select these galaxies penetrates to surface brightness limits below $\mu_{b_J} = 26.5$ arcsec^{-2}, it becomes hard to argue that the flat LF derived in earlier work is in error due to selection biases. The redshift distribution at $b_J = 24$ eliminates the possibility that the faint source counts are significantly contaminated by a population of low luminosity galaxies underrepresented in the original $b_J < 17$ surveys (Glazebrook et al. 1995a) and gives further support to the increase of a factor 2 in the Φ^* normalisation discussed in § 8.3. The best-fit local LF claimed by Ellis et al. (1996) is $< M_B^*, \alpha, \Phi^* >$ of $< -19.20, -1.09, 0.026$ Mpc$^{-3} >$ ($H_o = 100$ kms sec^{-1} Mpc^{-1}).

So where do the excess galaxies in the number counts come from? Figure 8.4 shows a highly suggestive steepening of the faint end slope of the LF with increasing redshift starting from the flat local determinations (Efstathiou et al, Loveday et al.).Formally, a change in *shape* with z is significant at the 99.9% level. However, as in § 8.3, there is no obvious brightening of the bright end of the LF to $z \simeq 1$. As noted before, it seems the LF is composed of two components—a luminous part evolving very slowly, if at all, over z <1-2 *plus* a sub-L* component which must evolve remarkably rapidly in order to satisfy the local LF with its flat faint end slope.

The results presented are qualitatively similar to those analysed in the impressive CFRS galaxy survey (Lilly et al. 1995). Lilly et al. likewise find evolutionary trends which are dominated by the bluest star-forming galaxies. The CFRS is I-band selected and probes a smaller but more uniform sample to higher redshift within a narrow apparent magnitude slice. The Autofib survey is b_J-selected with a lower mean redshift. However, a more detailed impression of the *shape* of the LF can be obtained at each redshift by virtue of the wider apparent magnitude range.

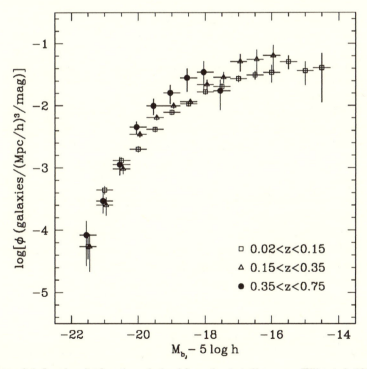

Figure 8.4: Luminosity functions derived from the Autofib survey (Ellis et al. 1995) in different redshift intervals.

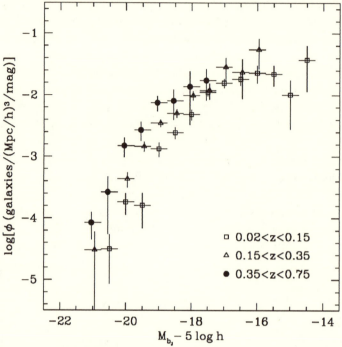

Figure 8.5: Luminosity functions at various redshifts from the Autofib survey for those galaxies whose rest-frame [O II] 3727 Å equivalent width exceeds 20 Å . The evolutionary trends for this class are sufficient to explain those seen in the overall population presented in Figure 4.

What physical property distinguishes the galaxies that lie in the evolving and non-evolving components? The missing clue appears to be related to the star-formation rate, as originally suggested by Broadhurst et al. (1988). Figure 8.5 shows how the galaxies with the strongest [O II] emission dominate the evolutionary trends. The luminosity density contributed by this class of star-forming galaxies has dropped by a large factor since $z \simeq 0.5$. Significantly, at the present epoch, such systems all lie at the faint end of the LF, whereas at modest redshifts they occupy a wide range of luminosities.

8.5   FAINT GALAXY MORPHOLOGIES FROM HST

The repair of HST is adding a new dimension to the study of faint field galaxies. Early Cycle 4 images (Griffiths et al. 1994) demonstrated the ability to recognise resolved morphological features in $I \simeq 22$ galaxies, the bulk of which lie at $0.4 < z < 0.7$ (Lilly 1993). A new set of questions can be addressed with such data: (1)

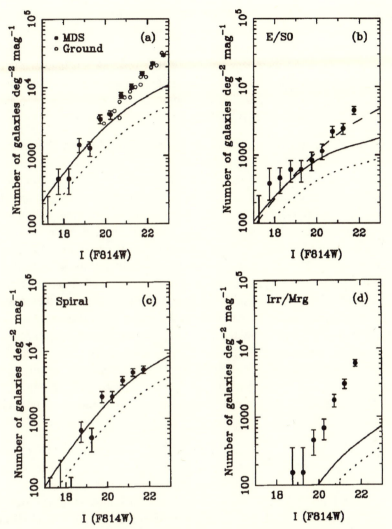

Figure 8.6: Morphological counts of 550 galaxies with $I < 22$ from the HST Medium Deep Survey. The sample is enlarged from that presented by Glazebrook et al. (1995b). Counts of 'regular' Hubble types are consistent with no evolution (solid and dotted lines depending on Φ^*) but those of irregular/peculiars show a steeper slope suggesting rapid evolution.

What is the morphological mixture of the high z field population? (2) Are the faint blue galaxies a distinct morphological population?

Ultimately a large sample of HST morphologies *with redshifts* is required. Meanwhile, a considerable amount can be determined from images alone. The largest collection is currently provided by the 'Medium Deep Survey' (PI: R. Griffiths). In this key project, WFPC-2 is used in parallel mode according to primary pointings defined by other observers. Redshifts for the galaxies surveyed have to be secured later. An increasing amount of HST imaging is being done in the reverse mode: i.e. primary WFPC-2 imaging of larger fields with existing multi-slit spectroscopy. First results of these programmes are also emerging (Schade et al. 1995, Cowie et al. 1995, Broadhurst, private communication).

The most significant claim to date has been made by the Medium Deep Survey. Over 300 I <22 galaxies were classified on a simple E/S0: Spiral: Irr/Pec scheme by Glazebrook et al. (1995b) and the number-magnitude counts determined as a function of type. Counts for an enlarged sample of 550 galaxies are shown in Figure 8.6. The results illustrate that the spiral and early-type classes show little evidence for evolution. Significantly, both classes fit the predictions only if the absolute normalisation is × ≃2 higher than that derived from the 17th magnitude surveys (e.g., Loveday et al., § 8.2). The possibility that the MDS fields are overdense by virtue of the parallel observing strategy can be eliminated by careful comparison to wider-field ground-based counts. On the other hand, the irregular and peculiar galaxies show a count slope much steeper than expected, consistent with significant evolution. Although the overlap with the spectral samples remains small, the limited data suggests that the [O II]-strong sources which decline dramatically in number since z ≃1 *are* the morphologically unusual examples in the HST samples. A morphologically-distinct population of star forming sources appears to be responsible for the well-established excess population of faint blue galaxies.

But can we really recognise galaxy morphologies readily at these faint limits? Is there not a danger that a faint galaxy whose signal to noise is poor in some respect or other is too readily cast into the 'peculiar/irregular' category? To quantify this possibility, Figure 8.7 shows a set of regular Hubble sequence galaxies observed (a) in the B-band at z ≃0 and (b) as simulated in the I-band viewed by HST at z=0.8. k-corrections are allowed for via the positioning of these two passbands and corrections have been applied to allow for the reduced HST night sky background and various other instrumental effects.

Such simulations suggest the trends seen in the MDS are reasonably robust although certain flocculent late type spirals could conceivably be misinterpreted as interacting/peculiar systems. Extending these trends, one might worry that star forming galaxies might ultimately break up into disparate H II regions of small apparent size (c.f. the 'train-wrecks' in Dickinson's image of 3C324 at z=1.2 and Dressler et al.'s (1993) 'nascent' galaxies supposedly at z ≃2).

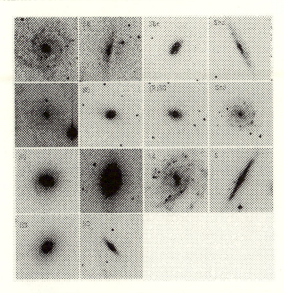

Figure 8.7a: A selection of regular Hubble sequence galaxies viewed at $z=0$ in the B band (courtesy of M. Pierce).

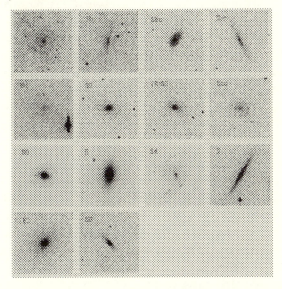

Figure 8.7b: As (a) simulated for WFPC-2 in the I band for the same sample viewed at $z=0.8$ taking into account various instrumental and background effects (see text for details).

8.6 CONCLUSIONS

Several different lines of enquiry converge on the identification of two very different galaxy populations. Massive regular galaxies (roughly defined as symmetrical spirals and ellipticals more luminous than $\simeq 0.1$ L^*) appear to have changed little since $z \simeq 1\text{-}2$ and the bulk of the star formation in the earlier types occurred very early indeed ($z > 3\text{-}5$). In contrast to this slow evolution, even the modest redshift ($z \simeq 0.5$) Universe was dwarf dominated and this dwarf population has rapidly declined in recent times. These systems are less well-formed and show intense star formation (possibly of a bursting type).

The interested graduate student could make a major contribution to the subject by addressing the physical origin of this dichotomy. Observationally, the trends seen in the HST morphological counts and in the faint redshift surveys need to be understood by combining the two currently disparate samples. Further, spatially-resolved ground-based spectroscopy is urgently required to understand the dynamical state of many of the irregular sources.

In the analytic and theoretical areas, I see two principal unsolved problems. In hierarchical theories of galaxy formation such as those that form the basis of simulations by Kauffmann et al. (1993) and Cole et al. (1994), dwarfs and globular clusters form early in great abundance around dark matter halos with low circular velocities. Merging reduces the number of dwarfs with time but steep faint end slopes to the local LFs are still expected today unless feedback and other processes are very effective. The end point of these simulations is closer to the observed Universe at $z \simeq 0.5$ (Fig. 8.4) rather than $z=0$. A further difficulty is the older ages for the massive galaxies which is, at first sight, counter to the expectations of recent hierarchical models.

The clue to solving this problem, as recognised by Babul & Rees (1992), lies in *preventing* a major component of the dwarf population from forming stars until late times. Babul & Ferguson (1996) show how star formation might be inhibited by the ionising UV background until recent epochs. A key question which has to be addressed is whether star formation controlled by feedback can occur at the rate required to account for the observed evolution. As the data improves with larger surveys, this will be an important area to study in more detail. One feature of this picture would be the sudden collapse of such systems over a narrow redshift interval when the associated gas neutralises (Efstathiou 1996). In this respect, the irregular morphologies of the HST images and the wide range in luminosity and redshift at which the evolving population is seen (Fig. 8.5) is puzzling. Although it is claimed the small physical sizes of the MDS sources (Im et al. 1995) is consistent with this picture, not enough work has been done to separate the evolving and quiescent components, or in understanding obvious biases which may render large sources as compact ones (Fig. 8.7).

The second unsolved problem concerns the fate of these star-forming dwarfs. Figures 8.4-8.5 demonstrate this problem observationally very clearly. Excess objects in the high z LF could fade to lower luminosities, but given the wide redshift range over which the [OII]-strong sources are seen, the decay would appear to be too rapid for conventional stellar mass functions. The fact that most of the present-day low luminosity systems are gas-rich (McGaugh 1996) argues that these are unlikely to be faded remnants of the higher redshift blue star-forming galaxies.

Alternatively, the dwarfs could disappear via minor merging into larger galaxies. Broadhurst et al. (1992) showed qualitatively how self-similar merging is consistent with both the optical and infrared photometric and redshift data. The picture is largely under attack from theorists who object to the high merging rates required and the likely effect this will have on the morphologies of the merger products. Observationally, it is exceedingly difficult to test the idea further. Although the angular correlation function, $w(\theta)$, shows no convincing statistical excess of pairs on small scales, a significant percentage (10-20%) of the MDS sample appear to be interacting systems supporting earlier investigations (Zepf & Koo 1989). Given the rapid timescale involved, even if only a fraction of these are genuinely merging, the phenomenon must be significant.

None of the proposed solutions to either unsolved problem is particularly attractive. On the positive side, the disparity between the two populations of galaxies revealed by a number of independent observations is a sound result which will hopefully lead to the understanding of a major new physical process in galaxy formation. On the negative side, I do worry about the absurdity of switching on a dominant optical component of the Universe in the most recent few Gyr only to allow it to fade rapidly into obscurity. This seems specifically contrived to fit the puzzling observations! However, "when you have eliminated the impossible, whatever remains, however improbable, must be the truth" (Conan-Doyle 1920).

ACKNOWLEDGMENTS

The *Autofib* survey includes Matthew Colless, Tom Broadhurst, Jeremy Heyl and Karl Glazebrook and the Medium Deep Survey is led by Richard Griffiths. The 'Morphs' project involves Harvey Butcher, Warrick Couch, Alan Dressler, Gus Oemler, Ray Sharples and Ian Smail. I thank all my co-workers in these projects and particularly Alfonso Aragòn, Karl Glazebrook and Ian Smail for their generosity in allowing me to present data prior publication.

BIBLIOGRAPHIC NOTES

- **Bower, R. G., Lucey, J. R., & Ellis, R. S. 1992, MNRAS, 254, 601. Rose, J. A., Bower, R. G., Caldwell, N., Ellis, R. S., Sharples, R. M., & Teague, P. 1994, AJ, 108, 2054.** These articles discuss the homogeneity of the local elliptical galaxy population and the possible degree to which their environmental location may have played a role in the star formation history. A key result is that the ellipticals in the cores of rich cluster represent a remarkably uniform population whose stars formed prior to redshift 2.

- **Dressler, A., & Gunn, J. E. 1992, ApJS, 78, 1. Aragón-Salamanca, A., Ellis, R. S., Couch, W. J., & Carter, D. 1993, MNRAS, 262, 764.** The advantage of look-back time is exploited here through the analysis of galaxies in distant clusters. The aim is to see whether the color evolution observed is consistent with the idea that most spheroidal galaxies formed at high redshift.

- **Broadhurst, T. J., Ellis, R. S., & Shanks, T. 1988, MNRAS, 235, 827. Colless, M., Ellis, R. S., Taylor, K., & Hook, R. 1990, MNRAS, 244, 408. Cowie, L., Songaila, A., & Hu, E. M. 1991, Nature, 354, 460.** The first deep magnitude-limited spectroscopic surveys of field galaxies found a surprising conflict between the absolute numbers of galaxies observed and the redshift distribution obtained. The articles provide empirical descriptions of the evolving luminosity functions.

- **Babul A & Rees, M. J. 1992, MNRAS, 255, 346. Kauffmann, G., White, S. D. M., & Guiderdoni, B. 1993, MNRAS, 264, 207. Cole, S., Aragón-Salamanca, A., Frenk, C. S., Navarro, J. F., & Zepf, S. E. 1995, MNRAS, 271, 781.** These important articles discuss the rate of galaxy formation and evolution in the context of standard hierarchical dark matter cosmologies. They provide a valuable interface with the observational data prior to the era of HST.

- **Phillipps, S., & Driver, S. P. 1995, MNRAS, 274, 832. McGaugh, S. 1994, Nature, 367, 538.** There are numerous possible pitfalls in the interpretation of field galaxy data. These include the inadequate determination of the faint end of the local luminosity function and surface brightness selection effects which may render comparisons difficult across deep and shallow surveys.

- **Ellis, R. S., Broadhurst, T. J, Colless, M. M, Heyl, J. S., & Glazebrook, K. 1996, MNRAS, (May/June edition). Lilly, S.J., Tresse, L., Hammer,**

F., Crampton, D.C. & LeFevre, O. 1995, ApJ, 455, 108. More comprehensive redshift surveys have been now been completed which provide a first glimpse on the evolution of the luminosity function as a function of color and spectral class. Important trends emerge in these two articles.

- **Glazebrook, K., Ellis, R. S., Santiago, B., & Griffiths, R. 1995b, MNRAS, 275, L19. Schade, D.J., Lilly, S.J., LeFevre, O., Crampton, D.C. & Hammer,F. 1995, ApJ 451, L1.** The impact of HST morphology is discussed in these articles both in terms of counts and connections with the ground-based spectra. The latter samples are currently small but important progress is expected in this area in the next few years.

BIBLIOGRAPHY

[1] Aragón-Salamanca, A., Ellis, R. S., Couch, W. J., & Carter, D. 1993, MNRAS, 262, 764.

[2] Aragón-Salamanca, A., Ellis, R. S., Schwartzenberg, J-M., & Bergeron, J. A. 1994, ApJ, 421, 27.

[3] Aragón-Salamanca, A., Ellis, R. S., & O'Brien, K. 1996, MNRAS, in press.

[4] Babul, A., & Ferguson, H. 1996, ApJ, in press.

[5] Babul, A., & Rees, M. J. 1992, MNRAS, 255, 346.

[6] Bergeron, J. A. & Boissé, P. 1991, A&A, 243, 344.

[7] Binggeli, B., Sandage, A., & Tammann, G. 1988, ARA&A, 26, 509.

[8] Bower, R. G., Lucey, J. R., & Ellis, R. S. 1992, MNRAS, 254, 601.

[9] Broadhurst, T. J., Ellis, R. S., & Shanks, T. 1988, MNRAS, 235, 827.

[10] Broadhurst, T. J., Ellis, R. S., & Glazebrook, K. 1992, Nature, 355, 55.

[11] Cole, S., Aragón-Salamanca, A., Frenk, C. S., Navarro, J. F., & Zepf, S. E. 1995, MNRAS, 271, 781.

[12] Colless, M., Ellis, R. S., Taylor, K., & Hook, R. 1990, MNRAS, 244, 408.

[13] Colless, M., Ellis, R. S., Taylor, K., Broadhurst, T. J., & Peterson, B. A. 1993, MNRAS, 261, 19.

[14] Colless, M. 1995, in Wide Field Spectroscopy, eds. S. J. Maddox, & A. Aragón-Salamanca (Singapore: World Scientific), p. 263.

[15] Conan-Doyle, A. 1920, The Sign of Four (Pan Publications).

[16] Cowie, L., & Lilly, S. J. 1990, in ASP Conference Series, Vol. 10, Evolution of the Universe of Galaxies, ed. R. Kron, p. 212.

[17] Cowie, L., Songaila, A., & Hu, E. M. 1991, Nature, 354, 460.

[18] Cowie, L., Hu, E. M., & Songaila, A. 1995, preprint.

[19] Dickinson, M. 1996, in Fresh Views on Elliptical Galaxies, ed. A. Renzini, et al., in press.

[20] Dressler, A., & Gunn, J. E. 1990, in ASP Conference Series, Vol. 10, Evolution of the Universe of Galaxies, ed. R. Kron, p. 200.

[21] Dressler, A., & Gunn, J. E. 1992, ApJS, 78, 1.

[22] Dressler, A., Oemler, A., Gunn, J. E., & Butcher, H. 1993, ApJ, 404, L45.

[23] Efstathiou, G. 1996, in Galaxies in the Young Universe, ed. H. Hippelein (Max Planck Publications), in press.

[24] Efstathiou, G., Ellis, R. S., & Peterson, B. A. 1988, MNRAS, 232, 431.

[25] Ellis, R. S. 1992, in IAU Symposium 149, The Stellar Populations of Galaxies, ed. B. Barbuy, & A. Renzini (Dordrecht: Kluwer), p. 297.

[26] Ellis, R. S. 1993, in ASP Conference Series, Vol. 43, Sky Surveys: Protostars to Protogalaxies, ed. T. Soifer, p. 165.

[27] Ellis, R. S., Broadhurst, T. J, Colless, M. M, Heyl, J. S., & Glazebrook, K. 1996, MNRAS, in press.

[28] Glazebrook, K., Ellis, R. S., Colless, M., Broadhurst, T. J., Allington-Smith, J., & Tanvir, N. 1995a, MNRAS, 273, 157.

[29] Glazebrook, K., Ellis, R. S., Santiago, B., & Griffiths, R. 1995b, MNRAS, 275, L19.

[30] Griffiths, R. E., et al. 1994, ApJ, 435, L19.

[31] Heyl, J. 1994, M.Sc. thesis, University of Cambridge, UK.

[32] Im, M., Casertano, S., Griffiths, R. E., Ratnatunga, K. 1995, ApJ, 441, 495.

[33] Kauffmann, G. 1996, preprint.

[34] Kauffmann, G., White, S. D. M., & Guiderdoni, B. 1993, MNRAS, 264, 207.

[35] Kennicutt, R. C. 1992, ApJ, 388, 310.

[36] King, C. R., & Ellis, R. S. 1985, ApJ, 288, 456.

[37] Kron, R. 1982, Vistas, 26, 37.

[38] Lanzetta, K.M., Bowen, D.V., Tytler, D. & Webb, J.K. 1995, ApJ., 442, 538.

[39] Lilly, S. J. 1993, ApJ, 411, 501.

[40] Lilly, S.J., Tresse, L., Hammer, F., Crampton, D.C. & LeFevre, O. 1995, ApJ, 455, 108.

[41] Loveday, J., Peterson, B. A., Efstathiou, G., Maddox, S. J., & Sutherland, W. J. 1992, ApJ, 390, 338.

[42] Maddox, S. J., Sutherland, W. J., Efstathiou, G., Loveday, J., & Peterson, B. A. 1990, MNRAS, 247, 1P.

[43] McGaugh, S. 1994, Nature, 367, 538.

[44] McGaugh, S. 1996, MNRAS, in press.

[45] Metcalfe, N., Shanks, T., Fong, R., & Roche, N. 1995, MNRAS, 273, 297.

[46] Parry, I. R., & Sharples, R. M. 1988, in ASP Conference Series, Vol. 3, Fiber Optics in Astronomy, ed. S. M. Barden, p. 93.

[47] Peterson, B. A., Ellis, R. S., Shanks, T., Bean, A. J., Fong, R., Efstathiou, G., & Zou, Z-L. 1986, MNRAS, 221, 233.

[48] Phillipps, S., & Driver, S. P. 1995, MNRAS, 274, 832.

[49] Rose, J. A., Bower, R. G., Caldwell, N., Ellis, R. S., Sharples, R. M., & Teague, P. 1994, AJ, 108, 2054.

[50] Schade, D.J., Lilly, S.J., LeFevre, O., Crampton, D.C. & Hammer,F. 1995, ApJ 451, L1.

[51] Schechter, P. L. 1976, ApJ, 203, 297.

[52] Smail, I., Ellis, R. S., & Fitchett, M. J. 1994, MNRAS, 270, 245.

[53] Steidel, C., & Dickinson, M. 1995, in Wide Field Spectroscopy, eds. S. J. Maddox, & A. Aragón-Salamanca (Singapore: World Scientific), p. 349.

[54] Steidel, C., Dickinson, M., & Bowen, D. V. 1993, ApJ, 413, L77.

[55] Stockton, A., Kellogg, M., & Ridgeway, S. E. 1995, ApJ, 443, L69.

[56] Tyson, A.J. 1988, AJ, 88, 1.

[57] Wolfe, A., Turnshek, D. A., Smith, H. E., & Cohen, R. D. 1987, ApJS, 61, 249.

[58] Zepf, S. E., & Koo, D. C. 1989, ApJ, 337, 34.

CHAPTER 9

QUASARS

MARTIN J. REES

Institute of Astronomy, Madingley Road, Cambridge, England

ABSTRACT

This paper focuses on two important questions: (i) How and when did supermassive black holes form, and what is their relation to galaxy formation? (ii) Do these holes, which seem to lurk in the centers of most galaxies and to be involved in the quasar phenomenon, have the exact properties predicted by general relativity?

9.1 QUASARS AND THE END OF THE 'DARK AGE'

The most remote quasar so far detected has $z = 4.89$ (Schneider, Schmidt, & Gunn 1991). The population genuinely seem to be 'thinning out' at redshifts above 2.5–3: the comoving density of quasars falls by at least 3 for each unit increase in z (see Shaver 1995 for a review). But what about redshifts larger than 5? What can quasars tell us about how and when cosmic structures emerged?

It took about half a million years for the Universe to cool down to 3000 K (corresponding to $z = 1000$). Thereafter, further expansion shifted the primordial radiation into the infrared; a 'dark age' began, which persisted until the first nonlinear perturbations developed into bound systems and released enough nuclear or gravitational energy to light up the universe again. The high-z quasars tell us that, after about one billion years, the dark age had certainly ended. The lack of complete 'Gunn Peterson' absorption in the spectra implies that there had by then been enough energy input to re-ionize the primordial material; the existence of the quasars implies that black holes of as much as $10^8 M_{\odot}$ had accumulated.

What happened during the timespan from a million to around a billion years? The answer depends on the relation between quasars and galaxies, and on when galaxy formation began—two important, and interlinked, unsolved problems. One need only recall three much-studied options for structure formation (none of which can yet be ruled out) to indicate the current level of uncertainty:

(i) The formation of cosmic structure depends primarily on how the dark matter behaves, since in gravitational terms it is ~ 10 times as important as the baryons. If the dark matter is in, for instance, heavy particles which are electrically neutral and weakly interacting, the random thermal motions are too slow to affect the dynamics and the dark matter is termed "cold." The 'cold dark matter' (CDM) model (Peebles 1982; Blumenthal et al. 1984) has been a 'template' for comparing with the real data for at least the last decade: Ostriker (1993) gives a comprehensive review. According to the simplest CDM model, the first structures—loosely bound systems of subgalactic scale—start to form at redshifts as high as 20 or 30. The galaxies themselves form more recently, but still early enough to have provided 'hosts' for the high-z quasars. More detailed studies suggest that the z-distribution of quasars can be fitted if the mass of the hole that forms within each dark halo depends in a plausible way on the depth of the potential well and the density. If the CDM model were correct, the quasar density should fall off steeply beyond $z = 5$; the intergalactic medium would have been originally ionized, perhaps as early as $z = 20$, by stars in shallow potential wells of subgalactic scale.

(ii) If at least one species of neutrino has a mass of a few eV, then the CDM model is modified by the presence of 'hot' dark matter. The light neutrinos would have thermal motions high enough to homogenize by free-streaming across the scales of protogalaxies and protoclusters. In the hybrid 'mixed dark matter' (MDM) model (e.g., Klypin et al. 1993), fluctuations on small scales have lower amplitude, relative to fluctuations on large scales, than is the case for 'pure' CDM; there is therefore less structure at early times. The existence of quasars at $z = 5$ is a severe constraint on the fraction of 'hot' dark matter. Indeed the inferred ionization of the IGM at high redshifts is itself a problem (even if it isn't due to quasars), since the fluctuation spectrum in MDM has less power on small scales, so not even subgalactic structures form early.

(iii) The so-called primordial isocurvature baryon (PIB) model (Peebles 1987)—which has been taken specially seriously here in Princeton—leads to non-linear structures soon after recombination. Bound systems that condense out early, and lose their angular momentum by Compton drag, could evolve directly into black holes (Umemura, Loeb, & Turner 1993), even before the virialisation of galactic-scale potential wells which seem a prerequisite for black hole formation in CDM and MDM.

Recent progress in the study of active galactic nuclei (AGNs) bring into sharper focus the question of how and when supermassive black holes formed, and how this process relates to galaxy formation. There is hope of learning more, not only by better quasar statistics but also by observing ordinary galaxies in the same redshift range. Quasar absorption line spectra offer another probe.

The next step is to push back further towards the 'dark age'. The primordial medium could have been reionized by the quasars themselves. However, in hierarchical structure formation scenarios like CDM, the first UV will come from stars forming in systems of subgalactic scale. So ionization is most likely achieved by the first stars. (The quasar spectrum has a flatter UV slope, and could nevertheless be dominant for ionizing He even if not for H.) Ionization could have occurred at redshifts of $\gtrsim 10$ in CDM models (Couchman & Rees 1986) and at even higher redshifts in PIB models. The epoch of reionization is important in two contexts: if ionization occurs very early, the opacity may affect the microwave background anisotropies by attenuating the number of photons travelling uninterruptedly from the recombination era ($z \simeq 1000$); second, it affects the temperature of the gas forming the clouds and filaments that produce Lyman-α absorption in high-z (Miralda-Escudé & Ostriker 1990; Miralda-Escudé & Rees 1994).

But can we probe the pre-ionization era, when the hydrogen was still in atomic form? The best hope here is the 21 cm line. This makes a much smaller contribution to the background continuum than either the relic radiation or the synchrotron emission from nonthermal sources, but it could be separately identified because it would display distinctive angular and spatial structure, due to incipient large-scale clusters, and perhaps also to patchiness in the original heating (Scott & Rees 1990; Madau, et al. 1996).

Another probe of the dark age would be 'secondary' effects on the microwave background. These could attenuate the fluctuations imprinted at recombination; also, they could imprint secondary fluctuations.

Whether there are quasars with $z \gg 5$ is, therefore, an important unsolved problem. Discovery of higher redshifts would push back our estimates of when galaxies formed, unless we adopt the radical view (which could be maintained if the PIB model were right) that these quasars are not closely connected with galaxies.

9.2   THE RELATION OF AGNs TO THE CENTRAL BULGES OF GALAXIES

I have discussed elsewhere (Rees 1993; Haehnelt & Rees 1993) the processes whereby, when the stellar bulge of a galaxy forms, part of the gas may collapse into a black hole. The formation and growth of the hole then manifests itself as a quasar. According to this picture, the quasar population peaked (i.e. redshifts in the range 2 - 3) during the era when large galactic bulges were forming. The mass of the hole would depend on that of its host galaxy, though not necessarily via an exact proportionality: the hole mass would depend on the depth of the potential well, the density profile, and the angular momentum.

This process involves complex gas dynamics and feedback from stars; we are still a long way from being able to make realistic calculations. I think the most compelling argument that a massive black hole forms comes from considering what would need to happen to prevent such an outcome. There are two conceivable ways of stopping the hole from forming.

(i) The formation of stars during the infall could be nearly 100 per cent efficient; moreover, the stars would all need to be of low mass (so that no material was expelled again), or

(ii) Gas may remain in a self-gravitating disc for hundreds of orbital periods, without the onset of any instability that redistributes angular momentum and allows the inner fraction to collapse.

Neither of these options seems at all likely—the first would require an IMF quite different from what is actually observed in bulges or the cores of ellipticals; the second is contrary to well-established arguments that self-gravitating discs are dynamically unstable. So it seems natural to expect that, when galactic cores form, a fraction of the baryons, some already processed through stars, condense into a massive central black hole. (Not all the matter, of course, would get incorporated in the central hole because some must carry away the angular momentum.)

The issue will not be definitively settled without detailed computer modelling of stars and gas in the central 100 pc of a newly-formed galaxy. We need to know at what stage in its progressive concentration towards the centre the gas stops being able to form stars (because of radiation pressure, magnetic fields, or whatever) and evolves instead into a supermassive object. A crude argument is the following. When a self-gravitating object releases its binding energy in a dynamical (or free fall) timescale, the luminosity is $(v_{\mathrm{virial}}^5/G)$. When v_{virial} gets high enough, this becomes comparable with the Eddington limit, at which the radiation pressure due to electron scattering equals the gravitational attraction. Fragmentation is then impossible: instead, the gas is 'puffed up' by radiation pressure into a supermassive object. The critical value of v_{virial} is then $\sim 300 M_6^{\frac{1}{5}}$ km/sec. This is of course a 'conservative' guess. Fragmentation may stop earlier if other types of opacity exceed electron scattering. Another possibility is that contraction and shear amplify magnetic stresses until they can prevent fragmentation. (The ionization level would be too high to allow ambipolar diffusion, which permits stars still to form, despite dynamically-important magnetic fields, within dense interstellar clouds).

The central hole's actual mass would depend on the angular momentum of the protogalaxy, the depth of the potential well, and no doubt other parameters as well.

It is hard to predict how long the quasar-level activity would last. However, a natural timescale is set by the 'Salpeter time'—the time it takes a hole to double its mass if it accretes at the rate necessary to supply the Eddington luminosity. This time is of course proportional to the efficiency e with which mass is converted into radiation. It is $(e\sigma_T c/4\pi G m_p)$. For e around ten per cent, the Salpeter time is 4×10^7 years.

If typical high-z quasars are indeed associated with this process (or its immediate aftermath), their lifetimes would be a few tens of millions of years. There would then have been 50 generations of quasars during the period over which the population density rises and falls. At redshifts of 2-3, where quasars were most common, their comoving space density was only a few per cent of the present density of bright galaxies. These 'demographic' arguments therefore tell us that every galaxy could have passed through a quasar phase, and that by $z = 2$ (2-3 billion years) most cores have formed with central holes of $10^6 - 10^9 M_\odot$. An important project for the future would be to refine this very rough argument to take account of the broad luminosity function for AGNs.

John Bahcall and his colleagues have recently made very interesting discoveries about the host galaxies of quasars at low redshift (Bahcall, Kirhakos, & Schneider 1994, 1995). Although some (including the 'prototype' 3C 273) are indeed in large ellipticals, some of the others are in much less luminous hosts. In some instances, most spectacularly in Pks 2349-014, there seems to have been a violent tidal interaction; whereas the host galaxies of others, for instance the Sb galaxy around PG 0052+251, seem undisturbed (Bahcall, Kirhakos, & Schneider 1996).

If it turns out that many low-z quasars really are in small galaxies, the implications could be very interesting. It is unlikely that all low-mass galaxies, with shallow potential wells, develop a $10^7 M_\odot$ hole soon after formation—if they did, it would be hard to understand why high-z quasars were not even more abundant than the number observed. There are then three options:

(i) Some special subset of small galaxies may form with massive central holes, and only these can become low-z quasars. (These might, for instance, be objects with specially low angular momentum, where gas can fall readily into the centre).

(ii) When activity is triggered at a recent epoch in a pre-existing galaxy, a black hole might form *ab initio* (or from a $\ll 10^7 M_\odot$ seed).

(iii) Low-z quasars may involve pregalactic black holes that find themselves transiently in a dense environment. (Such a population could be expected in the PIB model for structure formation)

9.3   QUASARS AND THEIR REMNANTS: PROBES OF
GENERAL RELATIVITY?

9.3.1   *Dead Quasars in Nearby Galaxies*

The arguments of the last section suggest that the total number of quasars there have ever been, summing over all the generations, could be comparable with the number of luminous galaxies. (Of course, a better estimate would require a specific model for the life cycle of an individual quasar, and also a proper integration over the quasar luminosity function.) Scott Tremaine (in his contribution to this book) has reviewed the evidence that there may indeed be central dark masses—dead quasars—in many nearby galaxies. This evidence comes from studying the spatial distribution and velocities of stars in the central core.

Much the most compelling evidence for a central black hole, however, was recently supplied by a quite different technique: probing gas motions by measuring the emission in the 1.3 cm line of H_2O in the nearby peculiar spiral galaxy NGC 4258 (Watson & Wallin 1994; Miyoshi et al. 1995). The spectral resolution in this microwave line is high enough to pin down the velocities with accuracy of 1 km/sec. Such observations, combined with VLBA mapping with a resolution of 0.5 milliarc seconds (100 times sharper angular resolution than the HST) have revealed, right in the galaxy's core, a disc with rotational speeds following an exact Keplerian law around a compact dark mass; this is hard to interpret as anything other than a black hole. The circumstantial evidence for black holes has been gradually growing for 30 years, but this remarkable discovery clinches the case completely. Whether supermassive black holes exist is now a *solved* problem.

Should we be surprised that these putative holes are so quiescent? Their environment could be almost free of gas, so that very little gets accreted; moreover, when the accretion rate is low it is also inefficient, in that the cooling is so slow (because of the low densities) that only a small fraction of the binding energy gets radiated before the gas is swallowed (see Narayan's paper in this book for some comments on this possibility). However, there is one unavoidable gas supply. Around these holes is a high concentration of stars—the stars whose motions are analysed by Kormendy, Richstone and others. The stars interact with each other gravitationally; every few thousand years one of them would get deflected onto an almost radial orbit that approaches the hole closely enough to be ripped apart by tidal forces. Some of the debris would then swirl inward, providing a high transient fuel supply.

The predicted flares offer a robust diagnostic of the massive holes in quiescent galaxies. They would attain high luminosity—the total photon energy radiated could be a thousand times more than the photon output of a supernova. They would, however, not be standardised—what is observed depends on the hole's mass and spin, the type of star, the impact parameter, and the orbital orientation relative to the hole's spin axis and the line of sight. What is observed may also depend on absorption in the galaxy. To compute what happens involves relativistic gas dynamics and radiative transfer, in an unsteady flow with large dynamic range, which possesses no special symmetry and therefore requires full 3-D calculations—a challenge to those who have many gigaflops at their disposal.

This 'stellar flare' phenomenon presents a worthwhile challenge to both observers and theorists. Any survey that accumulates several thousand 'galaxy years' of exposure should detect such a flare. Its duration would be, typically, a few months: this is determined by the timescale for the debris from a disrupted star to fall back onto the hole and be swallowed (Rees 1988; Kochanek 1994). There are two possible strategies. Searches of relatively nearby galaxies can be carried out using CCD detectors, which do not discriminate against the central regions that would be burnt out if photographic techniques were used. Alternatively, thousands of more distant galaxies can be monitored in just a small selected patch of sky. [The latter could be a byproduct of the ongoing searches for distant supernovae (S. Perlmutter, private communication) but has the disadvantage, compared to less deep searches, that it would be harder to check whether a galaxy exhibiting a flare was otherwise completely quiescent.]

The case for a black hole in our own Galactic Centre was until recently ambiguous, but is now strengthened by the discovery of stars within the central 0.1 pc (Krabbe et al. 1995). The case would be clinched if some of these stars turned out to be moving as fast as (say) 1000 km/sec. If there is indeed a hole of a few million solar masses in our Galactic Center, the most recent stellar-disruption event (even if it happened tens of thousands of years ago) may have left traces that could still be be detectable. Up to 10^{53} ergs of ionizing radiation could be released by accretion of the captured debris—more photons than would be emitted by steadier UV sources in the entire $\sim 10^5$ years interval between successive disruptions. Moreover the half of the star that is ejected may have left traces in some of the strange patterns of the gas within the central 2 parsecs. The mean energy input may be dominated by these rare events.

9.3.2   Do These Holes Have a Kerr Metric?

There is growing evidence that supermassive black holes exist. But do they have the exact properties predicted by general relativity? Chandrasekhar (1975) wrote "In my entire scientific life.....the most shattering experience has been the realisation that an exact solution of Einstein's equations, discovered by the New Zealand mathematician Roy Kerr, provides the absolutely exact representation of untold numbers of massive black holes that populate the Universe."

Chandra refers, of course, to the 'no hair' theorems, which prove that any stationary black hole is characterised just by the two parameters of the Kerr solution: mass and spin.

But I don't think anyone could yet claim that any observed features of AGNs offers a clear diagnostic of a Kerr metric. All we can really infer is that 'gravitational pits' exist, deep enough to allow several per cent of the rest mass of infalling material to be converted into kinetic energy, and then radiated away from a region compact enough to vary on timescales as short as an hour. General relativity has been resoundingly vindicated in the weak field limit (by high-precision observations in the Solar System, and of the binary pulsar) but we still lack quantitative probes of the strong-field domain.

The tidal disruption process mentioned above depends crucially on relativistic precession. When we model AGNs, we use the Kerr metric. Soon there will be data to compare with the kind of models Ramesh Narayan (see his contribution to this book) has discussed for flow around black holes. In particular, X-ray telescopes, foremost among them the Japanese ASCA satellite, have sufficient spectral resolution to reveal lines. The radiation coming from very near the hole has a high surface brightness—so high that the equivalent black body temperature is $\gtrsim$ 10^5+K. Any emission lines from the inner part of the accretion flow must therefore be in the X-ray band. Such lines should display substantial gravitational redshifts, as well as large Doppler shifts. There is already one convincing case (Tanaka et al. 1995) of a broad asymmetric emission line indicative of a relativistic disc viewed almost edge-on, and others should soon follow.

The fate of the debris from a tidally-disrupted star (mentioned above) depends crucially on relativistic precession. Many detailed attempts to model AGNs use specific features of the Kerr metric. For instance Blandford and Znajek's (1977) mechanism, which may create the energetic plasma jets in radio sources, involves direct extraction of a Kerr hole's spin energy. Corroborating these ideas—finding evidence as strong (albeit circumstantial) as the evidence that nuclear fusion powers ordinary stars—is a challenge for the future.

But gas dynamics is always messy and intractable. A much cleaner test would be a star orbiting close to the hole, whose orbit would precess. Could this happen?

Bahcall and Wolf (1976) analysed theoretically the 'cusp' that forms around a black hole. This extends inward only until the 'point mass' approximation breaks down for star-star collisions—in other words, it breaks down within $\sim 10^5$ gravitational radii. However a cluster of compact stars would extend inward until the stars either fell directly into the hole, or their orbits were ground down by gravitational radiation.

Ordinary solar-type stars would undergo physical collisions, rather than large-angle 'Coulomb' deflections, if their relative speed were more than 1000 km/sec; their orbits cannot therefore, by stellar-dynamical processes, achieve very high binding energies, corresponding to orbital speeds a substantial fraction of c. Neutron stars or white dwarfs, on the other hand, could exchange orbital energy by close encounters with each other until some got close enough that they either fell directly into the hole, or until gravitational radiation became the dominant energy loss. A solar-type star could achieve a very tight orbit if it lost orbital energy as the cumulative effect of successive impacts on a disc. (Such orbits could not be reached by tidal capture, because the star itself, rather than a disc or external resisting medium, has to radiate the orbital binding energy, and would puff up and disrupt before the orbit could circularise.)

There was a flurry of interest three years ago when X-ray astronomers detected an apparent 3.4 hour periodicity in the Seyfert galaxy NGC 6814. But it turned out that there was a foreground binary star, with just that period, in the telescope's field of view. But theorists shouldn't be downcast. It is more elevated to make predictions than to explain phenomena *a posteriori*, and that's all we can now do. There is a real chance that someday observers will find evidence that an AGN is being modulated by an orbiting star, which could act as a test particle whose orbital precession would probe the metric in the domain where the distinctive features of the Kerr geometry should show up clearly (cf., Karas & Vokouhlicky 1993).

But, for all that, the most impressive test of general relativity would be detecting *gravitational radiation*, which involves no physics other than that of spacetime itself. The most dramatic sources would be mergers of supermassive black holes. Some events of this kind are expected. Most galaxies may harbour black holes that formed at $z \gtrsim 2$. Moreover, many galaxies have experienced a merger since that time. When that happens, the holes in the two merging galaxies would spiral together. In the final stages of the inward spiral, almost all the energy lost is carried away as gravitational waves. In the few orbits leading up to final coalescence, up to 10 percent of their rest mass is emitted as a burst of gravitational radiation.

When the hole masses are $\gtrsim 10^6 M_\odot$, such bursts would be in a frequency range around a millihertz—too low to be accessible to ground-based detectors, which lose sensitivity below 100 Hz, owing to seismic and other background noise. Space-based detectors are needed. One such being proposed, by the European Space Agency, is the Laser Interferometric Spacecraft (LISA)—six spacecraft on solar orbit, configured as two triangles, with a baseline of 5 million km whose length is monitored by laser interferometry.

LISA could detect the mergers of supermassive holes, even whose occurring at high redshifts, with high signal-to-noise. The bad news is that the event rate is low. Even out to z = 5, there could be less than one event per decade involving holes above $10^6 M_\odot$, even if there were enough black holes altogether to account for all the quasars (Haehnelt 1994). This is of course a lower limit. There could be lower-mass holes in small galaxies that are more common and underwent more mergers.

The sensitivity of LISA is such that it could detect waves from a stellar-mass object orbiting a supermassive hole—this is a further reason for interest in the possibility that there may be stars in such orbits, even around holes that are electromagnetically quiescent. LISA is at the moment just a proposal—even if it is funded, it is unlikely to fly before 2015. Is there any way of learning, before that date, something about gravitational radiation? The dynamics (and gravitational radiation) when two holes merge has so far been computed only for cases of special symmetry. The more general problem—coalescence of two Kerr holes with general orientations of their spin axes relative to the orbital angular momentum— is one of the US 'grand challenge' computational projects.

When this challenge has been met (and it will almost certainly not take all the time until 2015), we shall find out not only the characteristic wave form of the radiation, but the *recoil* that arises because there is a net emission of linear momentum. This recoil could displace the hole from the centre of the merged galaxy—it might therefore be relevant to the low-z quasars that seem to be asymmetrically located in their hosts (and which may have been activated by a recent merger). The recoil might even be so violent that the merged hole breaks loose from its galaxy and goes hurtling through intergalactic space.

As data accumulates, it should be feasible to pin down the luminosity functions and lifetimes of quasars, and to correlate the masses of remnant holes in nearby galaxies to the morphology of their hosts. This subject should offer real opportunities for fascinating analysis and modelling, with the aim of understanding the mechanisms within quasars and how the quasar phase relates to the general process of galaxy formation.

Quasars also already pose several other questions that can only be answered computationally. Being myself one of those whose cerebration is mainly done at about a milliflop without electronic aids, I suggest these with some diffidence. However, as computers advance from gigaflops towards teraflops, people like me will be relegated to the role of marginalised cheerleaders unless we can change our ways. I list below some fundamental problems that can be addressed by the technically strong student. Among these suggestions are:

(i) *Stars and gas in the central 100 pc of a newly-formed bulge.* When does gas stop being able to form stars, and evolve instead into a supermassive object?

(ii) *Flow patterns and energy generation around a hole:* the effects of Lense-Thirring precession on the flow pattern, the role of magnetic fields, and the production of relativistic jets.

(iii) *Tidal disruption of a star by a massive black hole.* What is the fate (and observable signature) of the debris?

(iv) *Gravitational radiation from coalescing Kerr holes in cases with no special symmetry?* What is the recoil of the merged hole, etc?

An even greater challenge—perhaps not for a student—would be to compute the inter-related processes that happen in the region around a relativistic mass in an AGN, all the way from the gravitational radius out to scales 10^{10} times larger, where the energy sometimes manifests itself as relativistic plasma in radio source lobes.

It is a particular pleasure to thank John Bahcall for all the support and enlightenment he has generously provided over the more than 25 years I've known him. I'm grateful to him, and to Oleg Gnedin and Jeremy Goodman, for their comments on my draft manuscript.

BIBLIOGRAPHIC NOTES

- **Begelman, M. C., & Rees, M. J. 1995, Gravity's Fatal Attraction: The Discovery of Black Holes (W.H. Freeman, Scientific American Library).** Although this is a 'popular' book, at the level of articles in Scientific American, it introduces many of the issues, both observational and theoretical, and summarises the current evidence that black holes actually exist.

- **Cohen, M. H., & Kellermann, K. I. (eds.) 1995, Quasars and Active Galactic Nuclei: High Resolution Radio Imaging, Proc. Nat. Acad. Sci., 92, 11339-11450.** This is the proceedings of a conference dealing primarily with the radio continuum properties of compact sources; it also discusses the remarkable new evidence for a black hole in NGC 4258 obtained from molecular line spectroscopy.

- **Genzel, R., & Harris, A. I. (eds.) 1994, Nuclei of Normal Galaxies (Kluwer).** This contains good reviews of the data on nearby galaxies, including our Galactic Center.

- **Rees, M. J. 1995, Perspectives in Astrophysical Cosmology (Cambridge University Press).** This is the written version of a series of lectures aimed at physicists, which should be accessible to beginning graduate students. It outlines the relation of quasars to galaxies, and sets them in a cosmological context.

- **Robson, I. 1996, Active Galactic Nuclei (Cambridge University Press).** An up-to-date survey of the phenomenology of 'AGNs', with a basic description of the relevant physical processes.

BIBLIOGRAPHY

[1] Bahcall, J. N., Kirhakos, S., & Schneider, D. P. 1994, ApJ, 435, L11.

[2] Bahcall, J. N., Kirhakos, S., & Schneider, D. P. 1995, ApJ, 447, L1.

[3] Bahcall, J. N., & Wolf, R. A. 1976, ApJ, 209, 214.

[4] Blandford, R. D., & Znajek, R. L. 1977, MNRAS, 179, 433.

[5] Blumenthal, G., Faber, S. M., Primack, J., & Rees, M. J. 1984, Nature, 311, 517.

[6] Chandrasekhar, S. 1975, lecture reprinted in Truth and Beauty (Chicago UP 1987), p. 54.

[7] Couchman, H. M. P., & Rees, M. J. 1986, MNRAS, 221, 53.

[8] Haehnelt, M. 1994. MNRAS, 289, 199.

[9] Haehnelt, M., & Rees, M. J. 1993, MNRAS, 263, 168.

[10] Karas, V., & Vokrouhlicky, D. 1993, MNRAS, 265, 365.

[11] Klypin, A., Holtzman, J. A., Primack, J., & Regos, E. 1993, ApJ, 416, 1.

[12] Krabbe, A., et al. 1995, ApJ, 447, L95.

[13] Madau, P., Meiksin, A., & Rees, M. J. 1996, ApJ, in press.

[14] Miralda-Escudé, J., & Ostriker, J. P. 1990, ApJ, 350, 1.

[15] Miralda-Escudé, J., & Rees, M. J. 1994, MNRAS, 266, 343.

[16] Miyoshi, M., et al. 1995, Nature, 373, 127.

[17] Ostriker, J. P. 1993, ARA&A, 31, 689.

[18] Peebles, P. J. E. 1987, Nature, 327, 210.

[19] Peebles, P. J. E. 1982, ApJ, 263, L1.

[20] Rees, M. J. 1993, Proc. Nat. Acad. Sci., 90, 4840.

[21] Schneider, D. P., Schmidt, M., & Gunn, J. E. 1991, AJ, 102, 839.

[22] Scott, D., & Rees, M. J. 1990, MNRAS, 247, 510.

[23] Shaver, P. 1995, Ann. N.Y. Acad. Sci., 759, 87.

[24] Tanaka, Y., et al. 1995, Nature, 375, 659.

[25] Umemura, M., Loeb, A., & Turner, E. L. 1993, ApJ, 419, 459.

[26] Watson, W. D., & Wallin, B.K. 1984, ApJ, 432, L35.

CHAPTER 10

SOLAR NEUTRINOS: SOLVED AND UNSOLVED PROBLEMS

JOHN N. BAHCALL

Institute for Advanced Study, School of Natural Sciences, Princeton, NJ

ABSTRACT

This talk answers a series of questions. Why study solar neutrinos? What does the combined standard model (solar plus electroweak) predict for solar neutrinos? Why are the calculations of neutrino fluxes robust? What are the three solar neutrino problems? What have we learned in the first 30 years of solar neutrino research? For the next decade, what are the most important solvable problems in the physics of solar neutrinos? What are the most important solvable problems in the astrophysics of solar neutrinos?

10.1 WHY STUDY SOLAR NEUTRINOS?

Astronomers study solar neutrinos for different reasons than physicists. For astronomers, solar neutrino observations offer an opportunity to test directly the theories of stellar evolution and of nuclear energy generation. With neutrinos one can look into the interior of a main sequence star and observe the nuclear fusion reactions that are ultimately responsible for starlight via hydrogen burning:

$$4\,^1\text{H} \longrightarrow \,^4\text{He} + 2e^+ + 2\nu_e. \tag{10.1}$$

The optical depth of the sun for a typical neutrino produced by nuclear fusion is $\sim 10^{-9}$, about 20 orders of magnitude smaller than the optical depth for a typical optical photon.

Table 10.1 shows the principal nuclear reactions that accomplish equation (10.1) via the proton-proton chain of reactions. In what follows, we shall refer often to the reactions listed in this table.

Table 10.1: The Principal Reactions of the pp Chain

Reaction Number	Reaction	Neutrino Energy (MeV)
1	$p + p \rightarrow {}^2\text{H} + e^+ + \nu_e$	0.0 to 0.4
2	$p + e^- + p \rightarrow {}^2\text{H} + \nu_e$	1.4
3	${}^2\text{H} + p \rightarrow {}^3\text{He} + \gamma$	
4	${}^3\text{He} + {}^3\text{He} \rightarrow {}^4\text{He} + 2p$	
	or	
5	${}^3\text{He} + {}^4\text{He} \rightarrow {}^7\text{Be} + \gamma$	
	then	
6	$e^- + {}^7\text{Be} \rightarrow {}^7\text{Li} + \nu_e$	0.86, 0.38
7	${}^7\text{Li} + p \rightarrow {}^4\text{He} + {}^4\text{He}$	
	or	
8	$p + {}^7\text{Be} \rightarrow {}^8\text{B} + \gamma$	
9	${}^8\text{B} \rightarrow {}^8\text{Be} + e^+ + \nu_e$	0 to 15

Solar neutrino experiments constitute quantitative, well-defined tests of the theory of stellar evolution. The neutrino fluxes reflect directly the rates of nuclear fusion reactions in the interior, rates which cannot be measured otherwise. Nuclear fusion reactions are primarily responsible for stellar evolution. Since stellar evolution theory is used by most astronomers in some aspect of their work, direct tests of this theory are of importance to astronomers.

Solar neutrinos are of interest to physicists because they can be used to perform unique particle physics experiments. Many physicists believe that solar neutrino experiments may have already provided strong hints that at least one neutrino type has a non-zero mass and that electron flavor (or the number of electron-type neutrinos) may not be conserved. Neutrinos of finite mass may be related to the dark matter problem (see the discussion by Spergel in these proceedings).

For some of the theoretically most interesting ranges of masses and mixing angles, solar neutrino experiments are more sensitive tests for neutrino transformations in flight than experiments that can be carried out with laboratory sources. The reasons for this exquisite sensitivity are: 1) the great distance between the beam source (the solar interior) and the detector (on earth); 2) the relatively low energy (MeV) of solar neutrinos; and 3) the enormous path length of matter ($\sim 10^{11}\text{gm cm}^{-2}$) that neutrinos must pass through on their way out of the sun.

One can quantify the sensitivity of solar neutrinos relative to laboratory experiments by considering the proper time that would elapse for a finite-mass neutrino in flight between the point of production and the point of detection. The elapsed

proper time is a measure of the opportunity that a neutrino has to transform its state
and is proportional to the ratio, R, of path length divided by energy:

$$\text{Proper Time} \propto R = \frac{\text{Path Length}}{\text{Energy}}. \tag{10.2}$$

Future accelerator experiments with multi-GeV neutrinos may reach a sensitivity of $R = 10^2 \text{ km GeV}^{-1}$. Reactor experiments have already almost reached a level of sensitivity of $R = 10^2 \text{ km GeV}^{-1}$ for neutrinos with MeV energies. Solar neutrino experiments, because of the enormous distance between the source (the center of the sun) and the detector (on earth) and the relatively low energies (1 MeV to 10 MeV) of solar neutrinos involve much larger values of neutrino proper time,

$$R(\text{solar}) = \frac{10^8}{10^{-3}} \left(\frac{\text{Km}}{GeV} \right) \sim 10^{11} \left(\frac{\text{Km}}{GeV} \right). \tag{10.3}$$

Because of the long proper time that is available to a neutrino to transform its state, solar neutrino experiments are sensitive to very small neutrino masses that can cause neutrino oscillations in vacuum. Quantitatively,

$$m_\nu (\text{solar level of sensitivity}) \sim 10^{-6} \text{eV to } 10^{-5} \text{eV} \quad (\text{vacuum oscillations}), \tag{10.4}$$

provided the electron-type neutrino that is created by beta-decay contains appreciable portions of at least two different neutrino mass eigenstates (i.e., the neutrino mixing angle is relatively large). Laboratory experiments have achieved a sensitivity to electron neutrino masses of order one eV. Over the next several years, the sensitivity of the laboratory experiments should be improved by an order of magnitude or more.

Resonant neutrino oscillations, which may be induced by neutrino interactions with electrons in the sun (the famous Mikheyev-Smirnov-Wolfenstein, MSW, effect), can occur even if the electron neutrino is almost entirely composed of one neutrino mass eigenstate (i.e., even if the mixing angles between e and μ and between e and τ neutrinos are tiny). Standard solar models indicate that the sun has a high central density, $\rho(\text{central}) \sim 1.5 \times 10^2 \text{ gm cm}^{-3}$, which (at least in principle) allows even very low energy (< 1 MeV) electron-type neutrinos to be resonantly converted (to the more difficult to detect μ or τ neutrinos) by the MSW effect. Also, the column density of matter that neutrinos must pass through is large: $\int \rho dr \approx 2 \times 10^{11} \text{ gm cm}^{-2}$. The corresponding parameters for terrestrial, long-baseline experiments are: a typical density of 3 gm cm^{-3}, and an obtainable column density of $\sim 2 \times 10^8 \text{ gm cm}^{-2}$.

Given the above solar parameters, the planned and operating solar neutrino experiments are sensitive to neutrino masses in the range

$$10^{-4} \, \text{eV} \lesssim m_\nu \lesssim 10^{-2} \, \text{eV},\qquad\qquad(10.5)$$

via matter-induced resonant oscillations (MSW effect).

The range of neutrino masses given by equation (10.4) and equation (10.5) is included in the range of neutrino masses that are suggested by attractive particle-physics generalizations of the standard electroweak model, including left-right symmetry, grand-unification, and supersymmetry.

Both vacuum neutrino oscillations and matter-enhanced neutrino oscillations can change electron-type neutrinos to the more difficult to detect muon or tau neutrinos. In addition, the likelihood that a neutrino will have its type changed may depend upon its energy, thereby affecting the shape of the energy spectrum of the surviving electron-type neutrinos. Future solar neutrino experiments will measure the ratio of the number of electron-type neutrinos to the total number of solar neutrinos (via neutral current reactions) and will also measure the shape of the electron-type energy spectrum (via charged current absorption and by neutrino-electron scattering). These measurements (of the spectrum shape and the ratio of electron-type to total number of neutrinos) will test the simplest version of the standard electroweak model (in which neutrinos are massless and do not oscillate); these tests are independent of solar model physics.

10.2   WHAT DOES THE COMBINED STANDARD MODEL TELL US ABOUT SOLAR NEUTRINOS?

In this section, I will describe the combined standard model (standard solar model and standard electroweak theory) that is used to decide if solar neutrino experiments have revealed something unexpected. Then I will present the calculated solar neutrino spectrum as predicted by the standard model.

10.2.1   *The Combined Standard Model*

In order to interpret solar neutrino experiments, one must have a quantitative theoretical model. Unlike many other areas of astronomy in which one can make important discoveries by identifying new classes of objects (such as quasars, or x−ray sources, or γ− ray sources), solar neutrino research requires a reliable theoretical model for comparison with the observations in order to determine whether one has found something surprising. Our physical intuition is not yet sufficiently advanced to know if we should be surprised by 10^{-2}, by 10^0, or by 10^{+2} neutrino-induced events per day in a chlorine tank the size of an olympic swimming pool.

I will use the most conservative model for comparison with experiments, the "combined standard model," unless explicitly stated otherwise. The combined standard model is the standard model of solar structure and evolution and the standard electroweak model of particle physics.

A particle physics model is required to predict what happens to the neutrinos after they are created. I will assume, unless stated otherwise, that nothing happens to solar neutrinos after they are created in the interior of the sun. This assumption is valid if standard electroweak theory is correct. In the simplest version of standard electroweak theory, neutrinos are massless and neutrino flavor (electron type) is conserved. The standard electroweak model has had many successes in precision laboratory tests; modifications of this theory will be accepted only if incontrovertible experimental evidence forces a change.

A solar model is required in order to predict the number of neutrinos created in a given energy range per unit of time. On a fundamental level, a solar model is required in order to predict the rate of nuclear fusion by the *pp* chain (shown in Table 10.1 and discussed below) and the rate of fusion by the CNO reactions (originally favored by H. Bethe in his epochal study of nuclear fusion reactions).

In our discussion, I will assume a result common to all modern solar models, namely, that the CNO reactions contribute only a very small fraction of the luminosity of the sun. Although the dominance of the *pp* chain is often taken for granted in theoretical analyses of solar neutrino experiments, it is not an *a priori* obvious result. We shall come back to this fundamental prediction of solar models in the discussion given in §10.5.1, where we review what has been learned about astronomy from solar neutrino experiments.

A precise solar model is required to calculate accurately which nuclear reaction occurs more often at the two principal branching points of the *pp* fusion chain. Referring to Table 10.1, the branching points occur between reactions 4 and 5 and between reactions 6 and 8. If the *pp* chain is terminated by reaction 4, only low-energy ($< 0.4\,\mathrm{MeV}$) *pp* neutrinos are produced, whereas if the termination occurs via reaction 5, then higher-energy $^{7}\mathrm{Be}$ and $^{8}\mathrm{B}$ neutrinos are created. The ratio of the rates for reaction 6 and reaction 8 determines how often $^{7}\mathrm{Be}$ neutrinos (two lines: 0.86 MeV and 0.38 MeV) are produced rather than the rare, but more easily detected $^{8}\mathrm{B}$ neutrinos (maximum energy $\sim 15\,\mathrm{MeV}$) are produced.

The competition between reactions 4 and 5, and between reactions 6 and 8, determines the energies of the emitted solar neutrinos. The predicted rates in solar neutrino experiments depend sensitively upon the relative frequencies of these crucial reactions. Fortunately, the theoretical uncertainties in the predicted neutrino fluxes are not very large; they vary from $\sim 1\%$ to $\sim 17\%$, depending upon the neutrino source in question. The rates for individual detectors are determined by the energy spectrum, by the neutrino type of the incoming solar neutrinos, and by the interaction cross sections of the different detectors. As we shall see in the later discussion (cf. §10.4), one can largely avoid the dependence of the predictions upon solar models by comparing the results of experiments that have different energy sensitivities. These comparisons between experiments primarily test standard electroweak theory.

10.2.2   *The Solar Neutrino Spectrum*

Figure 10.1 shows the calculated neutrino spectrum for the most important neutrino sources from the sun. I will discuss briefly the *pp* (and *pep*) neutrinos, the ^{7}Be neutrinos, and the ^{8}B neutrinos. I will concentrate on the reliability of the predictions and will indicate the role of each of these neutrinos in the ongoing experiments.

The dominant source of solar neutrinos is the first reaction listed in Table 10.1,

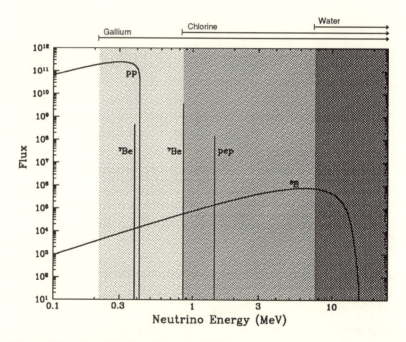

Figure 10.1: Solar Neutrino Spectrum. This figure shows the energy spectrum of neutrinos from the *pp* chain that is predicted by the standard solar model. The neutrino fluxes from continuum sources (*pp* and ^{8}B) are given in the units of number per cm^2 per second per MeV at one astronomical unit. The line fluxes (pep and ^{7}Be) are given in number per cm^2 per second. The arrows at the top of the figure indicate the energy thresholds for the ongoing neutrino experiments. The higher-energy ^{7}Be line is just above threshold in the chlorine experiment. For simplicity, CNO neutrinos are omitted.

the basic pp reaction ($p + p \longrightarrow D + e^+ + \nu_e$), which creates neutrinos with energies less than 0.4 MeV. Most of the nuclear energy that emerges as sunlight begins with this reaction. The theoretical uncertainty in the pp neutrino flux is less than 1%. Among the solar neutrino experiments that are currently operating or that are being constructed, only the GALLEX and SAGE gallium experiments have energy thresholds low enough to detect the pp neutrinos. (About 0.2% of the pp fusions are believed to occur via the pep reaction, the second reaction in Table 10.1. This reaction produces a neutrino line that contributes a small part of the calculated event rate in the chlorine and gallium experiments.)

The next most important source of neutrinos is from the ^{7}Be neutrino line at 0.86 MeV, which is produced by reaction 6 of Table 10.1. About 15% of the solar luminosity is produced by reactions which go through this channel; the uncertainty in the neutrino flux is $\sim$ 7%. The ^{7}Be neutrinos contribute significantly, according to standard model calculations, to the chlorine and the gallium experiments, but are too low in energy to be detected in the Kamiokande experiment. In an experiment under development called BOREXINO, ^{7}Be neutrinos will be detected by the unique signature they produce in scintillation light caused by neutrino-electron scattering.

The ^{8}B neutrino flux, produced by reaction 9 of Table 10.1, is tiny, $\sim 10^{-4}$ of the flux of pp neutrinos. However, the ^{8}B neutrinos are much of what the story is about for solar neutrino physics and astronomy. Because of their high energy ($\sim$ 10 MeV, which can excite a superallowed transition from the ground-state of chlorine to an excited state of argon), ^{8}B neutrinos dominate the predicted capture rate for the chlorine experiment. They are also the only significant source of neutrinos above the energy threshold in the water Cherenkov experiment, Kamiokande. The SNO and Superkamiokande experiments, both of which are under development and which should begin producing results in 1996, also observe only ^{8}B neutrinos. Unfortunately, the theoretical uncertainty in the predicted ^{8}B neutrino flux is relatively large, $\sim$ 17%. This relatively large uncertainty is due to the temperature sensitivity of the reaction that produces the ^{8}B nuclei and to a somewhat uncertain laboratory measurement of the cross section for reaction 8 of Table 10.1.

10.3   WHY ARE THE PREDICTED NEUTRINO FLUXES ROBUST?

The predicted event rates in the different solar neutrino experiments have been remarkable stable over the past 25 years. The published estimate in 1968, which accompanied the first report by Davis and his collaborators of measurements with the chlorine experiment, was 7.5 ± 1.0 Solar Neutrino Units (SNU); the most recent and detailed calculation gives in 1995 a predicted rate of 9.3 ± 1.3 SNU. A SNU is a convenient unit to describe the measured rates of solar neutrino experiments:

10^{-36} interactions per target atom per second. The theoretical errors are intended to be as close as possible to effective 1σ errors; they are obtained by carrying out detailed calculations using 1σ uncertainties on all the measured input data and, for the theoretical errors (which are less important), by taking the extreme range of theoretical calculations to be 3σ uncertainties.

There are three reasons that the theoretical calculations of the neutrino fluxes are robust: 1) the availability of precision measurements and precision calculations of input data; 2) the connection between neutrino fluxes and the measured solar luminosity; and 3) the measurement of the helioseismological frequencies of the solar pressure-mode (p-mode) eigenfrequencies.

Over the past three decades, many hundreds of researchers have performed precision measurements of crucial input data including nuclear reaction cross sections and the abundances of the chemical elements on the solar surface. Many other researchers have calculated accurate opacities, equations of state, and weak interaction cross sections. By now, these input data are relatively precise and their uncertainties are quantifiable.

The solar neutrino fluxes and the solar luminosity both depend upon the rates of the nuclear fusion reactions in the solar interior. Since we know experimentally the solar luminosity (to an accuracy of $\sim 0.4\%$), the calculated neutrino fluxes are strongly constrained by the fact that the standard solar models must yield precisely the measured solar luminosity.

Thousands of p-mode helioseismological frequencies have been measured to an accuracy of 1 part in 10^4. The standard solar models discussed here reproduce these p-mode frequencies to an accuracy of better than 1 part in 1,000. The standard solar models are in agreement with the measured helioseismological frequencies to a high level of precision without any special adjustments of the parameters. Helioseismological measurements show that standard solar models provide a reasonably accurate (much better than 1%) description of the sound velocity of the solar interior in the regions in which it has been studied precisely so far ($R \gtrsim 0.2\,R_{\odot}$).

The calculated solar neutrino fluxes are, after 30 years of intense study, known to relatively high accuracy because of the many precise measurements and calculations of input data, because of the strong constraint imposed on the models by the measured total solar luminosity, and because of the important tests of solar structure that are provided by helioseismological measurements.

10.4   WHAT ARE THE THREE SOLAR NEUTRINO PROBLEMS?

I will compare in this section the predictions of the combined standard model with the results of the operating solar neutrino experiments. We will see that this comparison leads to three different discrepancies between the calculations and the observations, which I will refer to as the three solar neutrino problems.

Figure 10.2 shows the measured and the calculated event rates in the four ongoing solar neutrino experiments. This figure reveals three discrepancies between the experimental results and the expectations based upon the combined standard model. As we shall see, only the first of these discrepancies depends sensitively upon predictions of the standard solar model.

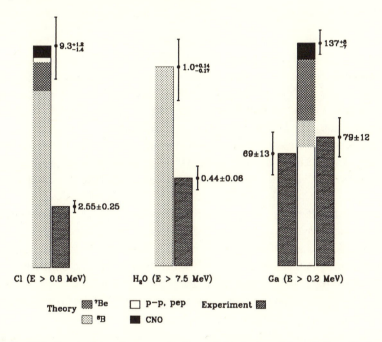

Figure 10.2: Comparison of measured rates and standard-model predictions for four solar neutrino experiments.

10.4.1 *Calculated versus Observed Chlorine Rate*

The first solar neutrino experiment to be performed was the chlorine radiochemical experiment, which detects electron-type neutrinos that are more energetic than 0.81 MeV. After more than 25 years of the operation of this experiment, the measured event rate is 2.55 ± 0.25 SNU, which is a factor ~ 3.6 less than is predicted by the most detailed theoretical calculations, $9.3^{+1.2}_{-1.4}$ SNU. Most of the predicted rate in the chlorine experiment is from the rare, high-energy ^{8}B neutrinos, although the ^{7}Be neutrinos are also expected to contribute significantly. According to standard model calculations, the *pep* neutrinos and the CNO neutrinos (for simplicity not discussed here) are expected to contribute less than 1 SNU to the total event rate.

This discrepancy between the calculations and the observations for the chlorine experiment was, for more than two decades, the only solar neutrino problem. I shall refer to the chlorine disagreement as the "first" solar neutrino problem.

10.4.2   *Incompatibility of Chlorine and Water (Kamiokande) Experiments*

The second solar neutrino problem results from a comparison of the measured event rates in the chlorine experiment and in the Japanese pure-water experiment, Kamiokande. The water experiment detects higher-energy neutrinos, those with energies above 7.5 MeV, by neutrino-electron scattering: $\nu + e \longrightarrow \nu' + e'$. According to the standard solar model (see also Table 10.1), ^{8}B beta-decay is the only important source of these higher-energy neutrinos.

The Kamiokande experiment shows that the observed neutrinos come from the sun. The electrons that are scattered by the incoming neutrinos recoil predominantly in the direction of the sun-earth vector; the relativistic electrons are observed by the Cherenkov radiation they produce in the water detector.

In addition, the Kamiokande experiment measures the energies of individual scattered electrons and therefore provides information about the energy spectrum of the incident solar neutrinos. The observed spectrum of electron recoil energies is consistent with that expected from ^{8}B neutrinos. However, small angle scattering of the recoil electrons in the water prevents the angular distribution from being determined well on an event-by-event basis, which limits the constraints the experiment places on the incoming neutrino energy spectrum.

The event rate in the Kamiokande experiment is determined by the same high-energy ^{8}B neutrinos that are expected, on the basis of the combined standard model, to dominate the event rate in the chlorine experiment. I have shown elsewhere that solar physics changes the shape of the ^{8}B neutrino spectrum by less than 1 part in 10^5. Using the known neutrino interaction cross sections for the chlorine and the Kamiokande experiments, and the standard neutrino energy spectrum, we can calculate the rate in the chlorine experiment that is produced by the ^{8}B neutrinos observed in the Kamiokande experiment. This partial (^{8}B) rate in the chlorine experiment is 3.2 ± 0.45 SNU, which exceeds the total observed chlorine rate of 2.55 ± 0.25 SNU.

Comparing the rates of the Kamiokande and the chlorine experiments, one finds that the net contribution to the chlorine experiment from the *pep*, ^{7}Be, and CNO neutrino sources is negative: -0.66 ± 0.52 SNU. The standard model calculated rate from *pep*, ^{7}Be, and CNO neutrinos is 1.9 SNU. The apparent incompatibility of the chlorine and the Kamiokande experiments is the "second" solar neutrino problem. The inference that is often made from this comparison is that the energy spectrum of ^{8}B neutrinos is changed from the standard shape by physics not included in the simplest version of the standard electroweak model.

10.4.3   Gallium Experiments: No Room for ^{7}Be Neutrinos

The results of the gallium experiments, GALLEX and SAGE, constitute the third solar neutrino problem. The average observed rate in these two experiments is 74 SNU, which is essentially fully accounted for in the standard model by the theoretical rate of 73 SNU that is calculated to come from the basic pp and pep neutrinos (with only a 1% uncertainty in the standard solar model pp flux). The ^{8}B neutrinos, which are observed above 7.5 MeV in the Kamiokande experiment, must also contribute to the gallium event rate. Using the standard shape for the spectrum of ^{8}B neutrinos and normalizing to the rate observed in Kamiokande, ^{8}B contributes another 7 SNU, unless something happens to the lower-energy neutrinos after they are created in the sun. (The predicted contribution is 16 SNU on the basis of the standard model.) Given the measured rates in the gallium experiments, there is no room for the additional 34 ± 4 SNU that is expected from ^{7}Be neutrinos on the basis of 13 standard solar models calculated by different groups using different input data and different stellar evolution codes.

The seeming exclusion of everything but pp neutrinos in the gallium experiments is the "third" solar neutrino problem. This problem is essentially independent of the previously-discussed solar neutrino problems, since it depends upon the pp neutrinos that are not observed in the other experiments and whose calculated flux is approximately model-independent (if the general scheme of the pp chain shown in Table 10.1 is correct).

The missing ^{7}Be neutrinos cannot be explained away by any change in solar physics. The ^{8}B neutrinos that are observed in the Kamiokande experiment are produced in competition with the missing ^{7}Be neutrinos; the competition is between reaction 6 and reaction 8 in Table 10.1. Solar model explanations that reduce the predicted ^{7}Be flux reduce much more (too much) the predictions for the observed ^{8}B flux.

The flux of ^{7}Be neutrinos, $\phi(^7\text{Be})$, is independent of measurement uncertainties in the cross section for the nuclear reaction $^7\text{Be}(\text{p}, \gamma)^8\text{B}$; the cross section for this proton-capture reaction is the most uncertain quantity that enters in an important way in the solar model calculations. The flux of ^{7}Be neutrinos depends upon the proton-capture reaction only through the ratio

$$\phi(^7\text{Be}) \propto \frac{R(e)}{R(e) + R(p)}, \tag{10.6}$$

where $R(e)$ is the rate of electron capture by ^{7}Be nuclei and $R(p)$ is the rate of proton capture by ^{7}Be. With standard parameters, solar models yield $R(p) \approx 10^{-3} R(e)$. Therefore, one would have to increase the value of the ^{7}Be$(p, \gamma)^8$B cross section by more than a factor of 100 over the current best-estimate (which has an estimated uncertainty of $\sim 10\%$) in order to affect significantly the calculated ^{7}Be solar neutrino flux. The required change in the nuclear physics cross section would also increase the predicted neutrino event rate by more than a factor of 100 in the Kamiokande experiment, making that prediction completely inconsistent with what is observed. (From time to time, papers have been published claiming to solve the solar neutrino problem by artificially changing the rate of the ^{7}Be electron capture reaction. Equation (10.6) shows that the flux of ^{7}Be neutrinos is actually independent of the rate of the electron capture reaction to an accuracy of better than 1%.)

I conclude that either: 1) at least three of the four operating solar neutrino experiments (the two gallium experiments plus either chlorine or Kamiokande) give misleading results, or 2) physics beyond the standard electroweak model is required to change the neutrino energy spectrum (or flavor content) after the neutrinos are produced in the center of the sun.

10.5   WHAT HAVE WE LEARNED?

No solar-model solution has been found that explains the results of the four existing solar neutrino experiments. Many particle-physics solutions have been proposed that can explain the existing data.

In this section, I will summarize the main lessons that have been learned from the first 30 years of solar neutrino research. I will first review the progress in astrophysics and then outline briefly the developments in physics.

10.5.1   *About Astronomy*

The chlorine solar neutrino experiment was proposed in 1964 as a practical test of solar model calculations in back-to-back papers (Physical Review Letters) that discussed the theoretical expectations and the experimental possibilities. The only motivation presented in those two papers for performing the chlorine experiment was "...to see into the interior of a star and thus verify directly the hypothesis of nuclear energy generation in stars."

What have we learned by direct experiments about nuclear energy generation in stars? How does our 1964 understanding compare with the results of the solar neutrino experiments? Table 10.2 summarizes the six principal predictions that

Table 10.2: Predictions versus Observations: 1964 vs. 1995

Property	Predicted	Observed
Direction	From the Sun	o.k.
Rates	Measurable	$\sim$ Predicted Rates (within factor of few)
Neutrino Energy	0–15 MeV	< 15 MeV
Time Dependence	Constant (except seasonal)	o.k.
pp not CNO	If CNO : $\begin{cases} \text{Cl}: & 28\,\text{SNU} \\ \text{Ga}: & 610\,\text{SNU} \\ \text{H}_2\text{O}: & 0.0 \end{cases}$	2.6 SNU 74 SNU 0.44 Standard Model
Central Temperature	16×10^6 K	$T(^8\text{B})/T_{\text{model}} \gtrsim 0.98$

were made (or which were implicit in the theory) in 1964 and compares those predictions with the results of the four ongoing solar neutrino experiments.

The neutrinos were predicted to originate in the solar interior; the direction of origin of the neutrinos has been verified by detecting neutrino-electron scattering (as was suggested in what seemed to be a futuristic paper in 1964) in the Kamiokande experiment. The rates of the four operating experiments are in semiquantitative agreement with the predictions; the ratios of the observed to the predicted rates are 0.3 (chlorine), 0.5 (water-Kamiokande), and 0.5 (gallium, average). This agreement is better than any of us dared hope for in 1964, especially since the dominant neutrino flux (from ^{8}B beta-decay) for the first two experiments depends upon the central temperature of the sun as approximately the 20th power of the central temperature. The energy range of the dominant neutrinos was predicted to be from 0 MeV to 15 MeV, which is consistent with the observations from the Kamiokande experiment. These observations do not yield detailed information about the incoming solar neutrino energy spectrum, but they do show that neutrinos exist in the expected energy range (at least above 7.5 MeV) and that no events have been observed at energies beyond the maximum energy resulting from ^{8}B beta-decay. Standard models predict that the neutrino fluxes are constant in time except for a small seasonal variation. (The Kelvin-Helmholtz cooling time for the solar interior is $\sim 10^7$ years.) The consensus, but not the unanimous view of the experimentalists, based upon the available data from the four pioneering solar neutrino experiments (all of which have rather low counting rates) is that the neutrino fluxes are indeed constant in time. Standard solar models predict that the sun shines almost entirely via the *pp* chain of nuclear fusion reactions, rather than the CNO reactions originally emphasized by Bethe. If CNO reactions were dominant, and no new neutrino physics occurs, then the event rates in solar neutrino experiments could be calculated precisely. These "all-CNO" rates differ, as shown in Table 10.2, from the observed rates by more than an order of magnitude. Finally, if we crudely characterize the rate of the ^{8}B neutrino emission by its approximate dependence upon the central temperature of the solar model, then the central temperature of the solar model agrees with the value obtained from the experimental rates to an accuracy of $\sim$ 2% or better.

The four ongoing solar neutrino experiments have shown directly that the sun shines by nuclear fusion reactions, thus achieving the original goal proposed in 1964. Quantitative improvements in the tests shown in Table 10.2 will occur with the next generation of experiments. But, the most important qualitative result has been established: **neutrinos have been observed from the interior of the sun in approximately the number and with the energies expected.**

In the three-decade long struggle to improve solar models in order to calculate more accurate solar neutrino fluxes, we have obtained a greater understanding of solar structure. The theoretical models have gradually been refined as improved input data, more accurate physical descriptions, and more precise numerical techniques have been employed. Perhaps most importantly, the complementary field of helioseismology has been developed and now provides precise data that determine the sound velocity over most of the solar interior; these beautiful measurements are used to test and to refine the standard solar model. Further improvements in the solar model are desirable and important, but the quantitative agreement, typically better than 1 part in 1,000, between the calculated eigenfrequencies of pressure modes and the measured (helioseismological) frequencies provides strong evidence for the basic correctness of the standard solar model.

10.5.2   *About Physics*

To perform more precise tests of the astrophysical predictions, we must first learn what happens to neutrinos after they are produced in the interior of the sun. Theoretical physicists have fertile imaginations; they have provided us with a smorgasbord of explanations based upon new particle physics, including vacuum neutrino oscillations, resonant oscillations in matter (the MSW effect), resonant magnetic-moment transitions, sterile neutrinos, neutrino decay, and violation of the equivalence principle by neutrinos. Most of these explanations can account for the existing experimental data if either two or three neutrinos are involved in the new physics beyond the standard electroweak model. All these particle physics explanations, and others that I have not listed, can account for the existing data from solar neutrino experiments without conflicting with established laws of physics or with other experimental constraints.

The number of proposed particle physics explanations exceeds the diagnostic power of the existing solar neutrino experiments. I think it is unlikely that the next generation of solar neutrinos experiments will be able to eliminate all but one possible particle physics explanation. But I hope that the powerful new experiments (SNO, Superkamiokande, and BOREXINO) will, together with the four operating experiments, point us in the direction of one of the previously-proposed explanations.

So, what have we learned about particle physics? We have learned that a number of particle physics explanations are consistent with the data obtained from the first four solar neutrino experiments. More specifically, we have learned that an elegant, conservative extension of the standard electroweak theory, the MSW effect, can describe all of the existing experimental information on solar neutrinos if the electron-type neutrino is mixed with another neutrino that has a finite mass,

$$m_\nu \approx 0.003 \, \text{eV}. \tag{10.7}$$

The key to the MSW solution of the solar neutrino problems is that neutrino oscillations cause the survival probability to depend upon energy for an electron neutrino created in the solar interior. This energy-dependent change in the electron neutrino spectrum invalidates the arguments, based upon standard model physics, that led to the three solar neutrino problems.

The MSW theory is not proven, but it is a beautiful idea. I think it would be a disgrace if Nature failed to make use of this marvelous possibility.

10.6 WHAT NEXT?

In this section, I will summarize the problems, first in physics and then in astronomy, that are likely to be solved in the next decade or two.

10.6.1 *Solvable Problems in Physics*

The fundamental goal of physics research with solar neutrinos is to measure the energy spectrum and flavor content (i.e., neutrino type: electron, muon, or tau) as a function of time of the solar neutrino flux. How many neutrinos reach the earth with a given energy and with a given flavor? Because of some exotic particle physics possibilities, we also want to know if the solar neutrino flux contains any anti-neutrinos.

The standard model predicts that the energy spectrum of neutrinos from any given neutrino source, e.g., from ^{8}B beta-decay, will be the same to high accuracy as the energy spectrum inferred from terrestrial laboratory measurements. In the simplest version of the standard electroweak theory, only massless electron-type neutrinos are created in nuclear beta decay or nuclear fusion reactions. Standard theory predicts that the solar neutrinos produced by nuclear fusion reactions are all ν_e, not ν_μ or ν_τ. (According to MSW and vacuum oscillation theories, neutrinos created in nuclear beta-decay or nuclear fusion reactions are linear combinations of different neutrino types and at least one neutrino type has a non-zero mass.) Finally, the total amount of thermal energy in the solar interior implies that the neutrino fluxes will be constant in time (for time scales less than 10^7 years) except for the seasonal dependences caused by the earth's orbital eccentricity.

Any departure from these expectations will be a signal of physics beyond the standard electroweak model. There are many suggested modifications of standard electroweak theory that are consistent with all existing laboratory experiments and with all four of the operating solar neutrino experiments, but which predict different results for future solar neutrino experiments. We have a lot to learn.

I will now list six specific problems in the physics of solar neutrino research that we can expect to be solved in the next decade or so.

• **The ratio of ν_e to ν_{total}.** The electron-type neutrinos, ν_e, that are created in the interior of the sun will all remain ν_e if the simplest version of the standard electroweak theory is correct. If neutrino oscillations occur, either vacuum or matter-induced oscillations (the MSW effect), then the total number of solar neutrinos observed at earth will exceed the number present as electron-type neutrinos only.

The ratio of ν_e to ν_{total} can be determined by measuring the ratio of the rates for two different reactions, one that occurs only with electron-type neutrinos and one that will occur with equal probability independent of the type of neutrino. This ratio will be measured for the first time in the SNO solar neutrino experiment[1], which utilizes a kiloton of heavy water, via the two reactions: $\nu_e + {}^2\mathrm{H} \longrightarrow p + p + e$ (only ν_e); and $\nu_{total} + {}^2\mathrm{H} \longrightarrow p + n + \nu_{total}$.

• **Shape of the ${}^8\mathrm{B}$ neutrino energy spectrum.** The shape of this spectrum is independent of any aspect of solar model physics to an accuracy of 1 part in 10^5. The shape is determined empirically from laboratory nuclear physics measurements.

The shape of the ${}^8\mathrm{B}$ neutrino energy spectrum can be determined by measuring the energy spectrum of electrons created in the reaction $\nu_e + {}^2\mathrm{H} \longrightarrow p + p + e$, which will be done in the SNO experiment. Important information about the energy spectrum will also be obtained by the Superkamiokande electron-scattering

---

[1]If non-interacting (sterile) neutrinos, $\nu_{sterile}$, exist in nature, they will not be detected in this experiment. In principle, the solution of the solar neutrino problems could involve $\nu_e \rightarrow \nu_{sterile}$.

experiment. Superkamiokande will measure accurately the energy spectrum of recoiling electrons produced by neutrino-electron scattering in pure water.

• **Flux of ^{7}Be neutrinos.** The simplest interpretation of the four operating solar neutrino experiments implies that the flux of ^{7}Be neutrinos is greatly reduced with respect to the value predicted by the standard solar model. I do not know of any proposed modification of solar physics that could explain a greatly reduced ^{7}Be neutrino flux and at the same time be consistent with only a factor of two reduction in the ^{8}B neutrino flux (as observed by the Kamiokande experiment). Therefore, a much reduced ^{7}Be neutrino flux would be a strong signal for new weak interaction physics.

A measurement of the ^{7}Be neutrino flux is required to interpret the chlorine and gallium experiments. Since the chlorine and gallium experiments are radio-chemical, they do not give any indication of the energy (above threshold) of the neutrino that initiated a reaction.

A first measurement of the flux of ^{7}Be neutrinos will be carried out by the BOREXINO collaboration in the Gran Sasso underground laboratory using electron-neutrino scattering in an organic scintillator.

• **Time Dependence of the Neutrino Fluxes.** All the proposed experiments have expected event rates that are much larger than the operating solar neutrino experiments. Typical event rates with existing experiments are in the range of 25 to 50 events per year, whereas the expected event rates in the planned experiments are typically of order a few thousand events per year. The tests for time-dependence with the existing data have yielded only marginally suggestive results.

With these future experiments, it will be be possible to test with high precision the standard model prediction that the solar neutrino event fluxes are independent of time. In addition, it will be possible to measure the 7% peak-to-peak seasonal dependence caused by the eccentricity of the earth's orbit. One can also search with high accuracy for the strong seasonal dependences predicted by explanations that involve vacuum neutrino oscillations. MSW theory also predicts, for certain choices of the parameters, that there will be a strong day-night effect. The different counting rate between day and night could occur if muon or tau neutrinos are reconverted to electron neutrinos as they pass through the earth at night on their way to neutrino detectors on the far side from the sun.

• **Proton-proton Neutrino Flux and Energy Spectrum.** The dominant neutrino flux created in the sun is the low energy flux of neutrinos from the basic *pp* reaction. Because of their low energy (< 0.4MeV), these neutrinos are unobservable in the chlorine, Kamiokande, Superkamiokande, SNO, BOREXINO, and Iodine experiments. Among the planned or operating experiments, only the gallium experiments, GALLEX and SAGE, have a sufficiently low threshold energy to detect

pp neutrinos. The gallium experiments detect neutrinos radiochemically; they do not measure the energy of the neutrinos that cause the conversion of ^{71}Ga to ^{71}Ge. Therefore, we have no experimental way at present of determining how much of the observed event rate in the gallium experiments is due to pp neutrinos and how much is due to ^{7}Be, CNO, or ^{8}B neutrinos.

Two experiments have been proposed recently that are potentially capable of detecting and measuring the energies of individual events caused by pp neutrinos. These experiments would both use cold helium. The HERON detector uses ballistic phonon propagation in liquid helium maintained in the superfluid state. The HELLAZ detector uses a high-pressure helium gas in a time-projection chamber.

The development of a detector that can measure the rate and the energy spectrum of the pp neutrinos is an exciting technical challenge and of fundamental importance to both physics and astronomy. All the proposed physics solutions make definitive predictions about what happens to the low-energy pp neutrinos.

• **Refined Nuclear-physics Parameters.** Over the past three decades, many precise, difficult, and beautiful nuclear physics experiments have been performed to determine the rates of solar fusion reactions with the accuracy required for solar-model calculations of neutrino fluxes. It is important to check these measurements with the improved experimental techniques that are now available.

The predicted rates in the Kamiokande, Superkamiokande, and SNO solar neutrino experiments are proportional to the low-energy rate of the ^{7}Be $+ p \longrightarrow$ ^{8}B$^* + \gamma$ reaction (reaction 8 of Table 10.1) and the dominant contribution in the standard model prediction of the rate of the chlorine experiment is also proportional to the rate of this reaction. There are a number of very beautiful experiments in which the rate of the $p(^7$Be$, \gamma)^8$B reaction has been measured by using a radioactive target of ^{7}Be and a beam of protons. It would be especially informative to reverse the usual experimental situation and use a gaseous target of protons and a beam of ^{7}Be; this reversal would involve different systematic uncertainties, which are often the most important source of errors in difficult experiments.

Measuring the cross section for the ^{7}Be$(p, \gamma)^8$B reaction with a ^{7}Be beam is, in my view, the most important experiment to be performed in nuclear astrophysics.

The fundamental goal of physics experiments with solar neutrinos is to measure, or set stringent limits on, the elementary properties of neutrinos, especially their masses and mixing angles. It seems likely that we will make important progress toward this goal in the next decade.

10.6.2 *Solvable Problems in Astronomy*

The fundamental goal of solar neutrino astronomy is to determine the rates of different nuclear fusion reactions in the solar interior. Neutrino fluxes created by the different nuclear sources are the signatures of the fusion reactions. We must know what happens to the neutrinos after they are created in order to infer the created neutrino energy spectrum from the measured neutrino energy spectrum.

With one exception, progress in solar neutrino astronomy is held hostage to progress in particle physics. As discussed in §10.6.1, it seems likely that we will learn enough about the particle physics in the next decade to permit accurate inferences about the rates of neutrino creation in the sun from the observed rates of neutrino arrival at the earth. The discussion in this subsection presumes that the required progress in understanding the properties of the neutrino will be achieved.

• **Completing Hydrogen Fusion.** Table 10.1 shows that the two principal ways of completing nuclear fusion in the sun are reactions 4 and 5, the so-called ^{3}He-^{3}He and ^{3}He-^{4}He reactions. Because of the slightly smaller reduced mass that exists for the ^{3}He-^{3}He reaction, Coulomb barrier penetration favors this reaction over the ^{3}He-^{4}He reaction at lower temperatures. According to the standard solar model, the ^{3}He-^{4}He reaction is dominant in the innermost region of the sun (where it is 1.5 times faster than the ^{3}He-^{3}He reaction), but overall occurs in only $\sim 19\%$ of the fusion terminations that are described by equation (10.1). That is, in the most detailed solar models, the ^{3}He-^{3}He reaction is on average more than 5 times faster in completing the nuclear fusion of protons into alpha particles than the competing ^{3}He-^{4}He reaction.

Is this prediction of the standard solar model correct? A determination of the pp and ^{7}Be neutrino fluxes (corrected for what non-standard particle physics has done to them after they were created in the sun) can answer this important question. The average ratio of the total number of ^{3}He-^{4}He reactions per unit time in the sun to the total number of ^{3}He-^{3}He reactions per unit time in the sun is

$$\frac{< \,^3\text{He}-^3\text{He} >}{< \,^3\text{He}-^4\text{He} >} = \frac{2\phi(^7\text{Be})}{[\phi(pp) - \phi(^7\text{Be})]}, \tag{10.8}$$

where $\phi(pp)$ and $\phi(^7\text{Be})$ are, respectively, the fluxes from the pp and ^{7}Be neutrinos.

Equation (10.8) is the most precisely-testable prediction that I know of that follows directly from the theory of stellar energy generation. The known theoretical uncertainties in the calculation of the average solar ratio of ^{3}He-^{4}He to ^{3}He-^{3}He reactions is 7%.

• **The ^{8}B Neutrino Flux.** The flux of neutrinos from ^{8}B beta-decay in the sun (see reaction 9 of Table 10.1) is, in principle, the simplest solar neutrino flux to measure. The higher energies of the ^{8}B neutrinos make them easiest to detect. For this reason, the Kamiokande, Superkamiokande, and SNO neutrino experiments will all concentrate on the ^{8}B neutrinos.

However, one must determine the total flux of ^{8}B neutrinos, including the more difficult to detect muon or tau neutrinos that may have been produced by neutrino oscillations from the originally-created electron-type neutrinos. The total number of neutrinos of all types will be measured directly in the SNO experiment via the neutral-current disintegration of deuterium and, less directly, via electron-neutrino scattering in Superkamiokande. (This statement presumes there are no sterile neutrinos, see footnote 1.)

The magnitude of the ^{8}B flux (all neutrino flavors) is a sensitive probe of the temperature, T, of the solar interior, varying approximately as T_{central}^{20}. Therefore, it is important to determine experimentally the total ^{8}B solar neutrino flux.

• **The Temperature Profile of the Solar Interior.** A precision test of the theory of stellar structure and stellar evolution can be performed by measuring the average difference in energy between the neutrino line produced by ^{7}Be electron capture in the solar interior and the corresponding neutrino line produced in a terrestrial laboratory. This energy shift is calculated to be 1.29 keV. The energy shift is approximately equal to the average temperature of the solar core, computed by integrating the temperature over the interior of a standard solar model with a weighting factor equal to the locally-produced ^{7}Be neutrino emission. The total range of values for the shift, calculated for a number of modern solar models (going back to 1982), is 0.06 keV.

A measurement of the energy shift is equivalent to a measurement of the central temperature distribution of the sun.

The calculated energy profile of the ^{7}Be line contains, analogous to line-broadening in classical (photon) astronomy, information about the distribution of solar interior temperatures. The theoretical shape of the ^{7}Be neutrino line is asymmetric: on the low-energy side, the line shape is Gaussian with a half-width at half-maximum of 0.6 keV, and on the high-energy side, the line shape is exponential, with a half-width at half-maximum of 1.1 keV. The low energy Gaussian shape reflects the Doppler shifts caused by the motion of the decaying ions away from the observer. The high energy exponential tail is produced by the center-of-momentum kinetic energies.

The calculated shape of the ^{7}Be neutrino line is not affected significantly by vacuum neutrino oscillations, the MSW effect, or other frequently discussed weak-interaction solutions to the solar neutrino problems. This is a key result: it implies that the astronomical information contained in the line shift and in the line profile is not dependent upon further progress in neutrino physics.

Detectors are available that have the resolution to measure the line shift. Unfortunately, their current sizes are too small to permit a full-scale solar neutrino experiment. However, proposals have been made in the literature for developing detectors that are sufficiently large to be able to measure well the average shift in energy of the solar neutrino line.

• **Ruling out 'Non-Standard' Solar Models.** In the first decade and a half following the initial report that the measured solar neutrino flux in the chlorine experiment was less than the calculated value, a number of authors invented imaginative non-standard solar models that were designed to "solve" the solar neutrino problem. The situation is now different. There are now three solar neutrino problems, and it does not appear possible to reconcile the four operating experiments with any modification of stellar physics.

I think that the neutrino fluxes from the nuclear fusion reactions in the sun are known as well as the neutrino fluxes in many of the best terrestrial laboratory experiments. In the solar context, we use different constraints on the theoretical calculations than we do with laboratory accelerators. However, the solar constraints (especially the measured solar luminosity and the helioseismological frequencies) provide powerful limits on the allowed values of the neutrino fluxes. These constraints are discussed in §10.3.

Nevertheless, many physicists are unfamiliar with stellar physics. They do not feel comfortable judging the plausibility of different solar models, even fanciful ones in which the solar model contains, for example, a central black hole or a non-Maxwellian energy distribution for the nuclei. Some physicists are willing to consider solar models that nearly all astrophysicists would dismiss as unworthy of discussion.

It would be instructive to calculate precise solar models based upon some of the more frequently discussed non-standard models (e.g., a low central heavy element abundance, iron precipitation, a very strong internal magnetic field, nearly complete element mixing, turbulent diffusion, large mass loss, or energy transport by Weakly Interacting Massive Particles, WIMPs). For each non-standard hypothesis, a solar model could be evolved using the best-available physics (opacities, equation of state, diffusion rates, and measured input data) while also imposing the *ad hoc* stellar structure hypothesis. The non-standard models computed in this way could be compared with the thousands of accurately-measured p-mode helioseismological frequencies.

I believe that the non-standard models that have been suggested as possible solutions of the solar neutrino problem would be ruled out by accurate and detailed comparisons with helioseismological data. Most of the suggested models would fail, I suspect, on a grosser level, predicting for example the wrong depth of the convective zone or the wrong dependence of sound velocity on depth within the sun. But it would be an important contribution to test the conjecture that the previously-suggested non-standard solar models that were concocted to solve the solar neutrino problem (when it was just one problem) all fail to account for well-established results of helioseismology.

• **Discover the g-Modes.** The most important discovery that one can anticipate being made in optical solar astronomy is the detection of the oscillations from gravity (g) modes. Unlike the many pressure-mode (p-mode) oscillations that have been studied so far, the largest amplitudes of the g-modes occur in the solar interior; they are expected to be damped heavily in the outer regions of the sun. This concentration toward the center is particularly desirable if one wants to learn about solar interior properties that are relevant to neutrino astrophysics. However, the interior concentration also makes the detection of g-modes difficult. The amplitudes of the g-mode oscillations are expected to be small on the surface of the sun, where they might be measured. They have not yet been detected convincingly in the sun.

New experiments are underway to attempt to detect the g-modes from space (with the SOHO satellite) and from an international network of ground-based telescopes (GONG). The results of these new experiments will be of great interest for solar physics even if they do not lead to the detection of g-modes, since they will provide refined observations of the p-modes. If the g-modes are detected, their discovery will be of epochal importance.

• **More Complete Models of the Sun.** The accuracy of the physical description that is currently achieved with one-dimensional (spherically symmetric) models of the sun that include diffusion is sufficient to permit excellent quantitative agreement with the measured p-mode oscillation frequencies. Numerical experiments and theoretical arguments also suggest that further improvements are unlikely to affect significantly the calculated neutrino fluxes.

Nevertheless, current models of the sun are incomplete. They are spherically symmetric and do not take account of the two-dimensional (or three-dimensional) nature of solar structure. They do not contain a self-consistent dynamical treatment of the effects of rotation, of magnetic fields, of mass loss, or of other possible effects that may violate the currently-used approximations of spherical symmetry and quasi-static evolution. We know observationally that the sun (at least near its surface) contains magnetic fields, that it is losing mass, and that it departs from spherical symmetry by ~ 1 part in 10^5.

There are both analytic and calculational challenges in including these complicated processes in a more complete physical description in the next generation of solar models. New self-consistent methods of calculating solar models (and stellar models) must be developed, and then the appropriate numerical techniques must be worked out, tested, and applied.

The goal of developing a more complete solar model is a challenge for the next decade and beyond. Fortunately, it is a challenge that could lead to important progress since computing power is much greater than it was in the past and there is an abundance of precision data with which to make detailed comparisons.

Solar astrophysics has a bright future.

10.7   SUMMARY

The first 30 years of solar neutrino research have verified experimentally that the fundamental predictions of nuclear energy generation in the sun are approximately correct. Obtaining direct evidence of nuclear fusion in the solar interior is now a solved problem.

The next 10 or 20 years of research will, I think, concentrate on using solar neutrinos to learn more about weak interaction physics. As the weak interaction questions are being resolved, it will be possible to carry out progressively more accurate tests of the theory of nuclear energy generation and of stellar structure.

In retrospect, the history of solar neutrino research is ironic. The research began with an effort to use neutrinos, whose properties were assumed to be well known, to study the interior of the nearest star. The project was an unconventional application of microscopic physics that was designed to carry out a unique investigation of a massive, macroscopic body, the sun. It now appears likely that a large community of physicists, chemists, astrophysicists, astronomers, and engineers working together may have stumbled across the first observed manifestation of physics beyond the standard electroweak model.

We may have been incredibly lucky.

Will the next generation of experiments show that physics beyond the electroweak model is unambiguously required to understand the solar neutrino experimental results? I do not know. But I am sure that we have already learned important facts about neutrino physics from the existing solar neutrino experiments and that we will learn additional things from the future experiments. This research may or may not lead to a consensus that physics beyond the electroweak model is implied by solar neutrino experiments. To me the marvelous lesson of solar neutrino physics is that work on the forefront of one field of science has the potential to lead to important and completely unanticipated developments in a different field of science. This seems to me both humbling and beautiful.

I am grateful to N. A. Bahcall, P. Goldreich, A. Gruzinov, E. Lisi, W. H. Press, M. Ruderman, and A. Ulmer for valuable comments and suggestions on how to make this manuscript more "student friendly." The first version of this talk was presented as the Dannie Heineman Prize for Astrophysics lecture at the annual AAS meeting in January 1995 (see ApJ, 467 (August 10, 1996). The Heinemann Prize for Astrophysics is jointly administered by the American Institute for Physics and the American Astronomical Society. This work was supported in part by NSF grant number PHY95-13835.

BIBLIOGRAPHIC NOTES

1. Progress in solar neutrino research requires developing and operating difficult experiments. Theorists, as well as experimentalists, must understand what has been measured and how, and what might be measurable in the future. If you want to do something relevant, you have to understand the experiments. Recent descriptions of the ongoing solar neutrino experiments are contained in the following papers. More detailed references are listed in these articles.

Davis, R. 1993, in Frontiers of Neutrino Astrophysics, eds. Y. Suzuki & K. Nakamura (Tokyo: Universal Academy Press), p. 47; Davis, R. 1994, Prog. Part. Nucl. Phys., 32, 13. This paper is a description of the chlorine experiment by the founder of observational neutrino astronomy.

Suzuki, Y. 1995, Nucl. Phys. B (Proc. Suppl.) 38, 54; Proc. of the 6th Int. Workshop "Neutrino Telescopes," Venice, February 22-24, 1994 (ed. M. Baldo Ceolin), p. 197. The Kamiokande pure water experiment was the first to demonstrate that the neutrinos come from the sun and to measure energies of individual events.

Abdurashitov, J. N., et al. 1994, Phys. Lett. B, 328, 234; Nico, G., et al. 1995, in Proceedings of the XXVII International Conference on High Energy Physics, July 1994, Glasgow, eds. P. J. Bussey and I. G. Knowles (Philadelphia: Institute of Physics), p. 965. The GALLEX gallium experiment was the first solar neutrino experiment to carry out a direct calibration with an artificial neutrino source (^{51}Cr).

Anselmann, P., et al. 1994, Phys. Lett. B, 327, 377; *ibid*, 1995, 342, 440. The SAGE gallium experiment is (like GALLEX) sensitive to the fundamental *pp* neutrinos and has also carried out a successful calibration with an artificial (^{51}Cr) neutrino source.

2. The Superkamiokande, SNO, and BOREXINO detectors will soon be operating. They will initiate a new era of precision, high statistics solar neutrino experiments. Descriptions of these experiments can be found in the following papers.

Superkamiokande:

Takita, M. 1993, in Frontiers of Neutrino Astrophysics, ed. Y. Suzuki and K. Nakamura (Tokyo: Universal Academy Press), 147.

Kajita, T. 1989, Physics with the SuperKamiokande Detector, ICRR Report 185-89-2.

SNO:

Chen, H. H. 1985, Phys. Rev. Lett., 55, 1534.

Ewan, G., et al. 1987, Sudbury Neutrino Observatory Proposal, SNO-87-12.

McDonald, A. B. 1994, Proceedings of the Ninth Lake Louise Winter Institute, ed. A. Astbury, et al. (Singapore: World Scientific), 1.

BOREXINO:

Arpesella, C., et al. 1992, BOREXINO proposal, Vols. 1 and 2, ed. G. Bellini, R. Raghavan, et al. (Milano: Univ. of Milano).

Raghavan, R. S. 1995, Science, 267, 45.

3. The standard model predictions used in this talk are taken from Bahcall, J. N., & Pinsonneault, M. 1995, Rev. Mod. Phys., 67, 781. The formulation of the neutrino problems is adapted from Bahcall, J. N. 1994, Phys. Lett. B, 238, 276. References to many independent solar model calculations are given in these papers.

4. For a general review of solar neutrino physics and astrophysics, the reader can consult two books devoted to the subject. *Neutrino Astrophysics* is a monograph by J. N. Bahcall that is published by Cambridge University Press (1989). *Solar Neutrinos: The First Thirty Years* contains reprints of 104 of the key papers in the development of the subject plus brief introductions and summaries of recent developments in the major subject areas: standard model expectations, solar neutrino experiments, nuclear fusion reactions, physics beyond the standard model, and helioseismology. This reprint volume is published by Addison-Wesley, Reading, Massachusetts (1995) and is edited by J. N. Bahcall, R. Davis, Jr., P. Parker, A. Smirnov, & R. K. Ulrich.

CHAPTER 11

PARTICLE DARK MATTER

DAVID SPERGEL

Department of Astrophysical Sciences, Princeton University, Princeton, NJ, and
Department of Astronomy, University of Maryland, College Park, MD

ABSTRACT

The nature of the dark matter that comprises most of the mass of our Galaxy and other galaxies is one of the great unsolved problems of astrophysics. Arguments that suggest that this dark matter is non-baryonic include: the absence of baryonic candidates; the need for dark matter to form galaxies; and the low baryon densities inferred from nucleosynthesis. Particle physicists have suggested plausible candidates for non-baryonic dark matter: massive neutrinos, axions and supersymmetric relics. This talk reviews the evidence for non-baryonic dark matter and discusses recent progress and future challenges in detecting dark matter candidates.

11.1   INTRODUCTION: THREE ARGUMENTS FOR NON-BARYONIC DARK MATTER

Several lines of evidence suggest that some of the dark matter may be non-baryonic: the non-detection of various plausible baryonic candidates for dark matter inferred, e.g., from galaxy rotation curves and from cluster of galaxy velocity dispersions, the need for non-baryonic dark matter for theoretical models of galaxy formation, and the large discrepancy between dynamical measurements implying $\Omega_0 > 0.2$ and the baryon abundance inferred from big bang nucleosynthesis, $\Omega_b h^2 = 0.015$. There are a number of well-motivated dark matter candidates: massive neutrinos, supersymmetric dark matter and "invisible" axions. Many of these dark matter candidates are potentially detectable by the current generation of dark matter experiments.

11.2   THE CASE FOR NON-BARYONIC MATTER

While there is a consensus in the astronomical community that most of the mass of our Galaxy and of most galaxies is in the form of some non-luminous matter [70], there is only speculation about its nature.

In his lecture, Charles Alcock (see the contribution by C. Alcock to these proceedings) presents a report of recent progress in efforts to detect baryonic dark matter. Here, I will focus on non-baryonic dark matter.

I will begin by presenting three arguments that suggest that the dark matter is non-baryonic. None of these arguments are definitive. John Bahcall has urged the speakers to identify interesting problems for graduate students. In addition to the grand challenge of detecting the dark matter, I believe that an easier problem is to make some of the arguments for dark matter more compelling.

11.2.1   *We've Looked for Baryonic Dark Matter and Failed*

Astronomers have already eliminated a number of plausible candidates for the dark matter. X-ray observations of galaxies imply that only a small fraction of the mass of a typical galaxy is in the form of hot gas [6, 44]. Even in rich clusters, hot gas makes up less than 20% of the total mass of the system [14]. Neutral hydrogen gas is detectable through its 21 centimeter emission: in most galaxies, neutral gas comprises only 1% of the mass of the system [56] and in only a handful of dwarf galaxies does the neutral gas mass exceed the stellar mass. Even in these systems (e.g., DDO 240 [17]), neutral gas does not account for more than 20% of the system mass. Molecular gas is detectable through dipole emission of CO and other non-homopolar molecules: in most galaxies, the molecular gas mass appears to be less than the neutral gas mass. Low luminosity (low mass) stars, M dwarfs, have often been proposed as a dark matter candidate but HST observations show that faint red stars contribute less than 6% of the unseen matter in the galactic halo [7].

If the dark matter is composed of baryons, then these baryons must be clumped into dense bound objects to evade detection. Gerhard and Silk [29] have proposed that the dark matter consists mostly of very dense tiny clouds of molecular gas. Their model, while provocative, is only marginally consistent with current observational limits. A more widely accepted proposal is that the dark matter consists of very low mass stars, called brown dwarfs. These brown dwarfs are not massive enough to burn hydrogen, so that their only energy source is gravitational energy.

While these brown dwarfs are difficult to detect through their own emission, they are potentially detectable through the gravitational effects. Paczynski [45] proposed gravitational lensing searches for these objects. Several groups have begun searching for these events in an effort to probe the nature of the dark matter.

So far, massive compact halo object (MACHO) searches are not finding as many events as predicted by spherical halo models [3]; however, they can not yet rule out MACHOs as the dominant component of the halo. The current experiment is limited by both small number statistics and by uncertainties in galactic parameters. Many important galactic parameters such as the circular speed, disk scale length and the local surface density are still quite uncertain. Because of these uncertainties, the local halo density is not certain to a factor of two.

It is particularly important to accurately determine the local circular speed as our estimates of the local dark matter density is very sensitive to its value:

$$\frac{\partial \log \rho_{\text{halo}}}{\partial \log v_c} = 2 \frac{v_{\text{tot}}^2}{v_{\text{tot}}^2 - v_{\text{disk}}^2} \sim 4 \, .$$

(Deriving this formula is a good exercise for a student new to dynamics. For an excellent introduction to the subject, see Binney & Tremaine [12]). Thus, a 10% uncertainty in local circular speed translates into a 40% uncertainty in the local dark matter density. Without more accurate determinations of v_c, it is difficult to definitively argue that MACHOs can not comprise much, if not all, of the mass of the dark halo.

There is also a need for better models of the large magellanic cloud (LMC) and more accurate measurements of its properties. Some of the lensing events reported by the MACHO and EROS collaborations may be due to "self-lensing" by the LMC [55] rather than dark matter in the halo.

11.2.2 We Can't Seem To Make the Observed Large-Scale Structure with Baryons

All of the most successful models for forming large-scale structure assume that most of the universe is composed of cold dark matter.

Models in which the primordial fluctuations are adiabatic and the universe is comprised only of baryons and photons are ruled out by cosmic background radiation (CBR) observations. The predicted level of fluctuations in these models exceed the observed level by more than an order of magnitude. Isocurvature models [47] fare better; however, these models also appear to be in conflict with CBR observations [18].

The current "best fit" models have $\Omega_0 \simeq 0.3$, $H_0 \simeq 0.75$, $\Omega_b \simeq 0.03$ and either a cosmological constant or space curvature (see Steinhardt's talk in these proceedings for a review). These models fit COBE observations; are consistent with age and H_0 determinations; are consistent with the large-scale structure (LSS) power spectrum, and are consistent with most large-scale velocity measurements. While they are in conflict with the large velocities detected by Lauer & Postman

[40], these large velocities are controversial [53]. Numerical simulations suggest that these models also agree with the properties of rich clusters [8].

Despite the success of structure formation models that assume non-baryonic dark matter, no one has proven a "no-go" theorem that rules out baryon-only models. It is an interesting challenge to determine what observations are needed to rule out these models.

11.2.3   Dynamical Mass Is Much Larger than
Big Bang Nucleosynthesis Allows

Measurements of the mass-to-light ratios in clusters suggest that Ω_{tot}, the ratio of the total density of the universe to the critical density, exceeds 0.2 [9]. This determination of Ω_{tot} is consistent with measurements based upon the large-scale velocity fields and the dynamics of the large-scale structure [67]. Values of Ω less than 0.2 are very difficult to reconcile with the 500 km/s random velocities seen in large scale structure surveys and even harder to reconcile with large-scale streaming motions.

The observed (presumed cosmological) abundances of deuterium, helium and lithium are only consistent with standard big bang nucleosynthesis if the baryon density is much less than Ω_{tot}. The best fit value for $\Omega_b h^2 \simeq 0.015$, which is nearly an order of magnitude below the dynamical values [72]. For example, if $H_0 = 75$ km/s/Mpc, $\Omega_b = 0.2$ implies that Y, the Helium/Hydrogen abundance ratio, is 0.262 and D/H, the Deuterium/Hydrogen abundance ratio, is 10^{-6} [72] while if $H_0 = 50$ km/s/Mpc, $\Omega_b = 0.2$ implies $Y = 0.253$ and $D/H = 5 \times 10^{-6}$. There are many extragalactic HII regions with $Y < 0.25$ and best estimates imply $Y \simeq 0.24$. These observations appear to require either a significant modification of our ideas about big bang nucleosynthesis or the existence of copious amounts of non-baryonic dark matter. (See, however, Goldwirth & Sasselov [30] for a dissenting view.)

All of the proposed modifications of big bang nucleosynthesis (BBN) appear to violate known observational constraints. For example, Gnedin & Ostriker [32] proposed that an early gamma-ray background photodissociated some of the primordial Helium. This model predicts a spectral distortion of $y > 7 \times 10^{-5}$ and a fully ionized universe. y describes the deviation of the observed spectrum from the thermal spectrum and is a measure of the energy injection in the early universe. COBE [42] found that the observed spectrum was consistent (within the experimental errors) with a thermal spectrum and constrained $y < 2.5 \times 10^{-5}$.

Inhomogeneous nucleosynthesis models have been studied extensively in the past few years. However, Thomas et al. [68] found that even models with large inhomogeneities imply $Y > 0.25$ for $\Omega_b h^2 > 0.05$. Thus, they are also not consistent with $\Omega_b = \Omega_{tot} = 0.2$.

While the theory of big bang nucleosynthesis is well developed, there is still uncertainty in converting the observed line ratios to abundances. Most of the abundances for external systems assume spherical clouds with constant rates of ionization. It would be interesting to study a nearby system such as the Orion nebula and estimate the error associated with this approximation in the analysis. Goldwirth & Sasselov [30] have made an important first step in studying the sensitivity of these element abundances to model uncertainties. There is a need for more work.

While none of these three arguments is incontrovertible, they all do suggest that most of the universe is in non-baryonic matter. The rest of this paper will review the most popular proposed candidates for non-baryonic dark matter and consider various schemes for detecting its presence.

11.3   NEUTRINOS AS DARK MATTER

In the standard big bang model, copious numbers of neutrinos were produced in the early universe. The universe today is thought to be filled with 1.7 K thermal neutrino radiation, the neutrino complement to the thermal radiation background. If these neutrinos are massive, then they can make a significant contribution to the total energy density of the universe:

$$\Omega_\nu h^2 \simeq \left(\frac{m_\nu}{100\text{eV}} \right) . \tag{11.1}$$

Recent results from solar neutrino experiments have revived interest in neutrinos as dark matter candidates. As John Bahcall has described in his talk (see these proceedings), recent experiments appear to be consistent with the Mikheyev-Smirnov-Wolfenstein (MSW) solution to the solar neutrino deficit. The MSW solution implies that the difference in mass squared between the electron neutrino and another neutrino family is of order 10^{-5} eV^2. While this mass difference is much smaller than the mass needed for neutrinos to be the dark matter, it does suggest that neutrinos are massive. It is thus certainly possible that the MSW effect is due to oscillations between electron and mu neutrinos and that the tau neutrino is much more massive and comprises much of the dark matter.

There are several astronomical problems for neutrino dark matter models. Because cosmic background neutrinos have a Fermi-Dirac distribution, they have a maximum phase-space density, which implies a maximum space density [69]. Dwarf irregular galaxies [17] have very high dark matter densities and dwarf spheroidals [28] have even higher dark matter densities: neutrinos can not be the dark matter in these systems. So, if neutrinos are the dark matter in our Galaxy, then there is a need for a second type of dark matter for low mass galaxies [28]. Neutrino plus baryon models have a difficult time forming galaxies early enough, and these models predict galaxy clustering properties significantly different from those observed in our universe.

There are, however, several modified neutrino models that appear more attractive. Cosmological models in which cosmic string seed fluctuations in the hot dark matter have several promising features for structure formation [2]. Mixed dark matter models in which neutrinos comprise 20% of the dark matter and the rest of the dark matter is comprised of cold dark matter also appear to be consistent with a number of observations of large-scale structure [51].

11.3.1   *Detecting Massive Neutrinos*

While it is very difficult to detect the cosmic background of neutrinos directly, there are several experimental approaches that might be able to measure the mass of the neutrino. As I noted earlier, the detection of a stable several eV neutrino would imply that neutrinos comprise a significant fraction of the mass of the universe.

The classical approach to measuring neutrino mass are measurements of the β decay endpoint. Current limits from these experiments imply that the electron neutrino is not the predominant component of the dark matter; however, these experiments cannot place astrophysically interesting constraints on the mass of the mu or tau neutrino.

If the neutrino is a Majorana particle, then it might be indirectly detected through the detection of a neutrinoless double beta decay. Deep underground experiments looking for rare decays have placed very interesting limits [34] on the electron neutrino mass: $m_{\nu_e} < 0.68$ eV. This is a limit on massive neutrinos *if* the most massive eigenstate contains a significant fraction of the electron flavor eigenstate and does not apply to all neutrino models.

Neutrino oscillation experiments are sensitive to mass differences, usually $\Delta m^2 = m_{\nu_\mu}^2 - m_{\nu_e}^2$ and sometime $m_{\nu_\tau}^2 - m_{\nu_e}^2$. Recent results from the Los Alamos experiment [5], which suggest a detection of neutrino oscillations, are controversial [35].

There is a possibility of an astronomical detection of neutrino mass using neutrinos from a supernova explosion. If the neutrinos are massive, then more-energetic neutrinos arrive earlier than less-energetic neutrinos. Thus, neutrino detectors would first see higher energy events and then see less energetic events. This effect was not observed in SN 1987A, which suggests that $m_{\nu_e} < 15$ eV [64]. Observations of a galactic supernova by the Sudbury detector, which is sensitive to

ν_μ, ν_τ could place interesting limits on their masses and possibly rule out neutrinos as cosmologically interesting.

11.4   WIMPs

There is broad class of particle physics candidates for the dark matter that are referred to as Weakly Interacting Massive Particles or WIMPs. This class includes several proposed particles [massive Dirac neutrinos, cosmions, supersymmetry (SUSY) relics] that have masses of order a few GeV to a few hundred GeV and interact through the exchange of W's, Z's, higgs bosons and other intermediaries. In this talk, I will give a brief introduction to WIMPs. I refer interested readers to recent, more detailed reviews [61, 50, 37].

The early universe is a wonderful particle accelerator. WIMPs could be produced through reactions such as $e^+ e^- \rightarrow X\bar{X}$, where X denotes the WIMP particle. WIMPs, of course, can be annihilated through the backreaction, $X\bar{X} \rightarrow e^+ e^-$. As long as $T > m_X$, the WIMP number density would be comparable to the number density of electrons, positrons, and photons. However, once the temperature drops below m_X, the WIMP abundance begins to drop. It will fall until the WIMP number density is so low that the WIMP mean free time for annihilation exceeds the age of the universe. This "freeze-out" occurs at a density determined by the WIMP annihilation cross-section and implies that

$$\Omega_x h^2 \simeq \left(\frac{\sigma_{\text{ann}}}{10^{-37} cm^2} \right)^{-1} .$$

The first proposed WIMP candidates were heavy fourth generation neutrinos [36, 41]. If the neutrino mass was of order 2 GeV, then its relic abundance would be sufficient for $\Omega_\nu = 1$. Experimental dark matter searches [1] ruled out these particles as dark matter candidates.

Supersymmetry is an elegant extension of the standard model of particle physics. It is the only so-far "unused" symmetry of the Poincare group and has the virtue of protecting the weak scale against radiative corrections from Grand Unified Theory (GUT) and Planck scale. Local supersymmetry appears to be an attractive route towards unifying all four forces and is a basic ingredient in superstring theory. Supersymmetry transforms bosons into fermions (and vice-versa). As supersymmetry has a new symmetry, R parity, it can imply the existence of a new stable particle. In much of the parameter space of the minimal supersymmetric model, this new stable particle (which we will refer to as the "neutralino") has predicted properties such that it would comprise much of the density of the universe [25].

11.4.1 *Searching for WIMPs*

While WIMPs interact weakly, they are potentially detectable [31, 73, 37]. The flux of WIMPs through an experiment is quite large: 10^6 (m/GeV)$^{-1}$ cm^{-2} s^{-1}. The difficulty lies in detecting the rare WIMP interactions with ordinary matter.

The challenge for dark matter experimenters is to design an experiment that is simultaneously sensitive to few keV energy depositions and has a large mass (many kilograms) of detector material. The experiment must also have superb background rejection as the expected event rate, less than an event/kilogram/day, is far below most backgrounds. There are two potentially experimental signatures that can aid in the WIMP search: a roughly 10% annual modulation of the event rate due to the Earth's motion around the Sun [24] and a large ($\sim 50\%$) asymmetry in the direction of the WIMP flux due to the Sun's motion through the galactic halo [63].

The first generation of WIMP experiments were rare-event experiments that were adapted to search for dark matter. The first set of experiments were ultra-low background germanium semiconductor experiments [1, 16, 54] that were developed as double beta-decay experiments and modified into dark matter detectors. In these experiments, a recoiling Ge nucleus produces e^--hole pairs that are detectable down to recoil energies ~ 5 keV. These experiments have been limited by microphonics, electronic noise, and by cosmogonic radioactivity.

We are now entering the era of second generation experiments that have been designed primarily as dark matter detectors. In this section, I will highlight several of the promising experimental technologies.

The Heidelberg-Moscow germanium experiment is a modification of the early germanium experiments. It consists of 6 kilograms of purified ^{76}Ge in a detector in the Gran Sasso Tunnel. Since it does not contain ^{68}Ge, it has a reduced cosmogonic background. In this experiment, electronics and microphonics are the dominant background. This experiment places the best current limits on the halo density of WIMPs more massive than 50 GeV [11].

Rather than detecting the electron-hole pairs produced by recoiling nuclei, the Stanford silicon experiment [76] detected the ballistic phonons produced by recoiling silicon nuclei. This experiment has been calibrated by neutron bombardment. The Munich group is developing a silicon detector that will detect the ballistic photons with an SIS junction [48].

At Berkeley, the Center for Particle Astrophysics (CfPA) group is developing a detector that is sensitive to both phonon and electron-hole pairs. This dual detection allows much better background rejection as electrons excited by radioactive decays have a different photon and electron-hole pair signature than nuclear recoils. Neutron bombardment experiments suggest that this dual detection technique can reject $\sim 99\%$ of radioactive background [58]. A more massive experiment that utilizes this technique has the potential to probe into an interesting region of parameter space in supersymmetric theories.

Several groups are developing scintillators that are potential WIMP detectors. There are several scintillator experiments currently under development: a 36.5 kg NaI experiment in Osaka that has begun to place interesting limits on heavy neutrinos [27, 26]; the Rome/Beijing/Saclay experiment [13], a smaller detector, with sensitivities similar to the Osaka experiment; and a Munich sapphire scintillator experiment that is designed to be sensitive to low mass ($m < 10$ GeV) WIMPs. This technology has several advantages over the germanium and silicon semiconductors; the material is sensitive to spin-dependent coupling (although, this is now thought to be less important for supersymmetric dark matter detection [37]), and it is relatively easy to build very large mass detectors. The challenge for these experiments is to improve their background rejection. Spooner & Smith [65] suggest that it might be possible to have some rejection of radioactive γ's in these NaI scintillators through measurements of UV and VIS signatures of recoils [65].

Gas Detectors

Time-Projection Chamber (TPC) detectors have been used extensively in particle physics experiments. While a gas detector with sufficient mass to be sensitive to neutralinos would have an enormous volume, this technology does offer the possibility of detecting the direction of WIMP recoil. Due to the Earth's motion around the Sun, the WIMP recoil events are expected to be highly asymmetric [63]. Buckland et al. [15] report their development of a 50 g H prototype detector. This detector, developed at UCSD, has been tested with neutron source and is potentially scalable to larger masses.

Superconducting Grains

Superconducting grains have an illustrious history in dark matter detection. Drukier & Stodolsky [23] proposed superconducting grains for neutrino detectors, and this work led Goodman & Witten [31] and Wasserman [73] to propose the development of WIMP detectors.

A superconducting grain detector would consist of numerous micron size superconducting grains in a meta-stable state. When one of these grains is heated by WIMP recoil, it would undergo a phase transition to the normal state. The resultant change in B field would be detected by a Superconducting Quantum Interference Device (SQUID). Most background events, due to radioactivity, would flip multiple grains in the detector. Since the events can also be localized in the detector, this can further enhance background rejection as background events should occur primarily near the outside of the detector. The challenge for superconducting grain detector development is the production of a large number of high quality grains. Recently, the Bern group [4] has been able to report significant progress in this direction: they been able to build a superconducting grain detector with several different types of grains (Sn, Al and Zn grains), which they have calibrated with a neutron source.

"Old" Mica

WIMP detection requires exposure times of ~ 100 kg-years. A novel approach is to replace the 100 kg detector with small amounts of material that has been exposed for nearly a billion years. Snowden-Ifft and collaborators [62] have looked for tracks produced by WIMP scatters off of heavy nuclei (such as cadmium) in ancient Mica. They identify these tracks by etching the Mica and have calibrated their experiment by bombarding the Mica with a neutron source.

Atomic Detectors

Recently, Glenn Starkman and I proposed searching for inelastic collisions of SUSY relics with atoms [66]. The cross-sections for these interactions are largest for $\delta E \sim 1$ eV. While the cross-section for atomic interactions are smaller than nuclear interactions, there is a wider range of material that could be used for detecting these atomic interactions. There are not yet any experimental schemes proposed to look for WIMP-atom scatterings. This proposal requires more experimental and theoretical study.

11.4.2   *Indirect WIMP Detection*

The Sun can potentially serve as an enormous WIMP detector. WIMPs streaming through the galactic halo would be gravitationally focused into the Sun, where they would be captured through collisions with atoms in the Sun's center [49]. Neutralinos are their own anti-particles; thus, the neutralinos in the Sun would annihilate each other. When neutralinos annihilate, they will produce high energy neutrinos that are potentially detectable in terrestrial experiments [59]. These few GeV neutrinos are much more energetic than the MeV solar neutrinos produced through solar nucleosynthesis. There is also the possibility of detecting WIMPs in the halo through their annihilation into protons and anti-protons, into electrons and positrons and into γ's. The predicted rates for these processes are unfortunately rather low [21].

There have been several experiments that have looked for WIMP annihilations in the Sun. Currently, there are limits from the Kamionkande, Frejus, and MACRO experiments. In the coming years, we can look forward to more sensitive searches by the DUMAND, AMANDA and NESTOR experiments. While these searches are worthwhile, Kamionkowski et al. [38] have argued that direct experimental searches may be a more effective technique than searches for neutrinos from annihilations of SUSY relics in the Sun. However, for the rarer models with predominantly spin interactions, the converse is most likely true They conclude that for most of parameter space, 1 kg of direct detector is equivalent to 10^5–10^7 m^2 of indirect detector.

11.4.3   *What Is To Be Done?*

Besides the challenge of helping to make any of the promising experiments discussed above work, there are a number of interesting open problems in the WIMP detection field for both theorists and experimentalists. Advances at LEP and at the Tevatron continue to place new limits on the properties of SUSY particles and may provide hints of their existence. We need an on-going reassessment of the viability of different experiment approaches (see, e.g., [38]). There is still much work to be done on the interactions of neutralinos with ordinary matter (see, e.g., [66]). In particular, it would be useful to consider the excitation of atomic levels through WIMP-nuclei collisions.

Advances in technology may enable new kinds of WIMP detectors. It would be very exciting to be able to build a detector composed of large numbers ($\sim 10^{31}$) of spin aligned nuclei. As this detector would have directional sensitivity, it would be sensitive to the large angular asymmetry in the WIMP flux [63]. The development of new purification techniques in the semi-conductor industry may help facilitate the construction of ultra-low background Silicon and Germanium detectors. It would be very exciting if an experiment such as DUMAND or AMANDA with their large detection volumes could be redesigned so that it was sensitive to SUSY relics scattering events. Because of their large active volumes, even lower event rate processes such as inelastic scattering are of potential interest for these experiments. Close collaborations between experimentalists, theorists and technologists are needed to advance the search for SUSY relics.

11.5   AXIONS

Axions are another well-motivated dark matter candidate. While axions are much lighter than the SUSY relics discussed in the previous section and are produced by a very different mechanism, they are indistinguishable to theoretical cosmologists studying galaxy formation and the origin of large-scale structure. Both axions and SUSY relics behave as cold dark matter (CDM) and cluster effectively to form galaxies and large-scale structure. (See Steinhardt's and Ostriker's articles on structure formation.)

Axions were proposed to explain the lack of CP violation in the strong interaction [74, 75]. They are associated with a new U(1) symmetry: the Peccei-Quinn symmetry [46]. As originally proposed, axions interacted strongly with matter. When experimental searches failed to detect axions, new models were proposed that evaded experimental limits and had the interesting consequence of predicting a potential dark matter candidate [39, 57, 77, 22].

In the early universe, axions can be produced through two very distinct mechanisms. At the QCD phase transition, the transition at which free quarks were bound into hadrons, a bose condensate of axions form and these very cold particles would naturally behave as cold dark matter. Axions can also be produced through the decay of strings formed at the Peccei-Quinn phase transition [19, 20]. Unless inflation occurs after the P-Q phase transition, string emission is thought the dominant mechanism for axion production. While Sikivie and collaborators [33] has argued that Davis and Shellard overestimated string axion production, recent analysis [10] confirms that strings are likely to be the dominant source of axions. Axionic strings will not produce an interesting level of density fluctuations as their predicted mass per unit length is far too small to be cosmologically interesting.

The properties of the axion are basically set by its mass, m_a, which is inversely proportional to the scale of Peccei-Quinn symmetry breaking, f_a. The smaller the axion mass, the more weakly the axion is coupled to protons and electrons. Raffelt [52] reviews the astrophysical arguments that imply $m_a < 10^{-2}$ eV. If the axion had a larger mass, then it would have had observable effects on stellar evolution and on the dynamics of SN 1987A. If we require that the energy density in axions not "overclose" the universe, then $\Omega_a h^2 < 1$ implies that $m_a > 1\mu$eV. If strings play an important role in axion production, then the cosmological limit lies closer to $m_a > 1meV$ and there is only a narrow window for the axion model [52].

Axions are potentially detectable through their weak coupling to electromagnetism [60]. In the presence of a strong magnetic field, the axionic dark matter could resonantly decay into two photons. The first generation of detectors consisted of experiments in Florida and at BNL that looked for this decay in a tunable resonant cavity. Since the Peccei-Quinn scale is not well determined, these experiments have to scan a wide range of frequencies in their search for the axion. These experiments were an important first step towards probing an interesting region of parameter space.

In the past few years, the search for axions has been revived by two new exper-

imental efforts. Karl von Bibber [71] and his group at LLNL have built a cryogeni-
cally cooled cavity; this detector should be able to reach into the cosmologically
interesting region of parameter space.In Kyoto, Matsuki [43] and his group plan
to use an atomic beam of Rydberg atoms as an axion detector. This detector would
detect an axion in the galactic halo through its excitation of a Rydberg atom in the
n-th energy state to the n+1 energy state. The Kyoto collaboration also promises
to probe the cosmologically interesting region of parameter space.

11.6 CONCLUSIONS

While there is no conclusive evidence for non-baryonic dark matter, there are
strong hints that it may comprise most of the mass of the universe. There are several well motivated particle physics candidates for non-baryonic dark matter. Most
excitingly, these candidates are potentially detectable in experiments currently under development.

ACKNOWLEDGMENTS

I would like to thank John Bahcall, Marc Kamionkowski, Chris Kolda, and Bill
Press for comments on an earlier version. I would also like to thank Bernard
Sadoulet and Karl von Bibber for loaning me slides and updating me on recent
experimental progress.

BIBLIOGRAPHY

[1] Ahlen, S., et al. 1987, Phys. Lett B, 195, 603. This paper presents the first
experimental limits on halo cold dark matter particles.

[2] Albrecht, A., & Stebbins, A. 1992, Phys. Rev. Lett., 68, 2121. This paper
computes the density power spectrum in a cosmic string seeded cold dark matter
cosmology.

[3] Alcock, C., et al. 1995, ApJ, 445, 133. This paper presents an analysis of
the first year microlensing data from the MACHO collaboration. See Alcock's
article in this book for a more recent review.

[4] Abplanalp, M., et al. 1994, cond-mat preprint 9411072. A report on recent
progress made by the Bern group in developing superconducting grain detectors.

[5] Athanassopoulos, C., et al. 1995, Phys. Rev. Lett., 75, 2650. A description of recent results from the Los Alamos experiment suggesting evidence for neutrino oscillations. See also, the article by Hill.

[6] Awaki, H., Mushotzky, R., Tsuru, T., Fabian, A., Fukazawa, Y., Loewenstein, M., Makishima, K., Matsumoto, H., Matsushita, K., & Mimara, T. 1994, PASJ, 46, 65. ASCA, the US-Japanese X-ray satellite, has enabled measurements of X-ray temperature profiles in galaxies. This paper discusses the gas and dark matter density distributions in elliptical galaxies as well as the chemical composition of the cluster gas. They conclude that elliptical galaxies are dark matter dominated at large radii.

[7] Bahcall, J. N., Flynn, C., Gould, A., & Kirhakos, S. 1994, ApJ, 435, L51.

[8] Bahcall, N., & Cen, R. 1994, ApJ, 426, L15. The formation of clusters and large-scale structure in a low Ω CDM dominated universe.

[9] Bahcall, N., Lubin, L. M. & Dorman, V. 1995, ApJ, 447, L81. Recent discussion of the evidence for dark matter in clusters.

[10] Battye, R. A., & Shellard, E. P. S. 1995, hep-ph preprint 9508301. The most recent analysis of axion production by cosmic strings. They conclude that this is likely to be the most important mechanism of axion production.

[11] Beck, M., et al. 1994, Phys. Lett. B, 336, 141. Limits on halo CDM matter from the Heidelberg-Moscow enriched Germanium experiment.

[12] Binney, J. & Tremaine, S. 1987, Galactic Dynamics (Princeton: Princeton University Press). An excellent graduate student text.

[13] Bottino, A., et al. 1992, Phys. Lett. B, 295, 330. Rome-Beijing-Saclay NaI experiment.

[14] Boute, D.A., & Canizares, C.R. 1996, ApJ, 457, 565. Hot gas comprises roughly $10 - 20\Omega_b = 0.05$, then this baryon/dark matter ratio implies that $\Omega_0 \sim 0.25 - 0.5$.

[15] Buckland, K., Lehner, M. J., Masek, G. E., & Mojaver, M. 1994, Phys. Rev. Lett., 73, 1067. San Diego TPC experiment which is sensitive to the direction of WIMP recoil.

[16] Caldwell, D., et al. 1988, Phys. Rev. Lett., 61, 510. Limits from Germanium semiconductor experiment on halo SUSY particles and 4th Generation Neutrinos.

[17] Carrignan, C., & Freeman, K. C. 1988, ApJ, 332, L33. DDO 240 is a gas and dark matter rich dwarf galaxy. The gas mass/stellar mass ratio in this galaxy is roughly 10:1 and the dark mass/(gas + stellar mass ratio) in the galaxy is also roughly 10:1.

[18] Chiba, T., Sugiyama, N., & Suto, Y. 1993, ApJ, 429, 427. This paper compares the primordial isocurvature baryon (PIB) model to then current experimental data. See also, Hu, W., & Sugiyama, N. 1994, ApJ, 436, 456. For more recent cosmic microwave background (CMB) data, see Bennett et al., astro-ph/9601067 and Netterfield et al., astro-ph/9601197.

[19] Davis, R. L. 1986, Phys. Lett. B, 180, 225. This paper argues that axions may be produced predominantly through the decay of cosmic strings.

[20] Davis, R. L., & Shellard, E. P. S. 1989, Nucl. Phys. B, 324, 167. Further exploration of axion production by cosmic strings.

[21] Diehl, E., Kane, G.L, Kolda, C., & Wells, J.D. 1995, Phys. Rev. D, 52, 4223. Theory, phenomenology and prospects for detection of supersymmetric dark matter.

[22] Dine, M., Fischler, W., & Srednicki, M. 1981, Phys. Lett. B, 104, 1955. One of the models for the "invisible axion."

[23] Drukier, A., & Stodolsky, L. 1984, Phys. Rev. D, 30, 2295. This paper proposes the use of superconducting grains as a solar neutrino detector. It stimulated Goodman and Witten's and Wasserman's proposals for searches for non-baryonic halo dark matter.

[24] Drukier, A., Freese, K., & Spergel, D. N. 1986, Phys. Rev. D, 30, 3495. This paper explores the use of a superconducting grain detector in dark matter searches. It shows how the Earth's motion around the Sun produces an annual modulation in the WIMP flux and in the detector signal.

[25] Ellis, J., et al. 1984, Nucl. Phys. B, 238, 453. This paper shows that minimal SUSY models predict the existence of stable neutralinos and that these neutralinos have cosmologically interesting densities.

[26] Ejiri, H., Fushimi, K., & Ohsumi, H. 1993, Phys. Lett. B, 317, 14. Osaka NaI experiment.

[27] Fushimi, K., et al. 1993, Phys. Rev. C, 47, R425. Osaka NaI experiment.

[28] Gerhard, O. E., & Spergel, D. N. 1992, ApJ, 389, L9. Phase space constraints imply that neutrinos can not be the dark matter in dwarf galaxies.

[29] Gerhard, O., & Silk, J. 1995, astro-ph preprint 9509149 and astro-ph preprint 9511036. They present a model in which halo dark matter is composed of a combination of low mass stars and very cold gas clouds.

[30] Goldwirth, D., & Sasselov, D. 1995, ApJ, 444, 15. This paper shows that the systematic uncertainties in estimating the Helium abundances in low metallicity external galaxies are much larger than previously estimated. While many of my colleagues who study the physics of the interstellar medium agree with the conclusions of this paper, its implications have not been fully absorbed by the cosmology community.

[31] Goodman, M. W., & Witten, E. 1985, Phys. Rev. D, 31, 3059. This seminal paper started the field of cold dark matter searches. *I recommend this paper as the first article that someone interested in this field should read.*

[32] Gnedin, N., & Ostriker, J. P. 1991, ApJ, 400, 1. This paper proposed that the γ-rays produced by accretion onto black holes ionized the primordial Helium. See Mather et al. (1992) for limits on this model.

[33] Hagmann, C., & Sikivie, P. 1991, Nucl. Phys. B, 363, 247. This paper argues that cosmic string production of axions has been overestimated in earlier papers.

[34] Heidelberg-Moscow Experiment hep-ex/9502007. Best limits on the neutrino mass. These limits are based on limits on the rate of neutrinoless $\beta\beta$ decays and apply only to Majorana neutrinos.

[35] Hill, J. E. 1995, Phys. Rev. Lett., 75, 2654. This paper discusses the reported detection of Neutrino Oscillations (Athanassopoulos et al. 1995).

[36] Hut, P. 1977, Phys. Lett. B, 69, 85. This paper shows how several GeV neutrinos could be the dark matter. These dark matter candidates are now experimentally ruled out (see Ahlen et al. 1987).

[37] Jungman, G. U., Kamionkowski, M., & Griest, K. 1995, to appear in Physics Reports. This is an excellent up-to-date review of cold dark matter detection. It also contains several new results.

[38] Kamionkowski, M., Griest, K., Jungman, G., & Sadoulet, B. 1995, Phys. Rev. Lett., 74, 5174. This paper compares the relative effectiveness of experiments that look for the decays of SUSY particles in the Sun and experiments that are sensitive to WIMP recoils.

[39] Kim, J.-E. 1979, Phys. Rev. Lett., 43, 103. An invisible axion model.

[40] Lauer, T., & Postman, M. 1994, ApJ, 425, 418. Lauer and Postman use the properties of the brightest galaxy in each cluster as a "standard candle" to probe the large scale distribution of matter. They find evidence for large-scale motions relative to the microwave background frame (but also see Reiss et al. 1995).

[41] Lee, B. W., & Weinberg, S. 1977, Phys. Rev. Lett., 39, 165. This paper shows how several GeV neutrinos could be the dark matter. These dark matter candidates are now experimentally ruled out (see Ahlen et al. 1987).

[42] Mather, J., et al. 1992, ApJ, 420, 439. The COBE FIRAS detector showed that the CMB spectrum did not deviate (within their experimental limits) from the predicted thermal spectrum. This experiment places important limits on any kind of energy release (winds from stars, particle decay, etc.) in the early universe.

[43] Matsuki, S., et al. 1995, to appear in Proceeding of the XVth Moriond Workshop: Dark Matter in Cosmology, Clocks and Tests of Fundamental Laws, Villars-sur-Ollon, Switzerland, January 21, 1995. This paper describes a detection scheme for axions that uses Rydberg atoms.

[44] Mulchaey, J. S., Davis, D. S., Mushotzky, R. F., & Burstein, D. 1993, ApJ, 404, L9. X-ray observations of a group of galaxies shows that baryons account for only 4% of the mass. The authors place an upper bound of 15% on the baryon content in this small group of galaxies. These observations are strong evidence that dark matter dominates in these small groups.

[45] Paczynski, B. 1986, ApJ, 304, 1. This seminal paper describes how microlensing observations can be used to probe the composition of the halo.

[46] Peccei, R. D., & Quinn, H. R. 1977, Phys. Rev. D, 16, 1791. An important paper for understanding the role of the axion in CP conservation.

[47] Peebles, P. J. E. 1987, ApJ, 315, L73. This paper introduces the baryon isocurvature model. See Peebles, P. J. E. 1994, ApJ, 432, L1 for a more recent discussion of the model.

[48] Peterreins, T., et al. 1991, J. Appl. Phys., 69, 1791. This paper discusses the use of SIS junctions in WIMP detection.

[49] Press, W. H., & Spergel, D. N. 1985, ApJ, 296, 679. This paper describes how the Sun will capture WIMPs. Once in the Sun, the WIMPs can annihilate (see Silk and Srednicki 1984).

[50] Primack, J. R., Seckel, D., & Sadoulet, B. 1988, Ann. Rev. Nucl Part. Sci., 38, 751. A nice review of cold dark matter candidates and dark matter detection. See Jungman et al. (1995) for a more recent discussion.

[51] Primack, J. R., Holtzman, J., Klypin, A., & Caldwell, D. O. 1995, Phys. Rev. Lett., 74, 2160. This paper discusses the galaxy and structure formation in the mixed dark matter cosmogony.

[52] Raffelt, G. 1995, to appear in Proceeding of the XVth Moriond Workshop: Dark Matter in Cosmology, Clocks and Tests of Fundamental Laws, Villars-sur-Ollon, Switzerland, January 21, 1995 (hep-ph 9502358). This is an excellent introduction to axion dark matter physics.

[53] Reiss, A., Kirshner, R. P., & Press, W. H. 1995, ApJ, 445, L91. This paper uses supernova as "standard candles" to probe the large-scale structure of the universe. It appears to contradict earlier work by Lauer and Postman.

[54] Reusser, D., et al. 1991, Phys. Lett. B, 225, 143. Best limits on few GeV WIMPs as halo dark matter.

[55] Sahu, K.C. 1994, PASP, 106, 942. This paper argues that the microlensing events in the LMC are better explained as being due to stars in the LMC than by MACHOs.

[56] Scodeggio, M., & Gavazzi, G. 1993, ApJ, 409, 110. A survey of 112 nearby galaxies that discusses the neutral gas content, and star formation rate as a function of environment.

[57] Shifman, M. A., Vainshtein, A. I., & Zakharov, V. I. 1989, Nucl. Phys. B, 166, 493. A model for the invisible axion.

[58] Shutt, T., et al. 1992, Phys. Rev. Lett., 69, 3425; ibid. 3531. This paper reports progress on the development of a background rejection scheme for halo cold dark matter.

[59] Silk, J., & Srednicki, M. 1984, Phys. Rev. Lett., 53, 624. This paper describes the annihilation of WIMPs in the Sun and the possibility of detecting their annihilation signature.

[60] Sikivie, P. 1983, Phys. Rev. Lett., 51, 1415.

[61] Smith, P. F., & Lewin, J. D. 1990, Phys. Rep., 187, 203. A nice review of cold dark matter candidates and dark matter detection. See Jungman et al. (1995) for a more recent discussion.

[62] Snowden-Ifft, D., et al. 1995, Phys. Rev. Lett., 74, 4133. This describes the use of Mica as a particle detector.

[63] Spergel, D. N. 1988, Phys. Rev. D, 37, 1353. This paper shows how the Sun's motion through the Galaxy produces an asymmetric WIMP flux.

[64] Spergel, D. N., & Bahcall, J. N. 1988, Phys. Lett. B, 200, 366. Limits on neutrino masses from SN 1987a.

[65] Spooner, N., & Smith, P. F. 1993, Phys. Lett. B, 314, 430. This paper reports progress in background rejection in a NaI detector.

[66] Starkman, G. D., & Spergel, D. N. 1995, Phys. Rev. Lett., 74, 2623. This paper presents a new proposal for WIMP detection using WIMP coupling to bound electrons.

[67] Strauss, M., & Willick, J. 1995, Physics Reports, 261, 271. A very nice review of efforts to probe the large-scale structure.

[68] Thomas, D., Schramm, D. N., Olive, K. A., Mathews, G. J., Meyer, B. S., & Fields, B. D. 1994, ApJ, 430, 291. Constraints on inhomogeneous nucleosynthesis models.

[69] Tremaine, S., & Gunn, J. E. 1979, Phys. Rev. Lett., 42, 407. This paper shows that neutrinos have a maximum phase space density and a maximum space density.

[70] Trimble, V. 1987, ARA&A, 25, 425. A nice review of the astrophysical evidence for the existence of dark matter.

[71] von Bibber, K., et al. 1995, to appear in Proceeding of the XVth Moriond Workshop: Dark Matter in Cosmology, Clocks and Tests of Fundamental Laws, Villars-sur-Ollon, Switzerland, January 21, 1995 (astro-ph 9508013).

[72] Walker, T. P., et al. 1991, ApJ, 376, 51. Standard big bang nucleosynthesis.

[73] Wasserman, I. 1986, Phys. Rev. D, 33, 2071. This paper independently showed that supersymmetric dark matter and 4th generation neutrinos may be detectable experimentally.

[74] Weinberg, S. 1978, Phys. Rev. Lett., 40, 223. This paper describes how the axion can solve the CP problem.

[75] Wilczek, F. 1978, Phys. Rev. Lett., 40, 279. This paper describes how the axion can solve the CP problem.

[76] Young, B. A., Cabrera, B., & Lee, A. T. 1990, Phys. Rev. Lett., 64, 2795. This paper describes the Stanford Silicon experiment.

[77] Zhitnitsky, A. R. 1980, Sov. J. Nucl. Phys., 31, 260. A model for the invisible axion.

CHAPTER 12

STARS IN THE MILKY WAY AND OTHER GALAXIES

ANDREW GOULD

Department of Astronomy, Ohio State University, Columbus, OH

ABSTRACT

As the 20th Century comes to a close, long-standing questions about star counts and the stellar mass function in our own galaxy are being resolved. In the next century, it should be possible to extend these successes and then to begin measuring the stellar mass function in other "nearby" galaxies in the Virgo Supercluster. We may also measure the mass function in high redshift galaxies and so probe the star-formation history of the universe.

12.1   INTRODUCTION

I will touch briefly on some recent results about the stellar mass and luminosity functions in the Milky Way and other galaxies and then focus on outstanding problems that will be attacked in the next century. You will see that there are lots of unsolved problems in this field where the technology that is required to obtain solutions is now available, or will be available in the foreseeable future.

12.2   RECENT STAR COUNT RESULTS

Last year Bahcall et al. (1994) used the Wide Field Camera (WFC2) on the newly repaired Hubble Space Telescope (HST) to search for halo red dwarfs. They found no stars at all redder than $V - I = 3$ and brighter than $I = 25.2$ in a single high-latitude field. Using this null result (plus the knowledge that other similar fields are also deficient in faint red dwarfs) Bahcall et al. were able to show that red dwarfs account for no more than 6% of the dark halo.

241

The HST star counts raise a number of important questions. First, what can HST tell us about the disk luminosity function (LF) and disk mass function? Is there a large population of disk M dwarfs and/or brown dwarfs? Second, is the halo composed of other stellar objects, such as brown dwarfs, white dwarfs, or "bluish" red dwarfs? Several people have pointed out that the Bahcall et al. color cut off $(V - I > 3)$ may exclude the most metal poor faint red stars. How can one search for these various halo candidates?

HST can indeed provide rich material on the disk. Gould, Bahcall, & Flynn (1996) have now analyzed 22 HST WFC2 fields plus 162 pre-fix HST Planetary Camera (PC) fields. They restrict attention to faint red dwarfs (M stars) with absolute V-band magnitudes $8 < M_V < 18.5$ (as determined from their $V - I$ colors and assuming a disk color-mag relation) with heights above the plane $|z| <$ 3200 pc. The first condition excludes spheroid giants (which are too blue to make the cut). The second condition excludes all but a few spheroid dwarfs (which, being $\sim$ 2–3 mag fainter than disk dwarfs, must be $|z| < 1$ kpc to make the cut). These cuts leave an almost pure sample of 257 disk M stars distributed over 10 mag and 3 kpc which can help resolve a long-standing conflict over the shape of the faint end of the LF: Using nearby (< 25 pc) parallax stars, Wielen, Jahreiss, & Krüger (1983) found a basically flat LF from $M_V = 12$ to $M_V = 16$. By contrast, Stobie, Ishida, & Peacock (1989) found a sharply falling LF for $M_V > 12$ based on a photometric survey of stars < 130 pc. The HST data are in nearly perfect agreement with the Stobie et al. results. A recent reanalysis of the parallax stars by Reid, Hawley, & Gizis (who also obtained much new data) shows that the LF does indeed fall for $M_V > 12$. There remains a mild conflict between the parallax stars on the one hand and the HST and ground-based photometric studies on the other, but the basic picture is clear: the mass function peaks at stars with mass $M \sim 0.45\ M_\odot$ when plotted against log mass (or at $M \sim 0.23\ M_\odot$ when plotted against mass) and continues falling until $\sim 0.1\ M_\odot$.

Two other recent studies both point to a low scale height population of faint red "stars." Reid, Tinney, & Mould (1994) find that if the faintest stars ($M_V > 16$) in the solar neighborhood are divided into proper-motion selected and photometry selected subgroups, the former have normal M star kinematics but the latter are extremely cold, with mean motion consistent with the Local Standard of Rest and dispersions of order 10 km s^{-1} in each direction. Such a finding could be explained if these "stars" were not stars at all, but young brown dwarfs ~ 300 Myr old. Being young, they would be expected to be kinematically cold. Older brown dwarfs, which had been kinematically heated up, would have grown too faint to be seen in the survey. Kirkpatrick et al. (1994) find an excess of nearby (< 50 pc) red "stars" toward the southern Galactic hemisphere relative to the northern hemisphere. Since the Sun is generally believed to be ~ 30 pc above the plane, these observations could be explained if the "stars" were in fact young brown dwarfs with a low scale height. Then the effective volume would be much smaller toward the north than the south. Taken together, the Kirkpatrick et al. and Reid et al. studies involve only 17 stars, so the results must be regarded as still only suggestive. What further probes are available to detect a population of disk brown dwarfs?

12.3   MICROLENSING AND STAR COUNTS

Microlensing searches toward the Galactic Bulge by MACHO (Alcock et al. 1995a, 1995b) and OGLE (Udalski et al. 1994) promise to provide a direct measure of the mass function of the disk and bulge. To fully realize this promise would require additional ground-based (Gould 1995c) and space-based (Refsdal 1966, Gould 1994b, 1995a, Han & Gould 1995) observations. However, it is already possible to ask whether the time-scale distribution of microlensing events is consistent with what is expected from the known populations of stars. Han & Gould (1996) examined three classes of mass functions: a Gaussian, a power law, and the empirical HST mass function (Gould et al. 1996) augmented by the observed population of white dwarfs. Somewhat surprisingly, they found that the best-fit power law for lenses along the bulge line of sight has a cut off at $0.04\,M_\odot$ and a power index $p = -2.1$, close to the Salpeter value, -2.3. The power-law mass function was preferred over the HST mass function at the $5\,\sigma$ level. This result would not have been surprising a few years ago when the standard guess for the low-mass behavior of the mass function was based on a simple extrapolation of the Salpeter power law observed for $M > 1\,M_\odot$. But the mass function has now been measured and is found to be falling rapidly, not rising. Hence, the microlensing data also seems to be telling us that there is an unseen population of low mass objects. More microlensing data and especially more auxiliary data on these exciting events should clarify the problem of low-mass disk objects.

12.4   DISK DARK MATTER: STILL A QUESTION

Bahcall (1984) did the first modern reanalysis of the question of whether the mass of Galactic disk is fully accounted for by the observed luminous material. He concluded that on the contrary there is a substantial amount of disk dark matter. Today, the general community opinion seems to be that the problem of disk dark matter is "going away," that the dynamically measured mass is reasonably accounted for by the observed material. This view is mistaken. The total column density of observed material is $\sim 40\,M_\odot\,{\rm pc}^{-2}$ (Gould et al. 1996). Bahcall, Flynn, & Gould (1992) estimated the column density at $\sim 80\,M_\odot\,{\rm pc}^{-2}$ with large errors. Kuijken

& Gilmore (1991) estimated it at $71 \pm 6 \ M_\odot \ \mathrm{pc}^{-2}$ below 1.1 kpc. They argued that perhaps 25 $M_\odot$ of this could be due to a spherical dark halo. This is possible but by no means demanded by the data. Several other workers have estimated the disk column density at $\sim 60 \ M_\odot$ (not including halo) which, with errors, would be consistent with the commonly accepted "observed" value of $\sim 50 \ M_\odot$. However, it would not be consistent with the most recent determination of Gould et al. (1996). Since several lines of evidence given above suggest that low mass objects are being detected, the question of disk dark matter remains very much alive and should receive active attention.

12.5   MYSTERY OF THE LONG EVENTS

As mentioned above, the HST mass function does a poor job of explaining the short microlensing events seen toward the bulge. These might be accounted for by an unseen low mass population. But Han & Gould (1996) also found that the HST mass function does a poor job of explaining the long events. Of the first 54 events detected by MACHO and OGLE, 3 are longer than 70 days. Of the 3, one has a "parallax" measurement which puts its projected transverse speed at $\sim 75 \ \mathrm{km \ s}^{-1}$ (Alcock et al. 1995b). For typical lenses, parallaxes can be measured only by observing the event from two different lines of sight separated by ~ 1 AU. The parallax is measured by comparing the light curves as seen by the two observers (Gould 1994b, 1995a). However, for long events, the Earth actually moves far enough during the event to simulate two distinct observations (Gould 1992). The parallax measured by Alcock et al. (1995b) is almost certainly inconsistent with the lens being in the bulge. Han & Gould (1996) speculated that the long events could be due to a kinematically cold nearby population of massive dark objects. To resolve the issue, one must measure the mass and distance of individual events. This in turn requires measuring both the parallax and the proper motion.

For long events, it is generally possible to measure the parallax from the ground because, as mentioned above, the Earth changes its position substantially during the event. However, the standard method of measuring the proper motion, i.e. observing events in which the lens transits the face of the star (Gould 1994a; Nemiroff & Wickramasinghe 1994; Witt & Mao 1994; Witt 1995; Loeb & Sasselov 1995; Gould & Welch 1996) can be carried out only very rarely because the typical angular size of the star ($< 10 \ \mu$as) is so much smaller than the size of the Einstein ring ($\theta_e \sim 1$–2 mas). The Einstein ring is not quite large enough to resolve the two images which are separated by $\sim 2\theta_e$ (although this may become possible in the future using ground-based interferometry), but it is large enough to resolve by lunar occultation.

The Moon passes through the bulge every month. As seen from any fixed position on Earth it occults a strip $\sim 0.5°$ in width. However, the width of the strip occulted from *some* position on Earth is $\sim 2°$. If a long event lies in this strip, the Moon will occult it several times during the peak of the event. Each time, one of the two images will be occulted slightly before the other, thereby making it possible to detect the finite size of the Einstein ring.

Simulations by Han, Narayanan, & Gould (1996) show that for long events with $\theta_e > 1.5$ mas, and for clump giant bulge sources $H < 13.5$, it is possible to measure the proper motion with $> 3\,\sigma$ accuracy using a 4 m telescope. Such large Einstein rings are to be expected if the events are indeed generated by relatively nearby < 2.5 kpc objects. Thus, it may soon be possible to unravel the mystery of the long events.

12.6 PROPER MOTIONS FROM EROS II

Not all the questions about stars are going to be resolved by microlensers looking for dark matter. Some will be resolved by microlensers looking for supernovae. The primary goal of MACHO has been to find Massive Compact Halo Objects (Machos) toward the Large Magellanic Cloud (LMC) (Alcock et al. 1993). A French collaboration, EROS (Aubourg et al. 1993, 1995) has been competing with them. MACHO used a two-band (dichroic) 0.5 $\deg^2$ CCD camera while EROS used Schmidt plates. In the end the MACHO approach proved superior. So much so that EROS is building a new two-band 1 $\deg^2$ CCD camera. Not only is the EROS II camera twice as big as MACHO's, but they expect to get $1''$ seeing as well, compared to MACHO's $2''$ seeing. In fact, the EROS II camera is so good that they expect to have 1/3 of their observing time free to look for distant supernovae. While the exact strategy for the supernovae search has yet to be worked out, it is already clear that the observations will be a tremendous boon for Galactic astronomy.

As an example, assume that EROS II will during its supernova program observe four 30 $\deg^2$ fields, each for 15 minutes, 12 times per year for 3 years. For sufficiently bright stars ($I < 22.5$ or $V < 23.9$), each measurement should yield a position accurate to 1/20 pixel or $0.''03$. (For fainter stars, the accuracy scales inversely as the flux.) For these brighter stars, it should be possible to measure proper motions accurate to 5.5 mas yr^{-1}, or 25 km s^{-1} kpc^{-1}. Thus all halo objects ($v \sim 200$ km s^{-1}) should be detectable at the $5\,\sigma$ level out to 1 kpc. This covers a volume $\sim 10^7$ pc^3, containing a halo mass $M \sim 10^5\,M_\odot$ (assuming a local halo density $\sim 10^{-2}M_\odot$ pc^{-3}). Hence, if the halo is composed of bluish red dwarfs ($V - I \sim 2.7$; $M_I \sim 11.8$, i.e., the tip of the metal poor sequence seen by Monet et al. 1992) then $\sim 10^6$ should show up in the EROS II survey. A standard spheroid (Bahcall & Soneira 1980) alone would yield $\sim 10^4$ such stars and a heavy spheroid (Caldwell & Ostriker 1981) would yield a factor ~ 10 more.

Old white dwarfs ($V - I \sim 2$; $M_I \sim 15$), could be detected to ~ 500 pc. If the halo were composed of these, there would be 2.5×10^4 in the survey. Again, even a standard spheroid would yield 250 old white dwarfs.

Other aspects of the EROS II survey will also produce important proper motion data. For example, the bulge observations will give proper motions of 10^7 bulge giants accurate to ~ 50 km s^{-1}. Measurement of the proper motions of A stars along the line of sight to bulge will give information about the rotation curve and asymmetric drift as a function of height above the plane.

In brief, microlensing searches will provide a powerful probe of Galactic structure, not only regarding the dark, but also the visible matter.

12.7   Pixel Lensing: Stellar Mass Functions in Other Galaxies

The stellar mass function is being probed to greater and greater distances. Only a few years measurements were restricted to the "solar neighborhood." Now HST has measured the mass function at ~ 1 kpc and microlensing is beginning to make measurements at several kpc. But will we ever be able to measure the mass function in other galaxies?

Surprisingly, we are on the verge of doing this already. Arlin Crotts (1992) and a second French microlensing group (AGAPE, Baillon et al. 1993) independently proposed that microlensing could be detected even when the lensed stars could not be resolved. The idea is as follows: suppose that a single pixel, or more precisely a single resolution element ("seeing disk") contains 1000 stars of equal brightness. And suppose that one of these stars is lensed by a factor $A = 1.34$. That is, the source just barely comes within the Einstein ring of the lens. Then the pixel as a whole becomes brighter by 0.034%, an amount which is generally too small to be measured. However, for 1/10 of lensing events, the source will come within the inner 1/10 of the Einstein ring and so will be magnified by a factor 10. And 1/100 of the time it will be multiplied by 100. In these latter cases, the pixel will brighten by 10%, which can be measured. Of course, such extremely high magnifications of an individual star are extremely rare, but then there are 1000 stars in every pixel. So there are far more pixel lensing events per unit angular area of a distant galaxy than there are ordinary lensing events of a nearby galaxy. Pixel lensing is potentially very powerful.

Crotts and co-workers and AGAPE have begun carrying out observations toward M31 in the hopes of finding Machos in the halo of M31 (and secondarily in our Galaxy as well). They may indeed find Machos, but since MACHO and EROS have found the Milky Way halo may be only partially composed of Machos, there may not be too many in the halo of M31 either. However, these experiments will certainly find stars in M31, particularly when they observe M31's bulge. Recall that the Galactic microlensing experiments also started out looking for Machos, but so far have found mostly stars in our own bulge. Although each pixel lensing event contains less information than a resolved event (even the time-scale of the event is known only statistically) it should still be possible to constrain the mass function of the M31 bulge with pixel lensing.

Our extreme example of 1000 stars per pixel does not actually apply to M31. A typical clump giant has an absolute mag $M_I \sim 0.5$ or an apparent mag $I \sim 25$ at the distance of M31. In good seeing, a resolution element $< 1\,\mathrm{arcsec}^2$, while the surface brightness of a typical bulge field is $\sim 19.5\,\mathrm{mag\,arcsec}^{-2}$. Hence 200 stars per pixel is a more realistic estimate for M31.

At first sight, the density of stars per pixel appears unimportant because the rareness of the high magnification (required for the event to be seen in spite of the large number of stars) seems to be exactly cancelled by that same large number of stars. However, the stars are not actually point sources and so cannot be magnified to arbitrarily high brightness. If the angular radius of the star is θ_s and the angular Einstein radius is θ_e, then the maximum magnification of an event is $A_{\max} \sim \min(\beta^{-1}, 2\theta_e/\theta_s)$, where β is the impact parameter in units of θ_e. For clump giants in other galaxies lensed by sources in the same galaxy,

$$A_{\max} \leq \frac{2\theta_e}{\theta_s} \sim 80 \left(\frac{M}{0.5\,M_\odot}\right)^{1/2} \left(\frac{D}{\mathrm{kpc}}\right)^{1/2} \left(\frac{r_s}{10\,r_\odot}\right)^{-1} \tag{12.1}$$

where M is the mass of the lensing star, r_s is the radius of the source star, and D is the distance between them. Hence, at 200 stars per pixel, it is still possible to see significantly magnified events in M31. This calculation shows, however, that for a galaxy 20 times farther away (with 400 times more stars per pixel), the fractional change in pixel light is far too small to be measured.

However, using HST it is in fact possible to detect pixel lensing in M87, 20 times farther than M31: the area of an HST resolution element is smaller than a ground-based resolution element by a factor ~ 25. Gould (1995e) proposed that HST be used to monitor M87 in order to detect Machos in the halo of the Virgo cluster, noting that if Machos made up 20% of the Galactic halo (as the results of MACHO and EROS seem to indicate), then the amount of baryonic material in the Milky Way disk and halo would be about equal. Within the context of reasonable scenarios for cluster formation, clusters should then contain about equal amounts of X-ray gas and Machos. Such Virgo cluster Machos should be detectable at a rate of $\sim 10\,\mathrm{day}^{-1}$ with continuous HST monitoring. However, regardless of whether Virgo contained any Machos, the experiment would certainly detect stars in M87. From the standpoint of this (and all) Macho-detection proposals, stars are an unwanted background. However, it is possible, at least statistically, to distinguish the M87 stars from the Virgo cluster Machos. The basic idea is to make use of the truncation in the magnification curve which I mentioned above and which fundamentally limits the peak magnification. These truncated light curves affect M87 stars much more than Virgo Machos because $D \sim 3\,\mathrm{kpc}$ for stars and $D \sim 250\,\mathrm{kpc}$ for Machos. Moreover, $\langle D \rangle$ is a function of angular position for stars, being approximately equal to the impact parameter relative to the center of M87. For Virgo Machos, $\langle D \rangle$ is independent of position.

From the present perspective of finding the stellar mass function, the "unwanted background" of M87 stars is a very good thing. Moreover, since it is possible to distinguish stars from Machos, an unambiguous measurement of the M87 stellar mass function is possible. All that is necessary is to obtain the proposed observations.

12.8   STAR FORMATION HISTORY OF THE UNIVERSE

We began with the mass function of parallax stars ($z = 2 \times 10^{-9}$), moved out to HST stars ($z = 2 \times 10^{-7}$), to Galactic bulge stars ($z = 2 \times 10^{-6}$), to M31 ($z = 2 \times 10^{-4}$), and finally to M87 ($z = 0.004$). In one sense we have left the "solar neighborhood" of 10 pc far behind. (At least for M87, it is no longer necessary to write the redshift z in scientific notation.) But in another sense M87 is still the "solar neighborhood", i.e., the nearby universe. Can we probe the stellar mass function at distances that are so great that they represent different epochs in the history of the universe?

Actually, this may not be so difficult as it would first appear. There are 100 quasars to $B = 22$ per square degree. One could monitor these for gravitational lensing over, say, $10{,}000\,\mathrm{deg}^2$, yielding 10^6 quasars. Known stars in galaxies would yield a lensing event rate of $\Gamma \sim 20\,\mathrm{yr}^{-1}$. If the density of Machos (in or out of galaxies) were $\Omega_{\mathrm{Machos}} \sim 1\%$, these would generate an additional $\Gamma \sim 200\,\mathrm{yr}^{-1}$. The stellar events would typically last $\sim 10\,\mathrm{yr}$, while the Macho events would last $\sim 3\,\mathrm{yr}$ (Gould 1995d). Because the events do last several years, it would be sufficient to monitor the quasars only several times per year. In fact it would be unnecessary even to identify the 10^6 quasars. The 10,000 square degrees could be monitored with say a 4 square degree field on a 1 or 2 m telescope. Lensed quasars would be found by the same "pixel lensing" technique used by Crotts (1992) and Baillon et al. (1992) or in supernova searches.

What could be learned about stars in such a survey? First, the star formation history of the universe. If most stars formed before the epoch of quasars, then the optical depth to $z = 4$ quasars would be much higher than to $z = 1$ quasars, because the path length to the former would be densely populated with stars. On the other hand, if most stars formed $z < 1$, then the distant quasars would have only a slight advantage in optical depth (Gould 1995d).

Second, one could learn, at least statistically, about the stellar mass function. Measuring the mass of individual lensing stars would be more difficult, although even this would be possible given a sufficient investment. For most stars, or at least the stars in galaxies, one could identify the host galaxy by following up the detection with deep imaging. Once the redshift of the galaxy were measured, one would have two of the three parameters of the lensing event. That is, one would have two measurements (distance and time-scale of the event) against three unknowns (the mass, distance, and transverse speed). In other words, the mass could be statistically estimated by assuming a distribution of transverse speeds of galaxies. It is in fact possible to measure the scale of the transverse speeds from the monitoring experiment by comparing the lensing rate toward quasars in directions parallel and perpendicular to the Sun's motion relative to the microwave background. But it is also possible to measure the transverse velocities of the lensing stars themselves, provided that one observes the event simultaneously from a satellite in a Neptune like orbit (Gould 1995b). Such parallax measurements would be very similar to the measurements proposed for Galactic lensing events (Gould 1994b, 1995a) but on a much larger scale.

In fact, the first microlensing by a star was discovered not in our Galaxy but in Huchra's Lens (in front of QSO 2237+0305) (Irwin et al. 1989; Corrigan et al. 1991). While only at $z = 0.04$, this serendipitous discovery may one day be regarded as the first step toward a measurement of the star-formation history of the universe.

One might wonder how one could distinguish genuine quasar microlensing events from ordinary quasar variations. In fact, this is not difficult in principle: spectroscopic follow up would reveal a light echo in the broad line region for any variation coming from the continuum. No such effect would be seen for microlensing. Unfortunately, the telescope time necessary to follow up all variations among 10^6 quasars would be prohibitive. However, in the original survey, one could identify a subset of relatively quiescent quasars or, more to the point, eliminate quasars that showed too much variation. Whether such a population of quiescent quasars exists is not presently known, but could be determined from the EROS supernova search discussed above.

12.9   CONCLUSIONS

Star counting and measurements of stellar masses have been around for many decades and perhaps for that reason are regarded as "old hat" in some quarters. I have tried to show that new observations and new technologies have opened new questions and new possibilities in this field. My principal advice to students entering this field is: be bold. Technology is advancing so fast that many ideas which were regarded as impossible when they were proposed are in fact already feasible or will be in a few years. Most often, the conservatism that smothers a new idea is not that of a disbelieving referee, but comes from within oneself. I came close to not publishing two ideas for ground-based parallax and proper motion measurements of Machos because I thought that the possibility of observing these effects was remote. In fact, both effects were observed within two years. I have outlined a few experiments aimed at revealing the properties of stars that could be carried out by some ambitious young observer. Many others are waiting to be thought of.

In brief: there is a bright future in stars ... even when we can't see them.

BIBLIOGRAPHIC NOTES

The following articles will acquaint the student with the basic methods that have been used to extract information about stars and Galactic structure as well as the new technique of microlensing which promises to revolutionize this area of study.

- **Bahcall, J. N., & Soneira, R. M. 1980, ApJS, 44, 73**. The original paper in which characteristics of the stellar distribution were inferred by direct comparison of the predictions of models with the observations, as opposed to using integral equations. Also the origin of the Bahcall-Soneira model.

- **Bessel, M. S. & Stringfellow, G. S. 1993, ARA&A, 433**. Modern review of work (by traditional methods) on the faint end of the stellar luminosity function (LF).

- **Bahcall, J. N., Flynn, C., Gould, A., & Kirhakos, S. 1994, ApJ, 435, L51**. This paper showed how HST observations could be used to measure the faint end of the LF outside the immediate neighborhood of the Sun.

- **Paczyński, B. 1986, ApJ, 304, 1**. In this fundamental paper, Paczyński proposed microlensing as a means to search for MACHO dark matter. The microlensing technique is now revolutionizing our understanding of stars.

- **Paczyński, B. 1991, ApJ, 371, L63**; and **Griest, K. et al. 1991, ApJ, 372, L79**. The first proposals to use microlensing to systematically study stars in the Milky Way disk.

- **Gould, A. 1992, ApJ, 392, 442**. The first analysis of how the diverse higher order effects in microlensing can be used to extract detailed information about individual lenses.

- **Gould, A. 1995, ApJ, 441, L21**. This paper presents a simple analysis showing that it is possible to routinely acquire additional information about lensing events with a single heliocentric satellite.

- **Crotts, A. P. S. 1992, ApJ, 399, L43**. This paper, the first to show that it is possible to observe lensing of unresolved stars, has opened the way to studying low mass stars in external galaxies.

BIBLIOGRAPHY

[1] Alcock, C., et al. 1993, Nature, 365, 621.

[2] Alcock, C., et al. 1995a, ApJ, 445, 133.

[3] Alcock, C., et al. 1995b, ApJ, 454, L125.

[4] Aubourg, E., et al. 1993, Nature, 365, 623.

[5] Aubourg, E., et al. 1995, A&A, 301, 1.

[6] Bahcall, J. N. 1984, ApJ, 276, 169.

[7] Bahcall, J. N., Flynn, C., & Gould, A. 1992, ApJ, 389, 234.

[8] Bahcall, J. N., Flynn, C., Gould, A., & Kirhakos, S. 1994, ApJ, 435, L51.

[9] Bahcall, J. N., & Soneira, R. M. 1980, ApJS, 44, 73.

[10] Baillon, P., Bouquet, A., Giraud-Héraud, Y., & Kaplan, J. 1993, A&A, 277, 1.

[11] Caldwell, J. A. R., & Ostriker, J. P. 1981, ApJ, 251, 61.

[12] Corrigan, R. T., et al. 1991, AJ, 102, 34.

[13] Crotts, A. P. S. 1992, ApJ, 399, L43.

[14] Gould, A. 1992, ApJ, 392, 442.

[15] Gould, A. 1994a, ApJ, 421, L71.

[16] Gould, A. 1994b, ApJ, 421, L75.

[17] Gould, A. 1995a, ApJ, 441, L21.

[18] Gould, A. 1995b, ApJ, 444, 556.

[19] Gould, A. 1995c, ApJ, 447, 491.

[20] Gould, A. 1995d, ApJ, 455, 37.

[21] Gould, A. 1995e, ApJ, 455, 44.

[22] Gould, A., Bahcall, J. N., & Flynn, C. 1996, ApJ, 465, 000.

[23] Gould, A., Welch, R. L. 1996, ApJ, 464, 000.

[24] Han, C., & Gould, A. 1995, ApJ, 447, 53

[25] Han, C., & Gould, A. 1996, ApJ, 467, 000.

[26] Han, C., Narayanan, V. K., & Gould, A. 1995, ApJ, 461, 587.

[27] Irwin, M. J., Webster, R. L., Hewett, P. C., Corrigan, R. T., & Jedrzejewski, R. I. 1989, AJ, 98, 1989.

[28] Kirkpatrick J. D., McGraw, J. T., Hess, T. R., Liebert, J., & McCarthy, D. W., Jr. 1994, ApJS, 94, 749.

[29] Kuijken, K., & Gilmore, G. 1991, ApJ, 267, L9.

[30] Loeb, A. & Sasselov, D. 1995, ApJ, 449, L33.

[31] Monet, D. G., Dahn, C. C., Vrba, F. J., Harris, H. C., Pier, J. R., Luginbuhl, C. B., & Ables, H. D. 1992, AJ, 103, 638.

[32] Nemiroff, R. J., & Wickramasinghe, W. A. D. T. ApJ, 424, L21.

[33] Reid, I. N., Hawley, S. L., & Gizis, J. E. 1995, AJ, 110, 1838.

[34] Refsdal, S. 1966, MNRAS, 134, 315.

[35] Reid, I. N., Tinney, C. G., & Mould J. 1994, AJ, 108, 1456.

[36] Stobie, R. S., Ishida, K., & Peacock, J. A. 1989, MNRAS, 238, 70.

[37] Udalski, A., Szymański, J., Stanek, K. Z., Kaluzny, J., Kubiak, M., Mateo, M., Krzemiński, W., Paczyński, B., & Venkat, R. 1994, Acta Astron, 44, 165.

[38] Wielen, R., Jahreiss, H., & Krüger, R. 1983, IAU Coll. 76: Nearby Stars and the Stellar Luminosity Function, eds. A. G. D. Philip & A. R. Upgren, p. 163.

[39] Witt, H. 1995, ApJ, 449, 42.

[40] Witt, H. & Mao, S. 1994, ApJ, 430, 505.

CHAPTER 13

SEARCHING FOR MACHOS WITH MICROLENSING

CHARLES ALCOCK

Lawrence Livermore National Laboratory, Livermore, CA

ABSTRACT

Baryonic matter, in the form of Machos (Massive Compact Halo Objects), might be a significant constituent of the dark matter that dominates the Milky Way. This article describes the experimental searches for Machos that exploit the gravitational microlens magnification of extragalactic stars. These surveys monitor millions of stars, in some cases every night, looking for magnification events.

The early results from the surveys have yielded some spectacular events, and pose a significant new puzzle for galactic structure: toward the Large Magellanic Cloud we see fewer events than anticipated for a standard dark halo dominated by Machos, but toward the galactic bulge, the event rate is much higher than anticipated.

This is a field of research that is ripe with opportunities for beginning (and senior) scientists.

13.1 INTRODUCTION

Most of the mass of the Milky Way and similar galaxies is in some presently invisible form (see, e.g., the review of Fich and Tremaine [12]). This "dark matter" cannot be in the form of normal stars or gas, which can readily be detected. Additionally, there is compelling evidence for much larger quantities of dark matter on larger scales in the universe.

Many candidates have been proposed to account for this dark matter. These fall into two main classes: the particle-physics candidates such as massive neutrinos, axions or other weakly interacting massive particles (WIMPs) [24], and the

253

astrophysical candidates, including substellar objects below the hydrogen burning threshold $\approx 0.08 M_\odot$ ('brown dwarfs'), or stellar remnants such as white dwarfs, neutron stars or black holes; these are generically known as massive compact halo objects (Machos). Such objects would be much too faint to have been detected in current sky surveys.

The searches for Machos described here (and all of the experimental searches for particle physics candidates) are looking exclusively for objects located in the dark halo of the Milky Way. The total amount of dark matter in all galactic halos (located say, within 50 kpc of spiral galaxies) is approximately known from rotation curve data, and contributes to the mean density of the universe $\Omega \sim 0.05$: if $\Omega = 1$, either the halos must extend far beyond 50 kpc or there must be intergalactic dark matter.

Paczynski [21] suggested that Machos could be detected by their gravitational 'microlensing' of background stars. This indirect technique does not depend upon light emitted by the Machos. (It should be remembered, though, that the lensing objects in question are not required to be dark, merely significantly fainter than the source stars.)

13.2   THE GRAVITATIONAL MICROLENS

The simple gravitational lens comprises a point-like source of light (typically a star), a point-like massive deflector (the Macho), and an observer. If the target star lies directly behind the Macho, its image will be a ring of light of angular radius, θ_E, called the Einstein radius. If the Macho is separated from the line of sight to the source by some finite angle, $\theta = b/D_d$ (where b is the physical distance of the Macho from the line of sight), the ring splits into two arcs. The combined light from the two images produced by the gravitational lens causes a net magnification

$$A(u) = (u^2 + 2)/u\sqrt{u^2 + 4} \qquad (13.1)$$

that depends only on the ratio of the angular separation to the Einstein radius, $u = \theta/\theta_E$. Note that $A > 1.34$ when $u < 1$, and $A \sim u^{-1}$ when $u \ll 1$. The Einstein radius is related to the underlying physical parameters by

$$\theta_E = \sqrt{\frac{4GM D_{ds}}{c^2 D_d D_s}} \qquad (13.2)$$

where M is the Macho mass, and D_d, D_s, and D_{ds} are the (observer-lens), (observer-source), and (lens-source) distances, respectively. The term 'microlensing' is used when θ_E is so small that the two images cannot be separated with current observing equipment, and the image doubling cannot be seen.

Frequently the related quantity $R_E = \theta_E D_d$ is referred to as the radius of the Einstein ring, where R_E is the physical size of the ring described above, measured at the location of the Macho. For a source distance of $50\,\text{kpc} \approx 10^{10}\,\text{AU}$ and a deflector distance of $10\,\text{kpc}$, the Einstein radius is $R_E \approx 8\sqrt{M/M_\odot}\,\text{AU}$. The coincidence between this scale and the orbit of the earth around the sun can be exploited in the study of microlenses, as will be described below.

The observable phenomenon, as the Macho moves at constant relative projected velocity v, is the varying magnification of the star. This magnification is given as a function of time by $A[u(t)]$, where

$$u(t) = \sqrt{\omega^2(t - t_0)^2 + \beta^2} \tag{13.3}$$

and β is the impact parameter in units of the Einstein radius, t_0 is the epoch of maximum magnification, and ω^{-1} ($\omega \propto v$) is the characteristic time (duration) of the event.

Of course many astronomical sources are variable. Fortunately, simple microlensing has distinctive signatures which can be used to discriminate it from intrinsic stellar variability:

- Since the probability that any given star will be significantly lensed is small, only one event should be seen in any given star (see discussion of microlensing optical depth).

- The deflection of light is wavelength-independent; hence, the star should not change color during the magnification.

- The events should have lightcurves well described by the theory [eqs. (13.1) and (13.3)].

All these characteristics are distinct from known types of intrinsic variable stars; most variable stars are periodic or semi-regular, and do not remain constant for long durations. They usually change temperature and hence color as they vary, and they usually have asymmetrical lightcurves with a rapid rise and slower fall.

In addition to these individual criteria, if many candidate microlensing events are detected, further statistical tests can be applied:

- The events should occur equally in stars of different colors and luminosities.

- The distribution of impact parameter u_{min} should be uniform from 0 to the experimental cutoff $u(A_{min})$.

- The event timescales and peak magnifications should be uncorrelated (after correcting for experimental selection effects).

When a microlensing event is detected and its light curve is measured, one determines three parameters, ω, t_0, and β. Of these, only ω is related to the physical parameters of the lens:

$$\omega = \frac{v}{D_d \theta_E}. \tag{13.4}$$

For microlensing of sources in the Large Magellanic Cloud (LMC), $\omega^{-1} \approx 70\sqrt{M/M_\odot}$ days. One of the principal limitations of present experiments in this area, one which cannot be resolved with purely ground-based work, is that our uncertain knowledge of the quantities D_d, D_{ds}, and v for an observed microlensing event means that the uncertainty in the inferred mass of the Macho, M, spans more than an order of magnitude [16]. The mass is our only clue to the true nature of the Macho.

Parallax, obtained by the simultaneous observation of a microlens event from telescopes separated by a distance of order R_E, allows one to measure a second parameter [14],

$$\tilde{v} = (D_s/D_{ds})v. \tag{13.5}$$

(Generally one obtains one vector component of $\tilde{v}$, and the absolute magnitude of another.)

The complete solution of a microlensing event requires, in addition, the determination of θ_E. This is very difficult to achieve, since typically $\theta_E < 10^{-3}\,''$. For a handful of cases in which the Macho passes directly in front of the source star (or nearly in front of the star), it will be possible to measure θ_E. This is possible because the magnification equation given above must be modified to take into account the finite angular size θ_S of the star. The modification depends principally upon the ratio θ_S/θ_E. (An important complication is introduced by the center to limb variation of brightness over the face of the star.) If θ_S can be estimated spectroscopically, one can in turn infer θ_E.

In the rare cases when one measures ω, $\tilde{v}$, and θ_E, the lens parameters mass, distance, and transverse velocity can each be determined. For example,

$$M = \frac{c^2 \tilde{v} \theta_E}{4G\omega}. \tag{13.6}$$

13.3   THE "MACHO FRACTION" IN THE GALACTIC HALO

The gravitational microlens 'optical depth' is the quantity that probes directly the Macho fraction of the dark matter, since it is 'proportional' to the density of microlensing objects along the line of sight to the target stars. If the distribution of the total mass density $\rho_{\text{total}}(D_d)$ is known along a given line of sight, then the experimental estimate of the optical depth τ along this line of sight yields the fraction of the total dark matter that is in the form of Machos.

The 'optical depth' τ for gravitational microlensing is defined as the probability that a given star is lensed with $u < 1$ ($A > 1.34$) at any given time, and is

$$\tau = \pi \int_0^{D_s} \frac{\rho(D_d)}{M} R_E^2(D_d) \, dD_d, \tag{13.7}$$

where ρ is the density in Machos. Since $R_E \propto \sqrt{M}$, while for a given ρ the number density of lenses $\propto M^{-1}$, the optical depth is independent of the individual Macho masses. Using the virial theorem, one finds that $\tau \sim (V/c)^2$, where V is the rotation speed of the Galaxy.

More detailed calculations [16] give an optical depth for lensing by Machos in the Galaxy of stars in the Large Magellanic Cloud of $\tau_{\text{LMC}} \approx 5 \times 10^{-7}$, under the assumptions that (1) all of the dark matter is in the form of Machos; and (2) the most naive model of the halo (spherically symmetric, small core radius) is correct. This very low value means that only one star in two million will be magnified by $A > 1.34$ at any given time. (Note that this estimate assumes that all of the dark matter is in Machos, and hence is a crude upper limit to the optical depth.)

Surveys for gravitational microlensing follow millions of stars photometrically in order to obtain event rates of a few per year against the Large Magellanic Cloud. The optical depth in principle can be estimated directly from the experimental data, once a statistically significant number of events has been recorded. This interpretation is complicated, in practice, by inefficiencies introduced by the irregular sampling, and by the very crowded star fields that must be observed in order to make the large number of photometric measurements.

The interpretation of microlens optical depths in terms of the fraction of the dark matter in the form of Machos is limited by our poor knowledge of $\rho_{\text{total}}(D_d)$. The dark halo may be spherical or flattened [25], and if flattened it may or may not be aligned with the plane of the disk [13]. The core radius of the halo is not securely known. Also important is a complete understanding of the contribution of the disk and bulge to the total mass interior to the solar circle. This has turned out to be more significant than expected.

Improving our knowledge of ρ_{total} is clearly important. Improvement can be obtained by measuring the optical depth along many well separated lines of sight, combined with the measured rotation curve of the disk. The variation of τ with location on the sky would provide us with a form of *tomography* of the dark matter distribution, which gives the missing shape information. The rotation curve data provide the normalization for ρ_{total}. To date measurements have only been attempted along two well separated lines of sight, towards the Large Magellanic Cloud and towards the central bulge of the Milky Way. The latter line of sight does not probe the dark halo unambiguously, but is very useful in probing the structure of the Milky Way, and thus helping to determine ρ_{total}.

13.4   THE EXPERIMENTAL SITUATION

Four groups have reported detections of microlens events: the MACHO Project [3, 4, 5], the EROS Project [7], the OGLE Project [27, 28], and the DUO Project [1]. The MACHO Project follows $\sim 10^7$ stars in each of the Large Magellanic Cloud and the galactic bulge. The MACHO Project has recorded ~ 90 events, of which eight are toward the Large Magellanic Cloud, and the rest toward the galactic bulge. The EROS Project has followed $\sim 4 \times 10^6$ stars in the Large Magellanic Cloud, and recorded two events (both of these are possibly due to intrinsic stellar variation). The OGLE Project follows $\sim 2 \times 10^6$ stars in fields near the galactic center, and has recorded ~ 18 events, and the DUO Project follows $\sim 10^6$ stars also in fields near the galactic center, and has recorded ~ 12 events.

Two example events are shown in Figure 13.1. The upper panel shows a high magnification event from the MACHO Project (the star is in the galactic bulge). The lower panel shows an event that was noticed well before it reached maximum magnification. (The epoch at which the event was recognized is marked on the figure.) This "alert recognition" is a new capability now routinely used by both MACHO and OGLE. The MACHO Project currently uses the alert process to select events for more careful photometric monitoring at Cerro Tololo Interamerican Observatory, and for spectroscopic follow-up. Spectra have been obtained during two alert events, and in each case spectra were taken at epochs spanning the maximum. The spectra showed no evolution during the events, confirming the gravitational lens interpretation.

The early results from these surveys are surprising. Preliminary estimates of the optical depth toward the Large Magellanic Cloud indicate that (1) the event rate is probably too high to be accounted for by a previously known population of objects; and (2) the event rate seems to be lower than what was expected for a spherical dark halo in which all of the dark matter is comprised of Machos [5]. The recent work of Bahcall et al. [8] has *almost* eliminated low luminosity stars as the objects responsible for the microlensing that is seen.

A quantitative examination of the MACHO observations towards the Large Magellanic Cloud [5] shows that the best-fit Macho fraction, assuming a standard model for the dark matter distribution, is $f \sim 0.2$, but with very large uncertainties. It is not possible to exclude the possibility that all of the dark matter is in the

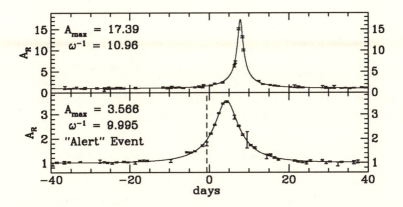

Figure 13.1: Two microlensing events seen by the MACHO Project, with the discovery date of the latter shown.

form of Machos (especially considering our poor knowledge of the true distribution of the dark matter), nor is it possible to exclude that essentially *none* of the halo dark matter is in Machos. These results do show that significant answers to the question of the Macho fraction are within reach of the current experiments, or by next generation experiments as described below.

The galactic bulge observations also have yielded substantial surprises. The estimated optical depths are $\sim 3\times$ higher than the upper limits that were estimated prior to the experimental results [29, 4]. (These upper limits [17, 22] were based upon standard models for the galaxy.) Suggestions have included (1) that the galactic bulge is a very massive bar, with the long axis pointing in our direction (a successful model fit to the OGLE and MACHO results using a pre-existing model bar was made by [30]); (2) that the mass of the disk is much greater than anticipated [4]; (3) that the dark halo has a low density or even hollow core [11]; and (4) that the dark halo is a very flattened structure, and the high rate toward the galactic center is due to halo Machos.

It is important to note that the interpretation of microlensing results toward the Large Magellanic Cloud is contingent upon understanding the galactic bulge results. In particular, the estimate of $\rho_{total}(D_d)$ in the halo, which is combined with τ to obtain the estimate of the Macho fraction, is derived from the mass model of the Galaxy. The model of the bulge is an important component of the mass model.

13.5   NEXT GENERATION EXPERIMENTS

The current experiments are far from completing their missions, and collectively they will clearly be able to detect and decipher many more microlensing events, probably tens of events toward the Large Magellanic Cloud and perhaps hundreds of events toward the Galactic bulge. It is probably more useful, however, to look at what can be achieved with more powerful or even substantially different experiments, based upon the experience of the current four projects.

13.5.1   *What Can Be Achieved from the Ground?*

The experience of the four groups allows us to project what could, in principle, be achieved from the ground. This assumes that next- generation experiments are exploited fully to realize the limits of ground-based microlens surveys.

It will be possible to measure the optical depth toward the Large Magellanic Cloud with reasonable accuracy.

It might be possible to do this also for the Small Magellanic Cloud. It will not be possible to obtain useful estimates of the microlens optical depth toward other targets that are high above the galactic plane. All other targets that contain adequate numbers of stars are either so distant that the observations are confusion limited (e.g., M33), or the stars are in globular clusters that are too faint to observe usefully from the ground. This limits the potential usefulness of the gravitational microlens tomography technique that was described above.

Two groups ([10, 9]) will attempt to measure microlens event rates toward M31. Their technique is sensitive only to Machos in the halo of M31, and will not in the end provide additional information about the nature and prevalence of Machos in our own galaxy.

Ground-based work alone will not be able to determine the Macho fraction of the dark matter. The uncertainties in the shape and radial profile of the dark halo will be the primary limitation. This limitation can be alleviated by using the Macho Parallax Effect, reducing the uncertainty about where along the lines of sight the individual Machos are, and adding some velocity information. This is discussed in the next subsection.

The situation with respect to the bulge observations is more promising. It is possible that ground-based observations will be able to distinguish between a large central bar, pointing in our direction, and a "maximal" disk (Han and Gould [18] have expressed doubts that this can be achieved). It will not be possible, however, to distinguish a "maximal" disk from an extremely flat halo; this would require either the parallax technique or the tomography technique.

Finally, there is an exciting new application of the gravitational microlens technique. It is possible to search for planetary companions around Machos by looking for the small distortions in the light curves that are produced [20, 15]. This search would require a latitudinally spread network of telescopes providing nearly continuous monitoring of events that are found in a survey such as MACHO. Two groups are attempting to put such a network together: MACHO and PLANET [2].

13.5.2 Observing Macho Parallax

Earlier we showed the great value of measuring the projected velocity $\tilde{v}$, especially in elucidating the masses of Machos. This can be accomplished by the parallax technique: measuring an event from two different lines of sight [14]. From the ground one can measure Macho parallaxes only for events of exceptionally long duration, so that the acceleration of the earth around the sun significantly changes the projected velocity during the event. These long duration events are uncommon.

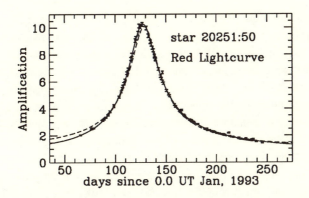

Figure 13.2: Lightcurve of an event from the MACHO Project showing the parallax effect. The dashed line is the best fit not including the motion of the earth around the sun, the solid line includes the motion of the earth around the sun.

Figure 13.2 shows the light curve of one of the more spectacular events discovered by the MACHO project. This is the only event for which the effect of parallax motion has been detected [6]. There is a significant asymmetry in the lightcurve of this star, which is well fit when the motion of the earth is taken into account. The best fit indicates that the velocity of the Macho with respect to the line of sight between the source star and the Sun is 75 ± 5 km/sec when projected to the position of the Sun. The knowledge of the projected velocity allows us to derive a relation between the mass of the Macho and its position along the line of sight as seen in Figure 13.3 (solid line).

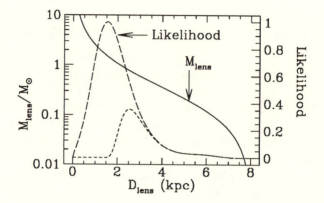

Figure 13.3: Mass of Macho versus distance to the Macho, and likelihood estimates assuming the lens is dark (long dash) or a main sequence star (short dash).

Also shown in Figure 13.3 are two likelihood estimates taken from [6]. The long-dashed line assumes a simple model for the velocity distributions of stars in the Galactic disk and bulge. The short-dashed line assumes additionally that the lensing object is a main sequence star, and that this star contributes the baseline flux observed (see [6] for details). From this figure, we can immediately conclude that the lens is either a dark object at $\sim 1.5\,\mathrm{kpc}$ or a star at $\sim 2.5\,\mathrm{kpc}$. If we had information like this for a large fraction of the microlensing events, then we would be able to tell whether the majority of the lensing objects are in the galactic disk, bulge, or bar. Similarly, toward the Large Magellanic Cloud, routine measurements of microlensing parallaxes would allow us to determine whether the observed events are Machos in the dark halo, the Milky Way disk, or perhaps in the Large Magellanic Cloud.

Parallaxes are difficult to measure from the ground; only for the longest duration events will the earth move far enough in its orbit during the event to make the parallax effect visible from the ground. This limits us to measurements made on events for which the Machos are in the galactic disk, and eliminates from useful study bulge and dark halo Machos (for which the events are of duration much shorter then six months, because of the high expected transverse velocities). For this vast majority of cases, the microlensing events will end before the earth has had a chance to move very far, so the parallax information will only be available if events can be observed simultaneously from the earth and a small satellite in a solar orbit of order $\sim 1\,\mathrm{AU}$ away from the earth [14].

Satellite parallaxes would greatly advance the study of Machos. For events seen toward the Large Magellanic Cloud, measurement of $\tilde{v}$ would distinguish between Galactic Machos ($\tilde{v} < 300\,\mathrm{km\,s^{-1}}$) and those in the Large Magellanic Cloud ($\tilde{v} \sim 2000\,\mathrm{km\,s^{-1}}$). For Galactic Machos, $\tilde{v} \sim v$, so that one could distinguish between Machos in the disk ($v \sim 50\,\mathrm{km\,s^{-1}}$), the thick disk ($v \sim 100\,\mathrm{km\,s^{-1}}$), and the halo ($v \sim 200\,\mathrm{km\,s^{-1}}$). It would be possible to measure the transverse direction for at least some of these Machos, yielding additional information about their distribution.

The situation is somewhat more complicated for Machos seen towards the bulge. Parallaxes can statistically distinguish between Machos lying in the disk versus those which lie in the bulge itself, and also allow one to estimate the mass of the former [18]. Moreover, for $\sim 10\%$ of bulge Machos and $\sim 4\%$ of disk Macho events it should be possible to measure θ_E from the ground and therefore use the parallax to determine M, D_d, and v individually.

13.6 Working on Gravitational Microlensing

Research on gravitational microlensing has become a well established sub-field of astrophysics, and there are abundant opportunities for young scientists. There is the special pleasure and reward that comes from working on a topic of great importance (in a period in which progress is being made rapidly), and there are few, if any, topics more important than determining what the universe (or at least, the Milky Way), is made of!

The experimental work differs in some important *cultural* ways from most contemporary astrophysics research. The most important difference is that the projects typically require teams that are much larger than the norm. (Some well-known astronomers have remarked to the author that they would not work in this area precisely for this reason!) A newcomer to this research should recognize that extremely demanding projects require larger than normal teams of scientists, and that group efforts are frequently necessary. In my view, the reward that comes from making an unusual contribution more than compensates for the reduced autonomy!

There is a notable exception to this remark: the DUO team is much smaller than the other successful microlensing survey teams. Keep in mind, though, that the only team that has repeatedly detected microlensing towards the Large Magellanic Clouds (MACHO) is relatively large.

Smaller groups and individuals dominate the theoretical and model building work which is needed to interpret the data and to suggest new directions. Prominent examples of this are the original paper in the field [21], and the many papers of Andy Gould.

I should make also a more general remark that comes from the history of the

MACHO Project. There was great skepticism among our colleagues about this project, and it was not widely believed that we would be successful. We decided that the scientific goal more than justified the technical risks we were taking (and fortunately some key sponsors agreed), and went ahead in spite of the best advice of our friends. The outcome has convinced me that it is important to take one or more large risks in a career.

13.7   SUMMARY

It is useful to summarize the situation in regard to research on gravitational microlensing in three parts: what we know, what we will likely learn from current experiments, and what we could learn with aggressive next generation experiments.

13.7.1   *What We Know Now*

- Gravitational microlensing has been observed.

- The event rate toward the Large Magellanic Cloud is lower ($\sim$ 20% but with large errors) than expected for a "standard halo" filled with Machos of substellar mass.

- The event rate towards the galactic bulge is 3 times higher than expected.

13.7.2   *What We Will Learn from Current Experiments*

- Reasonable statistics for the event rate towards the Large Magellanic Cloud (primary handle on halo Machos).

- Approximate event rate towards the Small Magellanic Cloud.

- Crude map of the event rate as a function of Galactic longitude and latitude near the galactic center (needed to disentangle galactic models).

13.7.3   *Next Generation Experiments*

- Develop photometric follow-up network to determine lightcurves better (to measure higher order effects such as finite angular size of star, which yields much more information per event, and to search for planetary companions).

- Macho Parallax Satellite (measure lightcurves simultaneously from earth and spacecraft 1 AU away to resolve the lens; this will determine where the Machos are).

13.8   Late Breaking News

Eight months after the *Unsolved Problems* meeting, the MACHO Project completed an analysis of a two year data set that included light curves for nine million stars. The new analysis differed from that described above and in [5] primarily in that new event acceptance criteria were developed following the project's experience in the Galactic bulge. Eight candidate microlens events passed the new acceptance criteria, of which six were of high quality. The best estimate of the Macho fraction in the Galactic halo increased in this new analysis from $\sim 20\%$ to $\sim 40\%$ (six events included) or $\sim 50\%$ (all eight events included), but still with very large error bars.

Bibliographic Notes

Two new and very good reviews have recently been written, by Paczynski [23] and by Roulet and Mollerach [26]. These provide a sound and broad introduction to the field. It is useful, in addition, to read some of the key early works, especially [21], which got all of this going. Paczynski himself is often very modest about this paper, pointing out (correctly) that the phenomenology of the point mass gravitational lens had been known already for decades! What was new, and important, was the connection he made to the dark matter problem.

On the experimental side, it is definitely worthwhile to read the first announcements from MACHO [4], from EROS [7], and from OGLE [27]. The papers are very cautious, but nevertheless there is a sense of barely contained excitement that marks a discovery paper!

Now that the field is maturing, new directions suggest themselves. Three papers have generated considerable discussion. In the first, Gould [14] pointed out the great value that satellite based parallax measurements would bring to this area of research. This paper opened up the investigation of parallaxes and the more general study of novel ways to add to the information we gain from individual events. In the other two, Mao and Paczynski [20], and later Gould and Loeb [15] described gravitational microlensing by planetary systems; this may lead to a major new initiative to search for extra-solar planetary systems using gravitational microlensing.

Research into gravitational microlensing makes use of the World Wide Web, and this is a useful way to learn up to date information. In particular, the MACHO and OGLE projects make real time event information available this way, so their locations are given here. Current information on the MACHO Project may be found at: http://wwwmacho.mcmaster.ca/, with a mirror copy at: http://wwwmacho.anu.edu.au/. Buttons on these pages direct the browser to real time information on events in progress, and in addition there are pointers to the other major projects (e.g., EROS).

Information on the OGLE Project is at: http://www.astrouw.edu.pl, with a mirror copy at: http://www.astro.princeton.edu/~ogle/.

Bibliography

[1] Alard. C., et al. 1995, The Messenger, No. 80, 31.

[2] Albrow, M., et al. 1995, preprint.

[3] Alcock, C., et al. 1993, Nature, 365, 621.

[4] Alcock, C., et al. 1995, ApJ, 445, 133.

[5] Alcock, C., et al. 1995, Phys. Rev. Lett., 74, 2967.

[6] Alcock, C., et al. 1995, ApJ, 454, L125.

[7] Aubourg, E., et al. 1993, Nature, 365, 623.

[8] Bahcall, J. N., Flynn, C., Gould, A., & Kirhakos, S. 1994, ApJ, 435, L51.

[9] Bouquet, A., Kaplan, J., Melchior, A., Giraud-Héraud, Y., & Baillon, P. 1993, preprint, astro-ph/931209.

[10] Crotts, A., & Tomaney, A. 1993, Columbia preprint.

[11] Evans, N.W. 1993, MNRAS, 260, 191.

[12] Fich, M., & Tremaine, S. 1991, ARA&A, 29, 409.

[13] Frieman, J., & Scoccimarro, R. 1994, ApJ, 431, L23.

[14] Gould, A. 1994, ApJ, 421, L75.

[15] Gould, A., & Loeb, A. 1992, ApJ, 396, 104.

[16] Griest, K. 1991, ApJ, 366, 412.

[17] Griest, K., et al. 1991, ApJ, 372, L79.

[18] Han, C., & Gould, A. 1994, preprint, astro-ph/9410052.

[19] Kiraga M., & Paczynski, B. 1994, ApJ, 430, 101.

[20] Mao, S., & Paczynski, B. 1991, ApJ, 374, L37.

[21] Paczynski, B. 1986, ApJ, 304, 1.

[22] Paczynski, B. 1991, ApJ, 371, L63.

[23] Paczynski, B. 1996, ARA&A, 34, in press.

[24] Primack, J., Seckel, D., & Sadoulet, B. 1993, Ann. Rev. Nuc. Part. Sci., 38, 751.

[25] Sackett, P., & Gould, A. 1993, ApJ, 419, 648.

[26] Roulet, E., & Mollerach, S. 1996, Physics Reports, in press.

[27] Udalski, A., et al. 1993, Acta Astronomica, 43, 289.

[28] Udalski, A., et al. 1994, ApJ, 426, L69.

[29] Udalski, A., et al. 1994, Acta Astronomica, 44, 165.

[30] Zhao, H. S., Spergel, D. N., & Rich, R. M. 1995, ApJ, 440, L13.

CHAPTER 14

GLOBALLY ASYMMETRIC SUPERNOVA

PETER GOLDREICH, DONG LAI, AND MIKAEL SAHRLING

California Institute of Technology, Pasadena, CA

ABSTRACT

Asymmetries in type II supernova explosions are the most plausible cause of the high space velocities of radio pulsars. The origin of these asymmetries is unknown. Recent work has stressed the potential importance of local instabilities which occur subsequent to the collapse of the stellar core. By contrast, relatively little attention has been paid to the possibility that substantial asymmetry might already be present during core collapse. The following line of argument suggests that this may not be farfetched. The presupernova star supports g-modes in which the core oscillates with respect to the stellar envelope. These modes gain excitation by modulating the release of nuclear energy. Asymmetries associated with these modes could then be amplified during core collapse. The most likely place for this to occur is in the supersonically collapsing outer core.

14.1 INTRODUCTION

14.1.1 *Preamble*

Our presentation is couched as a research proposal. The situation we describe may be similar to what a graduate student typically encounters when first approaching a thesis advisor. We describe some small steps taken to investigate the initiation of global asymmetries in presupernova stars and their amplification during core collapse. We advocate a particular scenario, fully aware that it is speculative and therefore fragile. Deeper analysis will be needed before it will be known whether our scenario contains even an element of the truth.

We are concerned with supernova of type II. These occur in massive stars ($M \gtrsim 8M_\odot$) which burn carbon nonexplosively and evolve to a configuration with a degenerate iron core overlaid by an "onion skin" mantle of lighter elements (e.g., Woosley & Weaver 1986; Nomoto & Hashimoto 1988; Bethe 1990). A dynamical instability in the iron core, initiated by electron capture and endothermic dissociation of iron, then leads to implosion.

The rapidly growing iron core of the presupernova star is encased in and being fed by shells burning silicon and oxygen, and the entire assemblage is surrounded by a thick convection zone. The core being nearly isothermal is stably stratified and supports internal gravity waves. These waves, in which surfaces of constant density oscillate with respect to those of constant gravitational potential, cannot propagate in the unstably stratified convection zone. Hence, they are trapped, giving rise to core g-modes in which the core oscillates with respect to the outer parts of the star.

Preliminary analysis suggests that core g-modes in presupernova stars may be overstable with driving provided by temperature sensitive nuclear burning in Si and O shells of the star before it implodes. If that is so, the number of e-foldings of the mode amplitudes prior to collapse is estimated to be large. A separate investigation indicates that the collapse of the outer core may be unstable with strain amplitudes growing by factors of order 10^2 for $l = 1$ modes (and less for higher degree modes). Together, these results hint that spherical symmetry may be at best a crude approximation in supernova explosions.

In our scenario, the recently-revised high space velocities of neutron stars might result from momentum exchange between the matter that ends up incorporated into the neutron star and that which just fails to do so and is ejected in the supernova explosion. This momentum transfer is likely to be mediated by pressure and gravitational forces acting during asymmetric core collapse and subsequent asymmetric mass ejection. Asymmetric neutrino emission might be a viable alternative provided the asymmetry is maintained for a time comparable to the neutrino diffusion time (see below).

14.1.2   *Evidence for Asymmetry*

Several lines of evidence suggest that supernova suffer from large scale asymmetries. Perhaps the best is the high space velocities of radio pulsars determined from proper motion studies (Harrison et al. 1993), interstellar scintillation measurements (Cordes 1986; Harrison & Lyne 1993), and associations with supernova remnants (Frail et al. 1994). Using the revised pulsar distance scale coupled with new measurements of proper motions and a better treatment of selection effects, Lyne & Lorimer (1994) give 450 km/s as the mean velocity of pulsars at birth, much larger than previously thought. Moreover, several individual pulsars may have velocities exceeding 10^3 km/s (Frail & Kulkarni 1991; Cordes et al. 1993; Harrison et al. 1993). A plausible explanation is that supernova explosions are asymmetric, and provide kicks to nascent neutron stars. Support for supernova kicks comes from the detection of geodetic precession in PSR 1913+16 (Weisberg et al. 1989; Cordes et al. 1990) and orbital plane precession in the PSR J0045-7319 system (Lai et al. 1995).

Other evidence for asymmetries comes from polarization measurements of SN1987A (and also, for example, SN1993J, Cropper et al. 1988, Trammell et al. 1993). Moreover, speckle interferometry of SN1987A (see, e.g., Papliolios et al. 1989) suggests that the envelope had an ellipsoidal shape when first resolved. Early X-ray and Gamma-ray observations were interpreted by McCray (1993) and Kumagai et al. (1989) as implying the existence of large-scale "clumps" of iron-group elements in the ejecta. Later analyses of emission-line profiles confirm their conclusion (Utrobin et al. 1995). Several oxygen-rich supernova remnants have large systemic velocities relative to the local ISM (Morse, Winkler & Kirshner 1995), implying that they were produced in asymmetric explosions. The discovery of ejected high-density clumps around the Vela supernova remnant from X-ray and radio observations (Aschenbach et al. 1995; Strom et al. 1995) is yet another indication that supernova explosions are not spherically symmetric.

In spite of the evidence cited above, there is another view. Iben & Tutukov (1996) suggest that only neutron stars born from presupernova stars in compact binaries spin fast enough to qualify as radio pulsars. They associate the high pulsar space velocities with the dissociation of these compact binaries.

14.1.3 *State of the Art*

Prior to 1987 almost all theoretical studies of supernova explosions focused on spherically symmetric models. In more recent attempts to explain observed asymmetries, several groups have investigated potential Rayleigh-Taylor, Kelvin-Helmholtz, and convective instabilities subsequent to the collapse of the stellar core (Burrows & Fryxell 1992; Burrows et al. 1995; Janka & Müller 1994; Herant et al. 1994; Herant & Woosley 1994). Numerical simulations indicate that these local, post core collapse instabilities may have difficulty accounting for kick velocities higher than 5×10^2 km/s (Janka & Müller 1994; Woosley, & Weaver 1992). Some authors attribute the large space velocities of neutron stars to an angular asymmetry, β, in neutrino emission. To obtain a kick $v = 10^3 v_3$ km/s requires $\beta = Mvc/E_\nu \simeq 0.08\, v_3 M_{1.4}/E_{\nu 53}$, where $M_{1.4}$ is the neutron star mass in units of $1.4 M_\odot$, and $E_{\nu 53}$ is the total neutrino energy emitted in units of 10^{53} erg. Note that β measures the angular asymmetry relative to inertial space. Since the bulk of the neutrino emission is spread out over several seconds, rapid rotation of the proto-neutron star tends to decrease this asymmetry.

Recent numerical simulations by Burrows and Hayes (1996) indicate that if the collapsing core is mildly asymmetric, the newly formed neutron star can receive a kick velocity comparable to the observed values. Asymmetric motion of the exploding material (since the shock tends to propagate more "easily" through the lower-density region) dominates the kick, although there is also contribution from asymmetric neutrino emission. The magnitude of the kick velocity is proportional the degree of initial asymmetry of the imploding core. Clearly, the important question remains: What is the origin of this initial asymmetry?

To date, little attention has been paid to the possibility that large scale asymmetries may already be present at the time of presupernova core collapse. As described below, we have carried out several reconnaissance investigations which have strengthened our belief that this line of research is worth pursuing. One irreducible difficulty such investigations face is that major aspects of presupernova stellar models are not well understood. Residual uncertainties are especially prevalent during the period of greatest interest to us, the final few hours prior to core collapse when O-Si shell burning is rapidly adding mass to the Fe core. Thus, at present, a complete picture of global asymmetries of supernova explosions is probably out of reach. Nevertheless, the study of simplified models, which are suitable projects for graduate students, seems likely to resolve whether global asymmetries are a generic feature of presupernova core collapse.

It is perhaps worth noting that the results of this research will bear on gravitational wave emission by supernova. Unfortunately, only a supernova within our galaxy seems likely to be detectable by instruments now under construction (Burrows & Hayes 1996; Müller & Janka 1996) unless significant angular momentum exists to induce a bar-mode instability in the nascent neutron star (Lai & Shapiro 1995).

14.2   INSTABILITY DURING CORE COLLAPSE

Using a parametrized polytropic equation of state, $P = K\rho^\gamma$ with $\gamma = 4/3$, Goldreich & Weber (1980) showed that the inner region of the collapsing core develops a self-similar structure. This homologous inner core has infall velocity proportional to radius and thus maintains a time invariant density profile. Time enters only through shrinkage of the overall scale. The mass M_{ic} of the homologous inner core is only slightly larger than the Chandrasekhar mass for a given K. Electron capture reduces the effective K below its precollapse value so the inner core typically has a mass $M_{ic} \sim 0.8M_\odot$, smaller than the initial iron core mass $M_c \sim 1.4M_\odot$. Yahil (1983) extended and improved upon this work obtaining self-similar solutions for general γ. His solutions match the subsonic inner-core to a supersonic outer core which collapses at about half the free-fall velocity. The bifurcation of the collapsing material into these two regions is a generic feature of core collapse, one that has been well corroborated by numerical simulations.

That the non-radial modes of the inner homologous core are stable (Goldreich & Weber 1980) is not surprising given the significant role played by pressure in its subsonic collapse. Pressure is less important in the supersonically collapsing outer region, making it more susceptible to large scale instability.

14.2.1   *Accomplishments*

We have studied a few simple models to explore potential nonradial instabilities of the outer core. One model involves a finite-mass, pressureless spherical fluid shell falling onto a central point mass. Here the spherical shell mimics the outer supersonic region, while the central mass represents the stable homologous core. Because it submits to analytic solution, such a simple-minded model is highly instructive. We find unstable nonradial modes associated with spherical harmonics $Y_{lm}(\theta, \phi)$ for all $l > 0$. In particular, there are two unstable $l = 1$ modes, both leading to the growth of the separation between the center-of-mass of the shell and the position of the central mass: (i) in one mode the surface density perturbation grows because one side of shell collapses faster than the other side, and the geometric center of the shell moves in opposition to the motion of the central mass; (ii) in the other mode the surface density perturbation grows due to the internal tangential flow in the shell, while the shell's geometric center suffers little displacement with respect to the position of the central mass. The first of these two modes is the more rapidly growing. Our results show that the perturbation $\delta R(\theta, t) \propto \cos \theta$ to the mean shell radius $R(t)$ grows as a power law in time, $\delta R(\theta, t)/R(t) \propto (-t)^s$ (s is negative), where t is measured from the time of complete collapse. For a core-shell mass-ratio of order unity, the power-law index $s \simeq -1$, corresponding to $\delta R/R \propto R^{-3/2}$. Thus an initial dipole perturbation can be amplified by a factor of order 10^3 while the shell collapses from white dwarf to neutron star size. Similarly, the quadrupole perturbation grows $\propto (-t)^s$, with $s \simeq -(0.8 - 0.9)$.

14.2.2   *Future Directions*

Several issues arise regarding the applicability to presupernova collapse of studies of the accretion of thin-shells onto a central mass. (i) Concentration of all the accreting mass into a thin-shell may lead to overestimates of perturbation growth rates. A preliminary analysis of the steady-state accretion of a collisionless fluid of test particles onto a central mass indicates that the fractional Eulerian density perturbation scales as $\delta\rho/\rho \sim 1/r$ for the fastest growing $l = 1$ mode. Thus the total amplification of dipole asymmetries are likely to be closer to 10^2 than to 10^3. The $l = 1$ modes are special because they involve a displacement of the central mass relative to the bulk flow. Under the assumption that the central mass maintains a spherical shape, the corresponding result for $l \geq 2$ is $\delta\rho/\rho \sim r^{-1/2}$. (ii) It may be misleading to refer to the growth of asymmetries in accreting thin-shells as instabilities. This behavior is primarily kinematic as evidenced by the fact that it persists even in the limit of zero shell mass (with $s = -1$). Feedback is present for finite shell mass, but does not affect the growth rates appreciably. (iii) The neglect of pressure also makes results obtained with the thin-shell models no more than suggestive. Pressure, which is not entirely negligible in the outer core, may weaken or even quench the instabilities. It is imperative to investigate the stability of more realistic models. A reasonable first step would be to incorporate tangential pressure into the thin-shell model in order to obtain a semi-quantitative understanding of its effect on instability growth rates. A recent analysis indicates that the more rapidly growing dipole mode is only slightly affected by pressure (chosen to correspond to Mach number $\simeq 2$), while the other mode is strongly modified.

A more ambitious project would be to determine the stability of Yahil's self similar collapse solutions. Perturbation analysis of these solutions will yield a set of ordinary differential equations from which instability growth rates can be obtained as eigenvalues. Depending upon the results of this study, a numerical simulation may prove worthwhile. This would require application of at least a 2D hydrodynamical code.

The rarefaction wave which initiates core collapse is another potential source of instability. Collapse induced by electron capture is triggered at core center where the pressure and density peak. Passage of an outward propagating rarefaction wave then initiates collapse throughout the entire iron core. The stability or instability of the rarefaction wave is largely unexplored territory. A reasonable approach might be to explore the stability of simple models, perhaps starting with that of the "inside-out" collapse of an isothermal sphere (Shu 1977).

14.3   Overstable Core G-Modes

Strain amplitudes of dipole and quadrupole instabilities probably grow by at most factors $\sim 10^2$ during core collapse. Thus perturbations $\delta R/R \gtrsim 10^{-3}$ are required at the onset of collapse in order that dynamically significant asymmetries, $\delta R/R \gtrsim 0.1$, be present when the core reaches nuclear density. The following line of argument suggests that this may not be farfetched.

By imposing pressure variations upon the temperature sensitive shell sources, core g-modes modulate the release of nuclear energy. As a result, some of these modes may be overstable. Preliminary estimates indicate that amplitudes of overstable modes might increase by large factors during the final day prior to core collapse. The "ϵ−mechanism" is commonly used to denote mode excitation through the modulation of nuclear burning (Cox 1980; Unno et al. 1989). That temperature sensitive shell burning might excite core g-modes has been recognized before. However, to our knowledge no one has considered this possibility in the context of presupernova stars.

14.3.1 *Accomplishments*

We have calculated core g-modes of arbitrary angular degree in a simple model consisting of a dense, rigid core centered inside a less dense, neutrally buoyant, incompressible envelope. The core mimics the iron core of the presupernova star, and the neutrally buoyant envelope represents the convective region which surrounds it. The relevant dipole g-mode frequencies are given by $\sigma^2 \sim (1 - \rho/\rho_c)$ $(1 - R_c^3/R^3)GM_t/R^3$. Here ρ and R are the shell density and outer radius, ρ_c and R_c are the core density and radius, and M_t is the total mass of the system. The corresponding oscillation periods range from a few seconds to tens of seconds.

We use eigenfunctions of these modes as input to calculations of mode growth rates, σ_I, in the quasi-linear approximation. In this approximation σ_I is obtained by evaluating the work integral (e.g., Unno et al. 1989) associated with nuclear energy generation in the shell source. A simple estimate gives $\sigma_I \sim \alpha \dot{E}_{nuc}/E_{th}$, where $\dot{E}_{nuc}$ is the total energy generation rate in the shell, α is the power-law index of the dependence of $\dot{\epsilon}_{nuc}$ (the nuclear energy generation rate per unit mass) on temperature ($\alpha \sim 50$ for Si burning), and E_{th} is the total thermal energy of the region into which the nuclear energy generated in the shell can spread in one oscillation period.

The ratio $\dot{E}_{nuc}/E_{th}$ may be replaced by the local quantity $\dot{\epsilon}_{nuc}/\epsilon_{th}$ (where ϵ_{th} is the thermal energy per unit mass) times a factor $R/\Delta R$ which specifies the spreading of the nuclear energy. These approximations yield $\sigma_I \sim (\alpha \dot{\epsilon}_{nuc}/\epsilon_{th})$ $(R/\Delta R)$. The nuclear lifetime of shell burning is $t_{nuc} \sim q_{nuc}/\dot{\epsilon}_{nuc}$, where q_{nuc} is the nuclear energy release per unit mass. Thus the number of e-foldings during shell burning is $n \sim (\alpha q_{nuc}/\epsilon_{th})(R/\Delta R)$. For Si fusion $\alpha \sim 50$ and $q_{nuc} \simeq 2 \times 10^{17}$ erg g^{-1}, so for a typical shell temperature $T \simeq 4 \times 10^9$ K and density $\rho \sim 10^7 - 10^8$ g cm^{-3} (e.g., Woosley & Weaver 1986), we find $n \sim 50 (R/\Delta R)$. Since ΔR is at most given by the convective speed, v_{conv}, times the oscillation period, and v_{conv} is less than the sound speed, we expect $\Delta R \lesssim R \sim 2 \times 10^3$ km. Thus the number of e-foldings of the oscillation amplitudes can be large. This is an encouraging result.

14.3.2 *Future Directions*

High priority should be given to calculations of core g-modes in sequences of pre-supernova stellar models of ever increasing reality. Ultimately, modes should be obtained for sophisticated models like those computed with stellar evolution codes by Weaver & Woosley (1993) and Nomoto & Hashimoto (1988). Those of the former pair of authors are publically available. Calculations of adiabatic eigenfrequencies and eigenfunctions could be carried out using standard programs. Handling composition discontinuities would require some care. Otherwise this part of the research should be fairly straightforward.

More difficult will be consideration of the potential overstability of the core g-modes. Initial efforts could again rely on quasi-adiabatic calculations of the work integral. Even at this level, several subtle issues will have to be addressed. Three of the more important ones are mentioned below.

Damping Due To Coupling to Acoustic Waves

Internal gravity waves are evanescent in the thick convection zones that surround the shells burning Si and O. This accounts for the existence of core g-modes. However, acoustic waves whose frequencies lie above the acoustic cutoff can propagate through convective regions. Each core g-mode will couple to an outgoing acoustic wave. The higher the mode frequency, the stronger will be the coupling. These outgoing acoustic waves will drain energy from the core g-modes. Consider again a rigid core (density ρ_c, radius R_c) oscillating with frequency σ in a uniform but compressible medium (density ρ). The damping rate due to dipole acoustic wave emission is given by $\sigma_d \sim \sigma(\rho/\rho_c)(\sigma R_c/c_s)^3$, where c_s is the sound speed in the medium. An overstable mode requires driving by the "$\epsilon-$mechanism" to exceed damping due to the outgoing acoustic wave, i.e., $\sigma_I > \sigma_d$.

The leakage of energy from g-modes due to outward propagating acoustic waves can be calculated using radiation boundary conditions. The eigenvalues and eigenfunctions become complex. Since the stability of a mode depends on competition between σ_I and σ_d, a proper treatment of the nuclear burning rate is necessary, especially for modes of low-order and low-degree.

Role of Neutrino Emission

Nuclear energy generated in the Si and O burning shells is balanced by pair neutrino emission, which is the main cooling mechanism until collapse begins in earnest; after that, electron-capture becomes the dominant mode of neutrino production. Since the nuclear energy generation rate depends more sensitively on temperature (power-law index ~ 47 for Si burning and ~ 33 for O burning) than pair neutrino emission (power law index ~ 9), cooling is never comparable locally to nuclear heating. Instead, thermal balance is mediated by the convective transport energy from the shells, where the rate of nuclear energy generation exceeds that of neutrino energy emission, to the cooler surroundings where the bulk of the neutrino emission takes place. Thus, nuclear heating, neutrino cooling, and convection are intrinsically coupled in the burning shells.

Initial estimates indicate that damping due to neutrino emission is always slower than growth due to nuclear burning because the bulk of neutrino cooling occurs in regions where the mode amplitude is small (if the mode is reasonably well trapped in the core). Detailed calculations including more careful treatment of neutrino energy loss are needed to confirm this result.

Role of Convection

For efficient driving, variations in energy generation associated with g-mode oscillations must produce corresponding variations in the temperature of the shell. Both radiative diffusion and heat conduction are too inefficient to smooth temperature variations on the time scale set by the oscillation periods. However, outward energy transport by convection, which is important in the steady state, deserves careful investigation. The nature of convective shell burning is rather subtle because the sound travel time, the convective turnover time, and the nuclear burning time are all of the same order of magnitude; the convective speed can be a significant fraction (~ 0.2) of the sound speed. Multi-dimensional hydrodynamical calculations of these fast evolutionary stages have recently been attempted (Arnett 1994; Bazan & Arnett 1994), but many uncertainties remain. These calculations (on O-burning shells) also indicate the presence of density perturbations of order 5% in the presupernova star.

14.3.3   *Turbulent Excitation of g-Modes*

These effects have undergone scrutiny in connection with the stochastic excitation of the solar p-modes (Goldreich, Murray, and Kumar 1994). By virtue of its intrinsic nonlinearity, stochastic excitation sets a floor to the level of excitation of core g-modes even in the absence of overstability.

Acknowledgments

I would like to thank John Bahcall, Jeremy Goodman, Pawan Kumar, and David Spergel for reading the manuscript and offering helpful suggestions. This research supported in part by the NSF and NASA.

BIBLIOGRAPHIC NOTES

- **Bethe, H. A., & Brown, G. 1985, Scientific American, 252 (#5), 60.** An exciting and simple description of the physics which causes stars to evolve and explode by two masters of the subject.

- **Bethe, H. A. 1990, Rev. Mod. Phys., 62, 801.** A more modern and detailed look at the physics of type II supernova.

- **Manchester, R. N. & Taylor, J. H. 1977, Pulsars (San Francisco: W. H. Freeman and Company)** The story of pulsars up to the time of its publication.

- **Lyne, A. G., & Graham-Smith, F. 1990, Pulsar Astronomy (Cambridge: Cambridge Univ. Press).** An almost up to date description of what we know about pulsars.

- **Cowling, T. G. 1941, MNRAS, 101, 367.** A clear original presentation of the normal modes of a star by the pioneer in the subject.

- **Cox, J. P. 1980, Theory of Stellar Pulsation (Princeton: Princeton Univ. Press).** An authoritative introduction to the theory of stellar pulsation.

- **Unno, W., et al. 1989 , Nonradial Oscillations of Stars (Tokyo: Univ. of Tokyo Press).** A more advanced treatise on stellar oscillations.

- **Goldreich, P., & Weber, S. V. 1980, ApJ, 238, 991. Yahil, A. 1983, ApJ, 265, 1047.** These papers derive analytic descriptions of supernova collapse.

BIBLIOGRAPHY

[1]   Arnett, D. 1994, ApJ, 427, 932.

[2]   Aschenbach, B., Egger, R., & Trumper, J. 1995, Nature, 373, 587.

[3]   Bazan, G., & Arnett, D. 1994, ApJ, 433, L41.

[4]   Bethe, H. A. 1990, Rev. Mod. Phys., 62, 801.

[5]   Burrows, A., & Fryxell, B. 1992, Science, 258, 430.

[6]   Burrows, A., Hayes, J., & Fryxell, B. A. 1995, ApJ, 450, 830.

[7]   Burrows, A., & Hayes, J. 1996, Phys. Rev. Lett., 76, 352.

[8]   Cordes, J. M. 1986, ApJ, 311, 183.

[9] Cordes, J. M., Romani, R. W., & Lundgren, S. C. 1993, Nature, 362, 133.

[10] Cordes, J. M., Wasserman, I., & Blaskiewicz, M. 1990, ApJ, 349, 546.

[11] Cox, J. P. 1980, Theory of Stellar Pulsation (Princeton: Princeton Univ. Press).

[12] Cropper, et al. 1988, MNRAS, 231, 695.

[13] Frail, D. A., Goss, W. M., & Whiteoak, J. B. Z. 1994, ApJ, 437, 781.

[14] Frail, D. A., & Kulkarni, S. R. 1991, Nature, 352, 785.

[15] Goldreich, P., Murray, N., & Kumar, P. 1994, ApJ, 424, 466.

[16] Goldreich, P., & Weber, S. V. 1980, ApJ, 238, 991.

[17] Harrison, P. A., & Lyne, A. G. 1993, MNRAS, 265, 778.

[18] Harrison, P. A., Lyne, A. G., & Anderson, B. 1993, MNRAS, 261, 113.

[19] Herant, M., Benz, W., Hix, J., Colgate, S. A., & Fryer, C. 1994, ApJ, 395, 642.

[20] Herant, M., & Woosley, S. E. 1994, ApJ, 425, 814.

[21] Janka, H.-Th., & Müller, E. 1994, A&A, 290, 496.

[22] Kumagai, S., et al. 1989, ApJ, 345, 412.

[23] Lai, D., Bildsten, L., & Kaspi, V. M. 1995, ApJ, 452, 819.

[24] Lai, D., & Shapiro, S. L. 1995, ApJ, 442, 259.

[25] Lyne, A. G., & Lorimer, D. R. 1994, Nature, 369, 127.

[26] McCray, R. 1993, ARA&A, 31, 175.

[27] Morse, J. A., Winkler, P. F., & Kirshner, R. P. 1995, AJ, 109, 2104.

[28] Müller, E., & Janka, H.-T. 1996, A&A, submitted.

[29] Nomoto, K., & Hashimoto, M. 1988, Phys. Rep., 163, 13.

[30] Papliolios, C., et al. 1989, Nature, 338, 565.

[31] Shu, F. H. 1977, ApJ, 214, 488.

[32] Strom, R., Johnston, H. M., Verbunt F., & Aschenbach B. 1995, Nature, 373, 590.

[33] Trammell, S. R., Hines, D. C., & Wheeler, J. C. 1993, ApJ, 414, L21.

[34] Unno, W., et al. 1989, Nonradial Oscillations of Stars (Tokyo: Univ. of Tokyo Press).

[35] Utrobin, V. P., Chugai, N. N., & Andronova, A. A. 1995, A&A, 295, 129.

[36] Weaver, T. A, & Woosley, S. E. 1993, Phys. Rep., 227, 65.

[37] Woosley, S. E., & Weaver, T. A. 1986, ARA&A, 24, 205.

[38] Woosley, S. E., & Weaver, T. A. 1992, in The Structure and Evolution of Neutron Stars, eds. D. Pines, R. Tamagaki, & S. Tsuruta (Reading, MA: Addison-Wesley).

[39] Yahil, A. 1983, ApJ, 265, 1047.

CHAPTER 15

IN AND AROUND NEUTRON STARS

MALVIN RUDERMAN

Physics Department and Columbia Astrophysics Laboratory,
Columbia University, New York, NY

ABSTRACT

There is not yet a consensus about the basic mechanisms for explaining many kinds
of pulsar observations. These include 1) coherent radio emission properties;
2) the origin of pulsar spin-period glitches and the dependence of their magnitude,
frequency, and healing fraction on pulsar period; 3) the effects of accretion induced
spin-up on a neutron star's magnetic dipole moment; 4) the apparent differences
between the magnetic field structures of the fastest millisecond-pulsars and those
of canonical radiopulsars. The relationship of some of these pulsar properties to
the microphysics of neutron star interiors and problems which remain to be solved
will be outlined.

15.1   INTRODUCTION

Efforts directed toward understanding the structure and dynamics of neutron star
interiors and magnetospheres were an important part of theoretical astrophysics for
most of the first two decades after the 1967 discovery of radiopulsars. Such work
then diminished very considerably, not because a consensus had been reached on
the answers to the many questions raised by observations of solitary neutron stars,
but because of a redirection of attention to other problems (e.g., modeling the genesis and physics of accreting neutron stars in binaries).

There is, for example, still no consensus on the description of important phenomena in regions of the rotating magnetospheres of solitary pulsars extending
from the stellar surface out to thousands of stellar radii. We first describe some

of the unsolved magnetosphere problems and then some of the problems of neutron star interior structure and dynamics. All are directly related to interpreting observations.

a) What is the mechanism for the coherent emission of a radiopulsar beam which accounts for the rich variety of pulse shapes, conal and core beams, polarization patterns, mode switching, subpulse drifting, nulling, giant pulses, and more? This primeval pulsar problem is still not resolved, although much more is now understood about how the needed coherence might be generated.

b) Most (perhaps all) radiopulsars with spin-down powers exceeding several times 10^{34}erg s^{-1} seem to be strong pulsed hard X-ray and/or γ-ray sources. A few of these γ-ray pulsars appear to be converting a very large fraction of their spin-kinetic energy to beams of 10^2 MeV–10 GeV γ-rays. Some authors propose that these γ-rays originate above the polar caps in the corotating near-magnetosphere of the neutron star. Others argue that these beams come from charge deficient regions of the outer magnetosphere. Among this group there is another dichotomy. Some propose emission from near the "light cylinder" but with the plasma-laden magnetic field very similar to the unladen retarded one of a rotating dipole in a vacuum. Others argue for an accelerator and X-ray/γ-ray emission region somewhat nearer the star where the magnetic field approaches that from a non-retarded dipole. The schism is considerable. Among its consequences are very different models for emission beam geometry. Who the "true believers" are has not yet been revealed.

Other unresolved pulsar issues which have observational significance involve the *interior* structure and dynamics of solitary strongly magnetized spinning neutron stars.

c) What is the equation of state of the matter near the center of observed neutron stars? This density greatly exceeds that in an atomic nucleus and there is not yet any laboratory testing of various proposals: we must rely on theoretical estimates of how much putative π-meson and K-meson condensation (the transition from usual neutron-proton-electron degenerate matter into one of baryons, each of which is a coherent combination of nucleon states, together with a boson-condensate of π^- or K^-) soften the equation of state? How much does such softening lower the maximum mass of a neutron star? How does it affect the division between residual neutron stars and black holes after supernova explosions?

d) What is the genesis of the low magnetic field (surface dipole $B \sim 10^9$ G) high spin millisecond pulsars? Are they all spun-up by accretion from low mass companions? Are many formed from white dwarfs which accrete enough mass to exceed the maximum which can be supported by the pressure of their relativistic degenerate electrons?

e) Do neutron star magnetic fields evolve? Do they, for example, grow during the very early post-birth cooling epoch ($\lesssim 10^3$ yrs), perhaps because of the thermoelectric generation of current in the stellar crust? Are they ultimately reduced during prolonged spin-down? Do canonical large magnetic field pulsars (surface dipole $B \gtrsim 10^{12}$ G) become small dipole millisecond radiopulsars after accretion induced spin-up in low mass X-ray binaries (LMXB's)? If so, what causes the dipole reduction? Is it the prolonged spin-up, the large total accretion or just the ohmic decay of currents in the crust of a neutron star?

f) What is the cause of the observed period "glitches" in spinning-down radiopulsars? How can we understand the large variety of observed glitch magnitudes, frequencies, and post-glitch healing within the now quite extensive family of radiopulsars with reported glitches?

g) After fitting the observed time evolution of some pulsar periods with the best values for P (period), $\dot{P}$, $\ddot{P}$ the residual errors in pulse arrival time sometimes oscillate about the best fit more than can be easily understood as an artifact of the fitting program. Evidence for such oscillations with a period of order 10^3 days has been proposed for PSR 1642-03 and the Crab pulsar. Free precession with such a long period is made difficult by the strong coupling between the nuclei in a neutron star's crust and the neutron superfluid vortex lines which must exist within the rotating neutron sea which fills the space between these nuclei. If the oscillations are indeed not artifacts how should they be understood?

h) How are canonical pulsar magnetic fields generated? Is it a coincidence that these pulsars have about the same magnetic flux as the most strongly magnetized white dwarfs? Can some pulsars have magnetic fields orders of magnitude larger than the canonical ones? Could the repeated release of such stored magnetic field energy in certain 10^4 year old neutron stars explain the observed family of "Soft Gamma-ray Repeaters"?

The answers to the questions in d)–f) should be among the easier unsolved neutron star problems: their solutions would seem to depend upon correctly describing neutron star structure and dynamics only above the inner core and up to the stellar surface. This is a region in which the conventional electron-proton-neutron sea description of neutron star matter is adequate and rather well understood quantitatively.

Below, I will sketch some ideas whose development could illuminate aspects of neutron star physics which may be relevant to these answers. A guide to some of the relevant literature is given at the end of this paper. (I would note, however, that if solutions to these problems were already clear the restriction of our agenda to "unsolved problems" would have removed them from further consideration.)

15.2   SUPERFLUID-SUPERCONDUCTOR INTERACTIONS
IN A NEUTRON STAR CORE

A neutron star is more like a giant atomic nucleus than a classical fluid. For many purposes this distinction is very unimportant, but there are phenomena for which the difference is crucial. A giant nucleus would consist of two coexisting, quantum superfluids, neutron and proton seas for which quantum mechanics plays an essential role in describing their dynamics. This does not mean that all of the classical magnetohydrodynamical descriptions are invalid. Rather, the infinite number of possible classical motions, one for each of an infinite number of possible initial conditions, is hugely constrained for these quantum fluids but the subset which is allowed is not increased by any new fluid modes which are unique to the quantum fluids. (The situation is somewhat analogous to the semi-classical Bohr-Sommerfield description of atoms: only those classical orbits which happen to possess appropriate values of the classical action are allowed, but no new orbits are introduced.) One can describe most of the dynamics of the neutron and proton fluids in spinning neutron stars without recourse to quantum mechanics except for the restrictions it brings in determining what classical magnetohydrodynamical motions are allowed. Ultimately, however, an understanding of the Bardeen-Cooper-Schrieffer (BCS) success in describing canonical electron superconductivity, and its application to neutral fermion systems, is needed for a complete description of what happens below the surface of a spinning, magnetized neutron star. It is not, however, at all essential for reading the sections below or the references given in the End Notes.

A neutron star is mainly a 10 km radius sphere of neutrons held at near nuclear densities by its own gravitational attraction. The same kind of BCS theory that is so successful in describing canonical electron superconductivity and He^3 superfluidity indicates quite unambiguously that at such densities the neutron sea of a neutron star must be a superfluid. Predicted energy gaps in the excitation spectra of such superfluids are confirmed in measurements of energy level spacings of excited states of heavy nuclei. Because the neutrons form a superfluid their allowed, incompressible, hydrodynamic motions are restricted to the irrotational ones. The neutrons within a neutron star mimic rigid rotation by establishing a dense array of vortex lines (each of which has the same quantized circulation) aligned parallel to the stellar spin (Ω). For a spin-period P they are spaced a distance $10^{-2} P$ (sec)$^{1/2}$ cm apart. When the star spins-down (up) each vortex line displaced by $r_\perp$ from the spin axis Ω moves outward (inward) with a velocity

$$v = r_\perp \dot{\Omega}/2\Omega. \qquad (15.1)$$

For the rapidly spinning-down Crab pulsar $v \sim 10^{-5}$ cm s^{-1}. Coexisting with the superfluid neutrons and their rotating quantized vortex array a neutron star are protons and a neutralizing sea of ultrarelativistic degenerate electrons whose abundances are typically a few percent that of the neutrons. The same robust arguments that imply the superfluidity of the neutron sea insist upon the superconductivity of the proton one. Because the effective potential between a neutron and a proton is velocity dependent (often approximated by assigning a proton in a nuclear density neutron sea an effective mass $m_p^* <$ the free proton mass $m_p : m_p^* \sim 0.7 m_p$) the motion of the rotating superfluid neutrons in a pulsar is closely coupled to that of the superconducting protons. One consequence is a superconducting proton current sheath around each neutron vortex line which encases the vortex line out to distance $\Lambda \sim 10^{-11}$ cm (the superconducting protons' "London length"). Inside that solenoidal current sheath there is a magnetic field exceeding 10^{15} G parallel to the neutron vortex line. Within a kilometer or so of the stellar surface the superconducting protons, together with some of the superfluid neutrons, clump into neutron rich nuclei which form the crystalline lattice of a pulsar's solid crust. The rotating neutron superfluid which fills the space between the nuclei still contains an array of quantized vortex lines. However, because of the absence of superconducting protons around them, there is no longer a strongly magnetized core along the center of each vortex line from induced proton currents.

A crucial question for understanding neutron star evolution is "What is carried outward (inward) by the expanding (contracting) superfluid neutron vortex array of a spinning-down (up) neutron star?" For example neutron vortex lines in the crust can be strongly pinned on crust nuclei. To move outward during spin-down do they unpin one at a time? Do they unpin collectively in bunches? Do they break the crust before unpinning? Could parts of the pinning crust move outward or inward with their pinned neutron vortex lines despite the strong constraints from crustal matter stratification? Most significantly what is the effect on the stellar magnetic field of changing the neutron star spin rate and, necessarily, the vortex spacing in the neutron superfluid vortex array?

A magnetic field which exists below the neutron star crust must somehow pass through the superconducting protons. But superconductivity and magnetic fields are largely incompatible. The same arguments that exclude incompressible superfluid neutron motions with $\nabla \times \underset{\sim}{v} \neq 0$ (except in central cores of vortex lines) allow only $\nabla \times \underset{\sim}{A} = \underset{\sim}{B} = 0$ wherever charged protons are superconducting (except within the cores of certain vortex-like structures which, when they exist, contain magnetic flux). A magnetic field passing through a superconductor will either form an array of very strongly magnetized "flux tubes" separated by spaces in which $B \sim 0$ ("type II" superconductor) or a mixed state ("type I" superconductor) in which B becomes large enough to quench superconductivity in some regions and vanishes in between them. In either case the field must become bunched into microscopic regions with huge $B(> 10^{15} \text{ G})$ separated by larger essentially field free ones. The necessarily strongly inhomogeneous structure of a magnetic field which passes through a proton superconductor is one reason, and probably the main one, why an outwardly or inwardly moving array of quantized neutron vortex lines sheathed by their own superstrong solenoidal fields, strongly pushes on the rest of the core's magnetic field structure. Details of that structure can determine when the neutron vortex array motion from spin-down or spin-up drags with it the stellar magnetic field below the crust.

Just below the stellar crust neutron star protons probably form a Type II superconductor: the magnetic field is organized into an array of quantized flux tubes whose structure, radius, and field strength are similar to those of the solenoidal sheath which forms around each neutron vortex line because of induced currents in those same protons. The quantized flux $\Phi_0 = 2 \cdot 10^{-7} \text{G cm}^2$ and the flux tube field $B_c \sim \Phi_0/\pi\Lambda^2 \sim 10^{15}$ G. The flux tube number per unit area ($n_\Phi = B/\Phi_0 = 5 \cdot 10^{18} B_{12} \text{ cm}^{-2}$) in a pulsar is typically 15 orders of magnitude greater than the area density of its neutron vortex lines. (And the total magnetic flux carried by these flux tubes is, similarly, over 15 orders of magnitude greater than that carried within the star's neutron vortex line sheaths.)

A moving neutron vortex array will push on the proton flux tubes which carry the stellar magnetic field through the neutron star core. The key questions which are far from fully answered concern how those flux tubes respond to that push. If the flux tubes are fairly uniformly distributed, the magnetic field complex behaves like a uniform field embedded in a conductor whose main ohmic dissipation comes from the scattering of the star's extreme relativistic electrons on phonons and other microscopic scattering irregularities.

There are several possibilities for flux tube response to the neutron vortex array push during stellar spin-down or spin-up.

1) Electron scattering (probably mainly on the departures from a smooth field from the flux tubes themselves) and proton flow very near the central axis of each flux tube where superconductivity effectively vanishes may give enough dissipation to allow very slow movement of the flux tube array with the vortex array. This is much more easily accomplished for the very slow ($\sim 10^{-9} \text{ cm s}^{-1}$) inward motion in neutron stars being spun-up slowly by accretion in LMXB's than in the relatively fast ($v \sim 10^{-6} \text{ cm s}^{-1}$) outward motion in young spinning-down solitary pulsars.

2) The charged ($e - p$) component of the stellar interior moves together with its flux tubes, so that no relative motion develops between them. The virtual incompressibility of this charged component (sound speed $\sim c/6$) and the essential

absence of $e + p \leftrightarrow n$ conversion tends to restrict such motions to the very limited allowed ones for incompressible flow on equipotential surfaces.

3) The magnetic flux tubes are bunched by the (flexible) vortex lines moving through them until there is back-flow between bunches.

4) The proton superconductor is Type I rather than Type II. The magnetic field is organized into regions within which B is large enough ($B > 10^{15}$ G) to quench superconductivity. This will occur where the $p - p$ attraction is not strong enough to give a radius for the superconductor's "Cooper pairs" less than the protons' London length (Λ). In the large superconducting regions between these "quenched" regions $B \sim 0$ so that $e - p$ flow there is not restrained by a magnetic field; $\boldsymbol{B}$ and the conducting $e - p$ seas still move together in the quenched "normal" regions but backflow compatible with 2) above takes place between such regions. The expected microscopic geometrical scale for alternating normal ($B > 10^{15}$ G) and superconducting ($B = 0$) regions is a geometric mean between the sub-microscopic $\Lambda(\sim 10^{-11}$cm) and the macroscopic length scale determined by that of the average B and the stellar structure ($\lesssim 10^6$ cm). (If, as seems quite possible, the superconducting protons are Type II just below the crust but become Type I far below it, there will already be some flux tube bunching in the Type II region to accommodate joining on to the Type I region below; cf. Fig. 15.1).

5) The flux tubes do not move significantly. The vortex line force on them then builds up until the neutron vortex lines cut through the flux tubes.

6) The presumed core magnetic field of order 10^{12} G is a huge overestimate: the core field is so small flux tubes hardly exist in the core.

Clearly the interior structure and dynamics of a pulsar are far from transparent.

Below we shall consider observable consequences of the coupling between a neutron star's moving neutron superfluid vortex array and the necessarily microscopically very inhomogeneous magnetic field below the stellar crust if that coupling does indeed move the magnetic field relative to the crust. At present this appears to me to be a reasonably well supported presumption for slow accretion induced spin-up of a neutron star but it needs more quantitative investigation for justification in rapidly spinning-down pulsars. The descriptions of §§ 15.4 and 15.5 below which are most emphasized should certainly not be represented as the consensus of those who work on this subject.

The magnetic field at the stellar surface (and beyond) directly affects observations. This surface is separated from the core by the strongly conducting stellar crust so that we must also understand some of its properties before relating core field motion to pulsar observations.

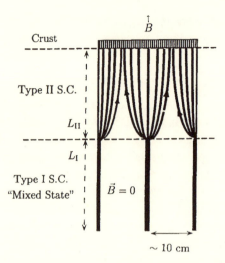

Figure 15.1: Schematic representation of the magnetic field organization expected to exist below a neutron star crust. In the Type II region the magnetic field is carried by flux tubes of radius $\Lambda \sim 10^{-11}$ cm initially separated by about 10^{-9} cm. In the Type I (mixed state) region slabs in which a magnetic field exceeding several $\times 10^{15}$ G quenches proton superconductivity are probably centimeters apart. The space between them is essentially field free. The lengths L_{II} and L_I are uncertain but likely to be of order several kilometers. In the crust the magnetic field spreads into a quasi-uniform one slightly squeezed by the diamagnetic nuclei there.

15.3   THE STELLAR CRUST

The one kilometer thick stellar crust is far below its melting temperature ($\sim 10^{10}$ K at the crust base) and for all except the very youngest pulsars forms a rather brittle shell around the neutron star's superconducting core. Because of its own extreme relativistic degenerate electron sea, the crust is an extremely good (conventional) electrical conductor. On time scales less than 10^7 years (and possibly very much longer depending upon the crustal impurities and temperature), spin-driven changes in the core magnetic field can cause similar changes at the stellar surface only when the shearing stress from moving core flux tubes whose fields are anchored in the crustal base exceed the crustal lattice yield strength. The flux tube pull must break the lattice since the crustal magnetic field which anchors and is fed from the flux tube array cannot move with respect to the local crustal matter. The crustal lattice motions which can result are more or less restricted to two-dimensional incompressible displacements on equipotential surfaces because of the strong stratification of crustal matter whose nuclear atomic number varies with electron pressure. The gravitational pressure is much greater than the field induced stress. The stresses from moving strong core fields do appear to be sufficient to move parts of the crust, accompanied by backflow on the equipotential surfaces of solitary spinning-down pulsars. Accreting spinning-up neutron stars can accomplish crust motion to accommodate that of moving core fields in other ways.

An additional source of crustal stress is the pinning by crust nuclei of the neutron superfluid vortices which must exist in the rotating superfluid neutrons which fill the interstitial spaces between these nuclei. As long as these vortices are pinned they are restrained from moving outward (inward) in response to crustal spin-down (up). This, at least temporarily, keeps the crustal neutron superfluid from spinning-down (up) with the crust (and the other components of the star which are strongly coupled to the crustal lattice and electrons). The resulting (Bernoulli) force on the crustal lattice from the flow of neutron fluid past its own pinned vortices stresses the crust in the same average direction as that from any spin-change induced core flux tube motion. The reinforced sum of these two crustal stresses is expected to be greatest where the surface field is greatest. This surface region should then be continually moved by this stress toward the spin-axis in spinning-up neutron stars. This is the simplest case of expected surface field evolution both for estimating the validity of the underlying microphysics modeling and its observational consequences.

15.4   Spun-up Neutron Stars

If millisecond pulsars are indeed initially canonical strongly magnetized neutron stars subsequently spun-up by prolonged accretion from a companion, then surface regions where magnetic flux enters or leaves should be squeezed to the spin axis for two reasons (cf. Fig. 15.2).

a) The core field is very slowly pressed to the spin axis by the large density of inward moving neutron vortex lines ($> 10^6$ cm^{-2} for a millisecond pulsar) even if mechanism 1) of § 15.2 is the only effective way of allowing such flux tube motion. A key circumstance is the long ($10^7 - 10^8$ yr) interval during which spin-up continues.

b) Because of the mass flow brought onto the surface by the accretion, the crustal lattice stress from pinned neutron superfluid vortex lines will be sufficient continually to move the crust and the magnetic field anchored in it toward the spin axis. Much more than 10^{-1} $M_\odot$ must be slowly accreted to spin-up a neutron star to a millisecond period. The crustal lattice will then be pushed down into the core and replaced over 10^2 times during the long accretion epoch. Simultaneously, the pinning force (toward the spin axis) on the lattice would continually displace new crust toward the spin axis. Crust motion is no longer constrained by stratification to essentially spherical shells.

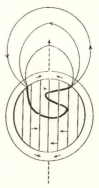

Figure 15.2: Schematic representation of structure in a spinning-up neutron star. The array of (magnetized) neutron superfluid vortex lines, with separation of order 10^{-2} cm, moves in toward the spin-axis. A small bundle of quantized flux tubes in the core is carried inward with them. In the crust, whose thickness is 10^{-1} the stellar radius, the flux tubes open and their fields merge.

Of special interest is the evolution of any initial magnetic field structure topologically similar to the sunspot-like configuration of Figure 15.3. During spin-up the surface dipole field is continually reduced as North and South magnetic poles are pushed toward each other and toward the spin axis (because of both core flux tube and crust motion). The more the spin-up, the smaller the dipole. But the smaller the dipole the faster the final spin of an accretion spun-up star. Through this symbiosis the fastest millisecond pulsars ($P \lesssim 2$ ms) can be formed with about the observed dipole field. Their resulting final magnetic field configuration is quite unique—an orthogonal dipole moment located on the spin axis very close to the

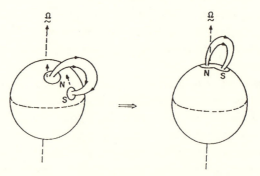

Figure 15.3: Evolution of the magnetic field of a spinning-up short-period neutron star when all the flux leaving a hemisphere reenters the same hemisphere ("sunspot" configuration).

neutron star surface as in Figure 15.4. This would give certain very special properties for the pulsar radiation for the fastest millisecond pulsars PSR 1937+24 and PSR 1957+20, both with periods $P \sim 1.6$ ms.

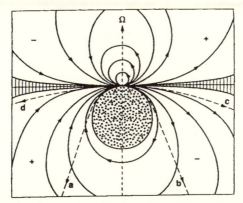

Figure 15.4: Magnetosphere of a strongly spun-up diamagnetic (because of core proton superconductivity) neutron star with the initial field geometry of Figure 15.3. One fan beam (cb) passes through electron filled magnetosphere, the other (da) through ions.

1) Such pulsars should have two subpulses of comparable intensity separated by an interval of half a period. Both have just this property (cf. Fig. 15.5) [Of 558 radiopulsars only 16 have large interpulses (relative amplitude $> 10^{-1}$) with a separation exceeding $P/3$. Of these 8 are millisecond pulsars with $P \lesssim 5$ ms.]

2) The subpulses should have anomalously narrow pulse widths. In most pulsar radiation models the minimum pulse width is determined by the angular spread of the magnetic field lines which leave the stellar surface and never return to it ("open" field lines). For a central dipole that angular width $w \sim (\Omega R/c)^{1/2}$, and the minimum observable pulse width is $w/\sin\theta$ with θ the angle between Ω and the dipole moment μ. For the field geometry of Figure 15.4 $\sin\theta = 1$ and the stellar radius R is replaced by the much smaller separation between the N and S poles which have been pushed almost to the spin axis. Observed widths in PSRs 1937 (Fig. 15.5) and 1957 are indeed anomalously narrow with about the expected magnitude.

3) Such millisecond pulsars seem to be much more frequently observed than would be the case if their radiation beams are narrow cones. However, gravitational bending of emitted beams (Fig. 15.4) is particularly important if such beams are emitted tangentially to the stellar surface as in the cases here. When the emitted beams originate close to the spin-axis the very narrow cone is stretched into a very

narrow fan, preserving the small pulse width but giving a high probability for beam detection.

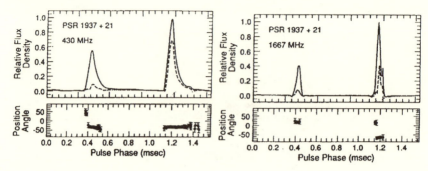

Figure 15.5: Pulse shape and polarization for PSR 1937+214 at 430 and 1667 MHz, respectively. The dashed line is the linearly polarized intensity; its position angle is shown below (Thorsett, E. 1991, Ph.D. Thesis, Princeton University).

4) The gravitationally bent observed subpulse beams from the solitary pulsar PSR 1937+214 pass through the plasma held on "closed" field lines near the neutron star. Because of the special geometry ($\Omega \cdot B$ has opposite signs for each beam) one beam passes through a magnetized electron plasma, while the second beam passes through a plasma of exclusively positively charged ions (Fig. 15.4). One consequence should be an almost complete linear polarization (with a constant polarization angle) for the first beam (but not the second) as long as the observing radiofrequency $\nu \lesssim 1$ GHz. In addition, the polarization angle of the second beam and also of the first, when $\nu > 1$ GHz, should be observed to have 90° "jumps." Again this is just what is observed for PSR 1937+214 (Fig. 15.5).

The pulsar PSR 1957+20 has a companion which eclipses it once *per* revolution. Therefore, this pulsar is probably being viewed by us from a direction very much closer to perpendicular to the pulsar orbital plane's normal and spin axis than is the case with respect to the spin axis of PSR 1937+214. Again observations of it do not conflict with what might be expected from a near orthogonal dipole on the spin axis near the stellar surface for this special viewing geometry.

Because properties of the two fastest radiopulsars agree so well with an orthogonal surface dipole model this description seems to have strong observational support. Then we have the problem of why they have this very strange and seemingly rare configuration. Would it be accomplished by the accretion induced collapse of a white dwarf to a neutron star? Could accretion alone somehow give it by surface field burial or thermoelectric destruction of a strong initial field by the accretion heat flow into the star at its polar caps? Could crustal field decay lead to it? In the absence of suggestions of how such models could give this unique field geometry, it is reasonable to take the implied magnetic field geometry as support for the model in which spin-up itself forces the stellar crust and field to move in just the way which leads to that configuration. If so, we should then consider the more difficult problem of what happens in a solitary spinning-*down* neutron star where there is a much shorter time scale for such motion.

15.5 SPINNING-DOWN RADIOPULSARS

Does the outward moving neutron vortex array within a spinning-down pulsar move core and crust magnetic field oppositely to that motion which seems to accompany the inward moving vortex lines in a spinning-up neutron star? Because the spin-down times (10^3 yrs for the Crab pulsar) can be so much shorter than accretion spin-up epochs, the role of mechanism 1) of § 15.4 in allowing such motion becomes more problematic and the other possibilities need further exploration. However, there are some clues that would seem to support the view (or at least not contradict it) that a spinning-down pulsar's magnetic field may evolve in just the "expected" way.

The spin-down rate of a spinning magnetic dipole moment (μ) in a vacuum is

$$\dot{\Omega} = -\frac{\mu^2 \Omega^3}{I c^3} f(\hat{\mu} \cdot \hat{\Omega}), \qquad (15.2)$$

with I the moment of inertia and f, of order unity, a simple function of the angle between the dipole direction ($\hat{\mu}$) and that of the spin ($\hat{\Omega}$). Of course the pulsar and especially the currents which flow into and out of it through its almost corotating magnetosphere, constitute a more complicated system but equation (15.2) generally survives. Its robustness is largely a consequence of dimensional analysis considerations. If μf and I are constant

$$\dot{\Omega} = -A\Omega^n \qquad (15.3)$$

with $n = 3$. The "spin-down coefficient" n is difficult to measure and has been meaningfully determined for only three pulsars, all young, relatively short period ones which are spinning down rapidly: the Crab pulsar has $n = 2.5$; its close relative PSR 0540 has $n = 2.0$; and PSR 1509 has $n = 2.8$; all are substantially less than 3. From equation (15.2) either I or μ is changing with Ω. Oblateness changes of spinning-down Crab-like pulsars are very much too small to give such large values for $3 - n$. It might be possible to achieve them by a redistribution of the differential rotation of neutron superfluid caused by changes in crustal vortex line pinning region locations. Such values for $3 - n$ are, however, just what would be expected for field changes from the spin-down of a sun-spot configured magnetic field if spin-down does indeed move core field and, especially, crust, as discussed in § 15.3. [It should be noted that expelling all or most of the magnetic field from the stellar core does not mean that it is then expelled from the star. The crust is a good but not perfect conductor. Field expelled from the core into the crust will slowly diffuse out of the crust—probably on a time scale exceeding several million years, when pulsar periods have slowed to of order a second. There is a competition between losing field by Eddy diffusion in the crust and building up field in the inner crust until the resulting stresses break the crust and allow reconnection. In either case the field does not remain indefinitely. If it did, solitary radiopulsars with periods of several seconds would tend to have orthogonal dipoles. The retention of magnetic field by very slowly rotating X-ray pulsars in accreting massive binaries (whose periods can approach 10^3 s) would be most simply attributed to an Eddy diffusion time out of the crust which exceeds the age of the neutron star.]

If the stellar crust does move in response to spin-down induced stresses—especially from the pull on it by the magnetic flux tubes which it anchors—we are led to consider in more detail just how the crust moves. Does it "flow" smoothly, or crumble, or does it crack with observable consequences?

15.6   GLITCHES OF RADIOPULSAR SPIN PERIODS

It is, of course, tempting to try to understand the sudden small radiopulsar spin-ups ("glitches") as a consequence of possible crust-cracking from the spin-down induced crust and field motions discussed above. There are now 34 published glitches and their number will undoubtedly increase greatly. The presently reported glitches are shown for pulsars with various spin periods in Figure 15.6. This collection has the following properties.

a) Among the youngest and shortest period pulsars (periods and ages less than those of the 10^4 yr old Vela pulsar with $P = 0.09$ s) glitches are rare and, when seen, small.

b) Large glitches ($\Delta\Omega/\Omega \gtrsim +10^{-6}$) begin with older pulsars but none have been observed for pulsars with $P > 0.7$ s.

c) Large and small glitches can occur in the same pulsar.

d) There is generally less spread among large (or small) glitches from a single pulsar than the differences between glitches from pulsars with very different periods.

The effect of glitches on the spin-period history of the Crab pulsar is shown in Figure 15.7. (The observed rotation frequency differences are plotted relative to a best fit for Ω, $\dot{\Omega}$, and $\ddot{\Omega}$ for the first few years of pre-glitch data.) Permanent changes in spin-down rate, $\Delta\dot{\Omega}/\dot{\Omega} \sim 3 \cdot 10^{-4}$, left after the larger Crab glitches, are the main consequence of such glitches. The sign of this change is just what would

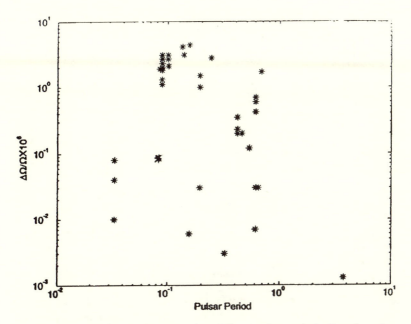

Figure 15.6: Pulsar glitch magnitudes ($\Delta\Omega/\Omega$) vs. pulsar periods. Data from Lyne, A., Pritcher, R., & Shemer, S. 1995, Proc. Indian Acad. of Sciences, in press (plus one anomalously small Vela glitch).

be expected if a warm Crab crust, moving as discussed in § 15.5, occasionally relaxes part of its stress by some form of cracking. Indeed, the estimated temperature for its crust (several $\cdot 10^8$ K) is also about that where a transition would be expected between plastic flow (creep) and brittle crust breaking. A sudden crust movement (Δs) would increase the dipole moment (and spin-down torque) enough for $\Delta\dot\Omega \sim 3 \cdot 10^{-4}\dot\Omega$ if

$$\Delta s \sim 3 \cdot 10^{-4}\, R. \tag{15.4}$$

If the cooler quite brittle crust of the Vela pulsar responds to stresses beyond its yield strength exclusively by such cracking, then the time between Vela glitches should be

$$\tau_g(\text{Vela}) \sim \left(\frac{\Delta s}{R}\right) \cdot \left(\frac{\Omega}{\dot\Omega}\right) \sim 3 \text{ years}.$$

This is about the observed interval between Vela's glitches.

Whether or not all or most substantial glitches in pulsars other than the Crab are accompanied by similar changes in their magnetic fields is still unknown. Either the glitches follow each other too closely to allow a simple determination of what

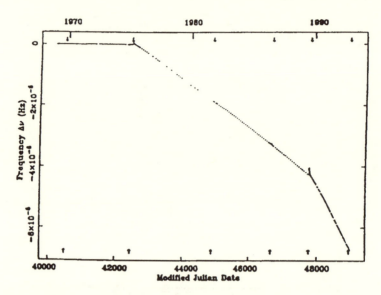

Figure 15.7: The rotation frequency of the Crab pulsar, PSR 0531+21, over a 23-year period, relative to a slow-down model fitted to the first few years of data (Lyne, et al., *loc cit*).

unhealed changes have occurred in $\Delta\dot{\Omega}/\dot{\Omega}$ from any one glitch (the case for Vela) or the post-glitch healing times are longer than present observation times.

Glitches are generally identified by their sudden period drops. There is considerable consensus on the cause for the observed spin-ups. Something causes a sudden release of some of the pinned vortex lines in the stellar crust. These then move outward to new equilibrium positions as the involved crustal neutron superfluid spins-down and consequently, to conserve angular momentum, the rest of the star spins up. One way to accomplish a simultaneous release of pinned crustal vortex lines may be by some kind of relatively fast crustal rearrangement ("cracking") which relieves some of the shearing energy stored in the crust from vortex and flux tube pinning. Another may be purely hydrodynamic—if a few released vortex lines can initiate the release of more to build up a vortex release avalanche. Both models have a similar problem. For the crust cracking it is why there is such an extreme event. Why does a crust not begin to "crack" locally on microscopic scales just as soon as its yield strength is reached locally? Why does it allow "global" cracks to develop instead of releasing stress earlier, slowly, bit by bit? Similarly for the hydrodynamic avalanche, why does a pinned vortex array allow such large potential instabilities to build up until they are released simultaneously in a relatively large global event?

In either case, as long as pinned crust vortex lines do not move from their pinning sites except during glitches, the "glitch activity parameter"

$$\frac{\Delta\Omega}{\Omega} \cdot \frac{1}{\tau_g} = \frac{I(n - SF \text{ in crust})}{I}\frac{\dot{\Omega}}{\Omega} \sim 10^{-2}\dot{\Omega}/\Omega, \tag{15.5}$$

where $I(n - SF)$ is the moment of inertia of the crustal superfluid whose angular velocity is determined by glitch released crustal vortex lines. (All pinned vortex lines need not be released in any one glitch.) Equation (15.5) gives a good fit to the glitch activity in the several pulsars with enough glitches to test the relation for them individually (e.g., Vela and PSR 1737). The observed glitch activity for all pulsars is shown as a function of pulsar age ($\Omega/2\dot{\Omega}$) in Figure 15.8. Again the fit to (15.5) is quite satisfactory except for the warmer youngest pulsars (age $\lesssim 10^4$ yrs) where plastic flow may be expected to relieve most crustal stress before cracking does and for the older pulsars ($P \gtrsim 0.7$ s, age $\gtrsim 3 \cdot 10^6$ yrs) where crust-stressing core flux tubes should have been expelled into the crust when spin-down has proceeded that long. The fact that glitch activity so well fits (15.5) seems a particularly compelling argument that crustal vortex line unpinning is the essential feature of at least the large glitches and perhaps of all glitches. So far, however, there has not been any observation which seems to discriminate definitively between possible causes for the unpinning. An outstanding problem of glitch modeling is to discover some characteristic, observable features that would distinguish them.

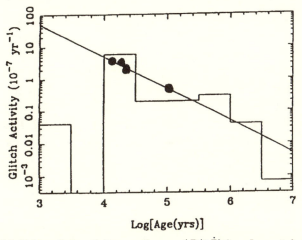

Figure 15.8: Pulsar glitch activity vs. pulsar age ($P/2\dot{P}$) from Lyne et al., *loc cit*. The individual pulsars designated by dots are PSRs 0833, 1758, 1737, 1823. The diagonal line is the glitch activity from (15.5).

Modeling glitches as unpinning events (or perhaps even as pinned vortices moved by the motion of cracking crust) may involve families of glitches with greatly differing magnitudes. The pinning parts of the crust are highly stratified into several layers with different nuclear species, pinning strengths, and neutron superfluid moments of inertia. If glitches occur independently in these layers a variety of glitch magnitudes similar to those observed seems possible.

Although there has been a considerable and impressive effort to describe exactly what happens to a neutron star when (for whatever reason) and after pinned crustal vortex lines are released, much remains to be explored. A particularly important unresolved glitch problem is the coupling between the core's superfluid neutrons and the rest of the star. In the many published analyses of post-glitch healing it is sensibly assumed that the crustal lattice and the core electron-proton sea spin-up or spin-down together because they are so strongly tied to each other by the strong magnetic field. A problem may arise in the additional assumption that the core's superfluid neutrons share in that motion without any resolvable time lag. The above discussion about core vortices pulling core flux tubes with them, or at least interacting strongly with these flux tubes when they move toward (away from) the spin axis in a spinning-up (down) neutron star raises a significant problem when these same neutron vortices are postulated to move relative to the spin-axis exceptionally rapidly in glitches. (For the Vela radiopulsar, which was actually observed during one of its glitches, the additional angular momentum acquired by core neutrons implies that their vortices must have temporarily acquired an inward velocity $\gtrsim 10^{-4}$ cm s^{-1} if the core superfluid neutron spin-up was indeed unresolved.) But there are 10^{15} magnetic flux tubes in the superconducting core protons for each neutron vortex line. How does the neutron vortex array move so rapidly (through a distance of about one cm) when it is so intimately surrounded by this very dense flux tube forest? Cutting through this dense flux tube array seems to take a much larger force than can be mobilized during the brief time available. Sufficiently rapid motion of flux tubes through the $e - p$ sea seems too limited by the small electron ohmic dissipation. Some conceivable ways of avoiding these alternatives were listed in items 2) 3) and 4) [or even 6)] of § 15.2. At this time all that is clear is that this important problem does not yet have a clear answer.

ACKNOWLEDGMENTS

I am grateful to J. Bahcall, P. Goldreich, A. Loeb, and A. Reisenegger for their very useful comments. This work was supported in part by NASA grant NAG5-2841.

BIBLIOGRAPHIC NOTES

So far, the only observable consequences of the structures and phenomena which are thought to exist within a neutron star are their small effect on pulsar spin periods. To exploit this connection it is necessary to interpret spin period data in terms of relatively detailed theoretical models. The construction of such models involved quite a variety of physics subfields. In addition to classical hydrodynamics and plasma physics these include quantum fluids, superconductivity, solid state physics, nuclear physics, general relativity, phase transitions, and some metallurgy and particle physics. The needed level of expertise is generally far less than that of a specialist. A graduate student in physics should probably have been exposed to most of the relevant background in courses, but this is less likely to be the case for a student in astronomy. This is an important reason for most of the investigations of what may be happening below the surface of a neutron star coming from a relatively small number of theoretical physicists who also work in astrophysics (and who usually work only occasionally on this problem). It is because the total effort is small and new observations alone seem unlikely to be decisive, that it is difficult to estimate a time scale on which an adequate description of the subsurface dynamics and structure of a neutron star will be achieved.

This brief survey is not meant for experts who already have much more complete lists of references than would be appropriate here. I have not, therefore, even attempted to reference the many topics mentioned as an abbreviated list would not do justice to all who should be on it. In addition, because the emphasis is on unsolved problems, much relevant work is not yet published. Included below are only some material familiar to the author which students might find useful as an introduction, and which generally do contain extensive references.

1. *Introduction.* a) What is known about radiopulsars is admirably summarized in *Pulsar Astronomy* by A. Lyne & F. Graham-Smith (Cambridge Univ. Press, 1990) and in the eagerly awaited sequel to J. Taylor & R. Manchester's *Pulsars* (W. H. Freeman, 1977) which will soon be available. A masterful short survey of mechanisms for coherent radioemission is given by D. Melrose in the proceedings of the Bangalore Discussion Meeting on Pulsars, Proc. Indian Acad. of Sciences, 1995. These proceedings, edited by G. Srinivasan, consist entirely of invited survey papers. Together they constitute a good collection of what workers in the field consider to be the important problems. b) Research on pulsar magnetospheres is well discussed by F. C. Michel in *Theory of Neutron Star Magnetospheres* (Univ. of Chicago Press, 1991). An excellent compact summary of high energy emission processes and their locations is given by P. Meszarós in chapter 8 of his *High Energy from Magnetized Neutron Stars* (Univ. of Chicago Press, 1992). Polar cap models for γ-ray pulsars are discussed in many papers by J. Daugherty, A. Harding, C. Dermer, S. Sturner, & others. Outer-magnetosphere accelerator authors include K. S. Cheng, C. Ho, M. Ruderman, A. Cheng, P. Sutherland, C. Chiang, & R. Romani. c) An introduction to neutron star structure including π-meson

condensates is *Black Holes, White Dwarfs, and Neutron Stars* by S. Shapiro & S. Teukolsky (John Wiley & Sons, 1983). G. Brown & J. Lattimer (SUNY at Stony Brook) and collaborators have more recently emphasized a possible extremely low mass upper bound for neutron stars when K-meson condensates are included in the equation of state of the central core. g) The evidence for such oscillations and the kinds of models needed to explain them may have been too cavalierly put aside. The Crab evidence is from Lyne, A., Pritchard, R., & Graham-Smith, F. 1990, MNRAS, 242, 370; that for PSR 1642-03 and other radiopulsars is summarized in Cordes, J., & Downs, G. 1985, ApJS, 59, 343. Cordes has also considered the relationship of such data to the detection of planets around pulsars.

2. *Superfluid-superconductor interactions.* A good survey is by J. Sauls (1989) in *Timing Neutron Stars*, ed. H. Ögelman & E.J.P. van den Heuvel (Dordrecht: Kluwer). The initial suggestion of core field expulsion by an expanding vortex array was by Srinivasan, G., Bhattacharya, D., Muslimov, A., & Tsygan, A. 1990, Current Sci., 59, 31. Additional discussion is in M. Ruderman 1991, ApJ, 382, 576.

3. *The stellar crust.* Some properties are summarized in M. Ruderman 1991, *loc cit* and ApJ, 382, 587. The original suggestion of neutron superfluid vortex pinning by the lattice in a neutron star crust was by P. Anderson, & N. Itoh 1975, Nature, 256, 25.

4. *Spun-up neutron stars.* This section is based upon K. C. Chen & M. Ruderman 1993, ApJ, 408, 179.

5. *Spinning-down radiopulsars.* Spinning-down neutron stars are discussed from this point of view in M. Ruderman 1991, *loc cit*, § 5.

6. *Glitches of spin periods of radio pulsars.* A current summary of all pulsar glitches (April 1995) and many glitch properties is in Lyne, A., Pritchard, R., & Shemar, S., Proc. Indian Acad. of Sciences 1995, in press (Jodrell Bank preprint CP 25/94). Details of a Vela glitch observed as it occurred were reported by, for example, C. Flanagan, 1991, IAU Circ. No. 5311. Discussion of consequences of sudden vortex line unpinning in the Vela radiopulsar crust and fits to observed glitch parameters are given in an extensive series of papers by Alpar, Chau, Cheng, Pines, & Shaham. A recent one with extensive references to their earlier work is Alpar, M., Chau, H., Cheng, K.S., & Pines, D. 1993, ApJ, 409, 345. Consequences of crust stratification for glitch families of different magnitudes in starquake models are considered in a doctoral thesis by T. Zhu (Columbia 1996). Radiopulsar data contain a great deal of additional information on microglitches and timing noise. The kinds of neutron star phenomena which underlie them have not yet been well understood.

CHAPTER 16

ACCRETION FLOWS AROUND BLACK HOLES

RAMESH NARAYAN

Harvard-Smithsonian Center for Astrophysics, Cambridge MA

ABSTRACT

Why do accreting black holes emit a substantial fraction of their luminosity in hard X-rays and γ-rays? This article reviews the constraints placed by observational data on theories of accreting black holes and summarizes the accretion processes. The principal features of hot accretion models are discussed, with special attention to the optically-thin advection-dominated model. Promising directions for future research are outlined. It might be fruitful to make a systematic search for stable equilibrium flows and to try and match theoretical models to the wealth of observational clues on X-ray binaries and active galactic nuclei.

16.1 INTRODUCTION

Relativistic stars such as black holes and neutron stars play a central role in modern astrophysics. The study of accretion flows around relativistic stars is an especially exciting branch of astrophysics, as it includes both black holes and neutron stars and spans objects ranging from stellar mass accretors to supermassive black holes. This subject is particularly appropriate for this collection, since there are basic unsolved problems in understanding the physics of accretion.

In this article we concentrate on accreting black holes. A small but rapidly growing subset of X-ray binaries (XRBs) has been identified as black hole candidates. A few of these candidates like Cyg X-1, A0620-00, V404 Cyg, Nova Muscae 1991 (Cowley 1994; van Paradijs & McClintock 1996) have good mass determinations which show that the compact accreting stars in these systems are too massive to be neutron stars and therefore must be black holes. The remain-

301

ing black hole candidates are identified purely on spectral similarity to the well-established candidates (e.g., White 1994). A second class of candidate black holes consists of active galactic nuclei (AGNs), which are believed to be accreting supermassive black holes (mass $\sim 10^5 - 10^9 M_\odot$, see the article by Martin Rees in this volume for a detailed discussion of AGNs). In a few cases like NGC4258 (Miyoshi et al. 1995), and Sgr A* at the center of our Galaxy (Genzel & Townes 1987; Genzel, Hollenbach & Townes 1994), the evidence for a black hole is based on dynamical information (see Scott Tremaine, this volume, for other examples). However, for the majority of AGNs, no such direct evidence exists, though extremely plausible theoretical arguments can be made (e.g., Rees 1984).

16.2   X-RAYS AND γ-RAYS FROM ACCRETING BLACK HOLES

The problem we focus on in this article is the fact that accreting black holes almost always emit a substantial fraction of their luminosity in hard X-rays and γ-rays, roughly in the energy range 10 keV to a few $\times 100$ keV. Their spectra in this energy range can be described as a power-law with a photon index α_N:

$$N_E dE \propto E^{-\alpha_N} dE, \qquad (16.1)$$

where $N_E dE$ represents the number of photons per second in the energy range E to $E + dE$. The quantity $E^2 N_E \propto E^{(2-\alpha_N)}$ is a useful way of characterizing the spectrum as it represents the energy emitted per logarithmic interval of E. Generally, spectra with $\alpha_N < 3$ may be considered to be hard spectra; spectra with $\alpha_N < 2$ are exceptionally hard since they have most of the energy coming out in the hardest photons.

The observational evidence suggests that all black hole XRBs spend at least part of their time in a hard state (the so-called Low State, see § 16.4.2) with $\alpha_N \sim 1.5 - 2.5$ (Grebenev et al. 1993; Harmon et al. 1994; Gilfanov et al. 1995). The power-law index α_N usually increases above $E \sim$ few $\times 100$ keV. AGNs too appear to have a more or less universal hard component in their spectra. The data are unfortunately less complete at this time (since bright AGNs, although $\sim 10^8$ times more luminous than bright XRBs, are $\sim 10^5$ times farther away and therefore $\sim 10^2$ times fainter.) The spectra of a number of quasars, Seyferts and other AGNs have been measured up to $E \sim 10 - 20$ keV using the Einstein Observatory, Ginga and other X-ray satellites (Williams et al. 1992; Elvis et al. 1994a). At these energies the spectra are power-laws with $\alpha_N \lesssim 2$. The quasar S5 0014+813, at redshift 3.384, has been observed with ASCA to have an $\alpha_N = 1.6$ power-law extending up to at least 40 keV in the rest frame of the source (Elvis et al. 1994b). The OSSE instrument on the Compton Gamma-Ray Observatory (CGRO) has been studying several AGNs at energies between 50 keV and 10 MeV and has found power-laws with $\alpha_N \sim 2$ extending above 100 keV in many cases (Maisack et al. 1993; Johnson et al. 1994; Kinzer et al. 1994).

These observations place strong constraints on theoretical models. The radiating gas has to be optically thin in order to produce a power-law spectrum, and the electron temperature has to be $\gtrsim 10^9$ K in order to produce photons up to 100 keV. These requirements are quite difficult to meet with accretion models.

The most commonly treated model of accretion flows is the thin accretion disk model developed by Shakura & Sunyaev (1973), Novikov & Thorne (1973) and Lynden-Bell & Pringle (1974). In this model, the accreting gas is assumed to be cool, with gas temperature T_{gas} much less than the virial temperature T_{vir} ($\sim 10^{13} K/r$, where r is the radius in Schwarzschild units). Since the gas is cool, (i) radial pressure forces are negligible and the angular velocity Ω of the gas is essentially equal to the Keplerian value, $\Omega_K = (GM/R^3)^{1/2}$, and (ii) the gas forms a thin disk with its vertical thickness H being much less than the radius. The energy equation, which describes the balance between the local energy generation due to viscous dissipation and the cooling due to vertical radiative transfer and radiation from the surface of the disk, simplifies considerably in the thin disk geometry and it is straightforward to calculate the density and temperature of the accreting gas as a function of r. The thin disk model has been applied successfully to accreting white dwarfs and pre-main-sequence stars (Frank, King, & Raine 1992).

The thin accretion disk model suffers from one major uncertainty which plagues all studies of accretion, namely the unknown nature of viscosity in the gas. Microscopic viscosity is much too small to produce significant accretion. Various macroscopic mechanisms such as shear-driven hydrodynamic turbulence, convection and MHD instabilities have been explored, the last process being considered particularly attractive (Balbus & Hawley 1991). In models of accretion flows it is usual to absorb the uncertainty in the viscosity mechanism into a dimensionless parameter α (Shakura & Sunyaev 1973) and to write the kinematic coefficient of viscosity ν as

$$\nu = \alpha c_s H \sim \frac{\alpha c_s^2}{\Omega_K}, \qquad (16.2)$$

where c_s is the sound speed of the gas. In certain circumstances, it is necessary to modify equation (16.2) to enforce causality (Popham & Narayan 1992; Narayan 1992; Syer & Narayan 1993; Narayan, Loeb, & Kumar 1994; Kato & Inagaki 1994). We will not discuss further questions related to viscosity, although they do represent major unsolved problems in accretion physics.

The thin accretion disk model, despite being plausible and self-consistent, is unable to explain the hard spectra observed in accretion flows around black holes.

Thin accretion disks around black holes are nearly always optically thick in the vertical direction and therefore produce blackbody-like spectra (e.g., Frank et al. 1992). This is of course in conflict with the power-law spectra observed. Further, the effective temperature is much too low:

$$T_{\text{eff}} \sim (6 \times 10^7 K)m^{-1/4}\dot{m}^{1/4}r^{-3/4}, \qquad (16.3)$$

where m is the mass of the black hole in solar units and $\dot{m}$ is the mass accretion rate in Eddington units ($\dot{M}_{\text{Edd}} = 2.2 \times 10^{-8}m\ M_{\odot}\text{yr}^{-1}$, assuming an efficiency of 0.1, cf. Frank et al. 1992). Equation (16.3) shows that with reasonable parameters it is impossible to obtain a temperature anywhere close to the observed $T \sim 10^9$ K. Moreover, the thin disk model predicts that T_{eff} should decrease with increasing black hole mass, whereas observations indicate that the characteristic temperature of the hard power-law radiation is $\gtrsim 10^9$ K for both stellar-mass and supermassive black holes.

Before proceeding to discuss other accretion models, two comments are in order. First, even though the thin disk model cannot explain power-law X-ray/γ-ray emission, it does explain softer components in the spectra of accreting black holes. Black hole XRBs in their "High" and "Very High" states (cf. § 16.4.2) have soft spectra with $kT \sim 0.7 - 1$ keV. This emission could very well be produced by a standard thin disk (e.g., Ebisawa 1994). Similarly, quasars have a "blue bump" in their spectra which is very likely due to thin disk emission (Malkan & Sargent 1982).

The second comment is that all the discussion in this article pertains to emission between 10 keV and 1 MeV. In addition to this emission, which is often thermal, a completely different kind of emission has been seen in blazars (a particularly active variety of AGNs). A number of these sources have been observed by the Egret detector on CGRO to produce tremendous amounts of radiation at GeV − TeV energies (Hartman et al. 1994). This emission appears to be relativistically beamed and seems to be produced by nonthermal mechanisms associated with relativistic jets. The description of the physical mechanism that produces this beaming is another unsolved problem, one that is not reviewed here.

16.3   HOT ACCRETION FLOW MODELS

In view of the inability of the thin accretion disk model to produce hard X-rays and γ-rays, various attempts have been made to develop alternate models of accretion. A common feature of these models is that the radiating electrons are relativistic, so that Compton-scattering is important (Rybicki & Lightman 1979 and references therein). In a seminal paper, Sunyaev & Titarchuk (1980, see also Titarchuk 1994) discussed Comptonization of soft photons by hot electrons and showed that, for optical depths $\sim$ a few, power-law spectra extending up to $E \sim kT$ are obtained. The soft seed photons for the Comptonization could come from a variety of sources — photons from a standard thin accretion disk, photons from the surface of the central star (in the case of accretion onto a neutron star), or synchrotron photons produced locally in the hot magnetized plasma. At the temperatures of interest, electron-positron pair processes can be important, and there is an extensive litera-ture on calculating these effects (Svensson 1982, 1984; Zdziarski 1985; White & Lightman 1989). Most discussions are in terms of thermal models, but nonthermal models have also been considered at times (Svensson 1994). Recent observations have confirmed earlier indications that most black hole XRBs and Seyferts have turnovers in their spectra between $E \sim 100$ keV and 1 MeV. This turnover poses severe difficulties for purely nonthermal models (e.g., Zdziarski et al. 1993; Fabian 1994). Therefore, the focus presently is on thermal models, with perhaps a fraction of the emission coming from nonthermal processes.

Three main categories of hot accretion flow models have been considered in the literature.

16.3.1 Corona Models

In this model, the hot gas is postulated to be in a corona above a standard thin accretion disk. The corona is characterized by its density (or equivalently optical depth) and temperature (or Compton y-parameter). The disk provides the soft pho-tons for Comptonization. X-ray data on AGNs and black hole XRBs show spectral features such as iron emission lines and a bump between 10 and 30 keV which indicates X-ray reflection in cool gas (Pounds et al. 1990; Matsuoka et al. 1990). These features are explained in the corona model as due to reflection of the hot coronal radiation off the underlying cool disk (Guilbert & Rees 1988; Lightman & White 1988; Fabian 1994 and references therein).

Haardt & Maraschi (1991, 1993) have developed an improved corona model where the temperature is not a free parameter but is solved for by balancing the viscous heating rate of the gas and the Compton cooling rate by soft photons from the disk. The disk emission is calculated self-consistently by including irradiation from the corona. The model assumes that most of the viscous energy dissipation occurs in the corona (as argued by Field & Rogers 1993) rather than in the disk, though most of the mass accretion is assumed to occur in the cool disk.

Coronal models though promising are still incomplete at this time since the dynamics of the coronal gas has not been treated in detail. The density and tem-perature of the corona are often adjusted freely, and even in the approach of Haardt & Maraschi the density is left unspecified.

16.3.2   *SLE Two-Temperature Model*

In an important paper, Shapiro, Lightman & Eardley (1976, SLE) showed that, in addition to the standard thin disk model, a second solution to the accretion flow equations is permitted at somewhat sub-Eddington accretion rates. The gas in this solution is optically thin and quite hot. SLE suggested that the accreting gas may be in the form of a two-temperature plasma close to the accreting star, with the ions being significantly hotter than the electrons. The electrons, which are cooled primarily by Comptonization of soft photons, achieve temperatures $\sim 10^8 - 10^9$ K.

The optically thin nature of the gas in the SLE solution, plus the high temperature, are precisely the features one needs to explain the high energy spectra of accreting black holes. Therefore, the SLE solution has been widely applied in models of XRBs and AGNs (e.g., Kusunose & Takahara 1985, 1989; White & Lightman 1989; Wandel & Liang 1991; Luo & Liang 1994; Melia & Misra 1993).

A major problem with the SLE solution is that the gas is thermally unstable (Piran 1978; Wandel & Liang 1991; Narayan & Yi 1995b). If the temperature of the electrons in an equilibrium flow is slightly perturbed, the gas either cools catastrophically to the thin disk solution or heats up catastrophically without bound. The instability operates on the thermal timescale ($\sim 1/\alpha\Omega_K$), which is much faster than the accretion timescale, calling into question the above-mentioned applications to XRBs and AGNs.

16.3.3   *Optically-Thin Advection-Dominated Model*

Recently, a new class of optically thin hot solutions has been discovered (Narayan & Yi 1994, 1995ab; Abramowicz et al. 1995; Chen 1995; Chen et al. 1995) which have all the advantages of the SLE solution without suffering from a violent thermal instability. A crucial feature of these solutions is that most of the viscously dissipated energy is advected with the accreting gas as stored entropy, with only a small fraction of the energy being radiated. The flows are therefore described as being *advection-dominated*. The accreting gas in these solutions is optically thin and is in principle thermally unstable. However, because of advection-domination the radiative cooling is only a small perturbation on the overall energetics of the gas, and further the accretion flow timescale is quite short. For both reasons, the thermal instability does not have time to grow to any significant amplitude and the flow is effectively stable to thermal perturbations (cf. Kato, Abramowicz, & Chen 1996). The flows are, however, convectively unstable (Narayan & Yi 1994, 1995a; Igumenshchev, Chen, & Abramowicz 1996), and it has been suggested that part of the viscosity could be due to the turbulence associated with the convective motions.

The idea of advection-dominated accretion goes back to Begelman (1978) and Begelman & Meier (1982) who suggested that at very high $\dot{m}$ radiation may be trapped in the accreting gas and be unable to escape. Solutions corresponding to this regime were obtained by Abramowicz et al. (1988), who considered flows that are optically thick and not very hot, and which are therefore not of interest for the discussion in this article.

Optically thin advection-dominated flows operate in a regime where the gas density is so low that the radiative efficiency is very poor. The physics here is very different from that of the optically thick, high $\dot{m}$ disk solutions of Abramowicz et al. (1988). In this case, energy is advected not because the radiation is trapped, but rather because the heated gas does not have enough time to radiate its internal energy. The net effect is the same, however, namely that most of the binding energy is retained in the gas as thermal energy and is carried into the black hole. Early work on these flows was done by Rees, Begelman, Blandford & Phinney (1982, who called the gas configuration an "ion torus," see also Phinney 1981), Matsumoto, Kato, & Fukue (1985), and Narayan & Popham (1993, who found optically-thin advection-dominated flows in accretion disk boundary layers at low $\dot{m}$). However, it is only recently that the properties of these hot flows have been studied in detail (Narayan & Yi 1994, 1995ab; Abramowicz et al. 1995; Chen 1995; Chen et al. 1995). Assuming the plasma becomes two-temperature, Narayan & Yi (1995b) found that the electron temperature lies in the range $10^9 - 10^{10}$ K, depending on $\dot{m}$. The ion temperature on the other hand is nearly virial ($\sim 10^{12} \text{K}/r$) because of the low radiative efficiency. The accreting gas is therefore almost spherical in shape rather than disk-like or toroidal (Narayan & Yi 1995a). Despite this, the gas does rotate (with sub-Keplerian Ω) and has angular momentum transport and viscous dissipation just as in regular thin disks. These features, plus the subsonic radial velocity (except close to the horizon), distinguish this model from pure spherical accretion models (e.g., Ipser & Price 1983; Melia 1994; Ostriker et al. 1976).

The optically-thin advection-dominated model requires the gas to be fairly optically thin, which means that the solutions exist only for relatively low accretion rates, $\dot{m} \lesssim 0.3\alpha^2$. In addition, because of the strong advection, the radiative efficiency is low. For both reasons, these solutions work best when applied to low luminosity sources. For such sources, the model has been quite successful.

Figure 16.1 shows an advection-dominated model (Narayan, Yi, & Mahadevan 1995) of Sagittarius A* (Sgr A*), the putative supermassive black hole at the center of our Galaxy (Genzel & Townes 1987; Genzel et al. 1994). This model,

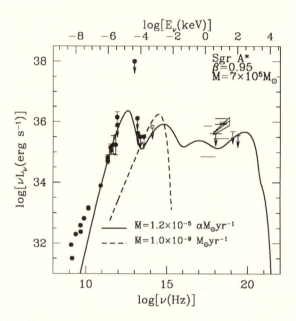

Figure 16.1: The solid line shows the spectrum of Sgr A* corresponding to an advection-dominated accretion flow with $M = 7 \times 10^5 M_\odot$ and $\dot{M} = 1.2 \times 10^{-5} \alpha M_\odot yr^{-1}$ (Narayan et al. 1995). The spectrum fits well the observed measurements and upper limits. The dashed line corresponds to a thin accretion disk model for the same M and with $\dot{M} = 10^{-9} M_\odot yr^{-1}$. The thin-disk spectrum is nearly blackbody in shape and does not explain either the radio/mm or X-ray observations.

which corresponds to a black hole mass of $7 \times 10^5 M_\odot$ and an accretion rate of $\dot{M} \sim 10^{-5} \alpha M_\odot yr^{-1}$, provides a plausible explanation for the observed spectrum of the source over nearly ten decades of frequency. Furthermore, the model is able to explain the extremely low X-ray luminosity of the source, $\sim 10^{37} erg s^{-1}$, despite making use of a fairly substantial $\dot{M}$ (see also Rees 1982). This is welcome because there is evidence for substantial amounts of gas in the vicinity of Sgr A* (Genzel & Townes 1987; Melia 1992), and until now it has been a mystery (Genzel et al. 1994) why the source is as dim as it is. An unusually low luminosity is of course natural with an advection-dominated flow since most of the energy disappears into the black hole. Models similar to that described here for Sgr A* have also been developed for NGC 4258 (Lasota et al. 1996) and to explain (Fabian & Rees 1995) the unusually low luminosities of supermassive black holes at the centers of nearby bright elliptical galaxies (Fabian & Canizares 1988).

Figure 16.2 shows another application, this time to the black hole soft X-ray transient A0620-00 in its quiescent state. In this case, the model consists of a

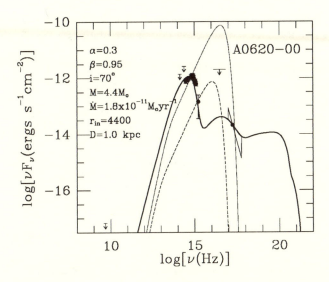

Figure 16.2: The thick solid line is a model spectrum of the soft X-ray transient source A0620-00 in quiescence (Narayan et al. 1995b). The model assumes a thin disk down to $r = 4400$ and an advection-dominated flow from $r = 4400$ to $r = 1$ with $\alpha = 0.3$. The black hole mass is $M = 4.4 M_\odot$ and the accretion rate is $\dot{M} = 1.8 \times 10^{-11} M_\odot \text{yr}^{-1}$. The dotted and dashed lines are two pure thin disk models with $\dot{M} = 3 \times 10^{-12}$, 3×10^{-14} $M_\odot \text{yr}^{-1}$ respectively. These models are unable to fit the data.

standard thin accretion disk on the outside, extending down to a radius $r = 4400$, and an optically thin advection-dominated flow on the inside extending down to the horizon (Narayan, McClintock, & Yi 1996). The model is able to explain the spectrum of the source quite well, fitting both the X-ray and the optical/UV observations.

The models described in Figures 16.1 and 16.2 make explicit use of the fact that the accreting objects are black holes. It is important that the advected energy be able to pass through a horizon and disappear rather than be re-radiated from the surface of a normal star. The successful application of these models could be optimistically viewed as "proof" that the accreting stars are indeed black holes (Narayan, McClintock & Yi 1996).

The hot advection-dominated solutions discussed here are currently the only known thermally stable accretion flows which achieve the kind of relativistic temperatures needed to explain the X-ray and γ-ray spectra of accreting black holes. For this reason, there is considerable effort under way to explore the full potential of these models. A currently open question is whether this model, which by its low $\dot{m}$ and low radiative efficiency is best suited for low-luminosity sources, can be applied also to objects with near-Eddington luminosities. Figure 16.3 shows the limit of validity of the hot advection-dominated solutions in the $r\dot{m}$ plane. For

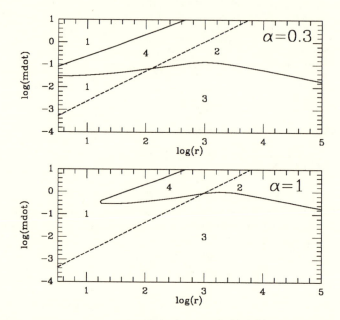

Figure 16.3: Based on Chen et al. (1995). The upper panel shows, for $\alpha = 0.3$, the regions in the $r\dot{m}$ plane where the various known equilibrium flows are allowed. The calculations were done using the two-temperature model of Narayan & Yi (1995b) which includes cooling due to synchrotron, bremsstrahlung and Comptonization. (Note: The corresponding Figure 2 in Chen et al. was calculated with a simpler one-temperature model which included only bremsstrahlung cooling.) The different regions are numbered $1-4$ as in Chen et al. (1995). The thin disk solution is allowed below the dashed line (note that above the dashed line the accreting gas is dominated by radiation pressure and suffers from the Lightman-Eardley (1974) instability), the optically thick advection-dominated solution is allowed above the upper solid line, and the hot optically-thin advection-dominated solution, which is the focus of this article, is allowed below the lower solid line. The lower panel corresponds to $\alpha = 1$. Note that the hot optically-thin advection-dominated solution exists only for $\dot{m} \lesssim 0.3\alpha^2$. In both panels, region 3 allows two distinct equilibrium solutions, while region 4 allows no solution at all.

reasonable values of α, say $\lesssim 0.3$, the hot solutions are present only for $\dot{m} \lesssim 0.03$ (the lower solid line in the upper panel of Figure 16.3). Only if $\alpha \sim 1$ (Fig. 16.3, lower panel) can these solutions be pushed to interesting values of $\dot{m}$ appropriate to luminous sources (Narayan 1995; Narayan, Yi, & Mahadevan 1996). An unsolved issue is whether α can be this large. Another problem is that only solutions which are very close to the upper limits shown in Figure 16.3 have high efficiencies. Therefore these models will need some measure of fine-tuning if they are to be applied to bright sources. It is always possible to modify the model of viscous dissipation and radiation in such a way as to achieve higher radiative efficiency even at low values of $\dot{m}$, but it remains to be seen if such a modification can be done in a physically plausible way. Other unresolved issues center on whether two-temperature plasmas are allowed in hot low density flows (see Begelman & Chiueh 1988, and Appendix A of Narayan & Yi 1995b), on what effect electron-positron pair production will have on the nature of the solutions, and on the role of the inner boundary condition near the horizon.

16.4   DIRECTIONS FOR FUTURE RESEARCH

The ultimate goal in the study of black hole accretion is easily stated:

Given the mass M of an accreting black hole, the mass accretion rate $\dot{M}$, and a few additional parameters (say the disk inclination and the specific angular momentum of the black hole), we would like to be able to calculate all the details of the flow and be able to predict the spectrum and variability of the source.

The thin accretion disk model does satisfy this goal in the sense that it gives a clear prediction for the spectrum of a source as a function of system parameters. Further, there has been interesting work on thermal instabilities which seems to explain X-ray nova outbursts in soft X-ray transients (Huang & Wheeler 1989; Mineshige & Wheeler 1989; but see Lasota 1995). Unfortunately, as we have seen, the thin disk model does not produce any hard radiation.

The most promising direction to pursue at this time is to try and develop the hot advection-dominated solutions further and to look for variants of these solutions which may be better suited to explain high luminosity sources. Orienting such theoretical work toward explaining the rich phenomenology that X-ray and γ-ray satellites are currently providing could be rewarding.

16.4.1   *Unresolved Theoretical Issues*

Perhaps what is needed most urgently is a systematic search for all stable accretion flow equilibria. Figure 16.3, based on Chen et al. (1995), shows the interrelationship among the three stable equilibria which are currently known, viz. the standard thin disk model (Shakura & Sunyaev 1973), the optically-thick advection-dominated model (Abramowicz et al. 1988), and the optically-thin advection-dominated model (Narayan & Yi 1994, 1995b; Abramowicz et al. 1995; Chen 1995). An important question is whether there could be other equilibria if we relax some of the assumptions made by Chen et al. (1995). For instance, in addition to the two-temperature models described by Narayan & Yi (1995b), are there single-temperature models where the ions and electrons are strongly coupled by non-thermal processes? How would these models differ from two-temperature models? (See Esin et al. 1996 for a recent analysis of this problem.) If we allow the electrons to have some degree of nonthermal energy distribution, how much will the results change? Can there be important nonlocal effects, such as Comp-tonization of radiation emitted at one radius by gas at another radius (cf. Ostriker et al. 1976), and can this open up new global equilibria which go beyond the local analyses done so far? These are important questions to which current research is being directed.

Once a fairly complete set of equilibria has been identified, two important issues will need to be resolved:

(1) It is possible to have multiple equilibria for given values of $\dot{m}$ and r; as Figure 16.3 shows, this happens already just with the three known equilibria. Consider a system at relatively low $\dot{m}$ in region 3 of Figure 16.3. Such a system has to choose between the standard thin disk solution and the hot advection-dominated solution. Which equilibrium does it choose? There are clearly several possibilities. Some systems may exist purely in the cool disk state, while others (for example Sgr A*, Fig. 16.1) may exist entirely in the hot state. Another possibility is that a system may exist in the cool state at some radii and in the hot state at other radii; our model of A0620-00 in quiescence (Fig. 16.2) is such an example. (See Meyer & Meyer-Hofmeister 1994, Narayan & Yi 1995b, and Honma 1996 for discussions of possible mechanisms which might drive a cool disk into the hot state.) Yet another possibility is that both solutions may be simultaneously present at the same radius; for example, part of the accretion may happen along an equatorial thin disk, and part may happen in a hot advection-dominated corona. Such a model would be very close in spirit to the corona models described earlier except that when fully developed this model would presumably include a full calculation of the vertical structure of the gas and would allow us to calculate how much mass goes into the corona and how the gas dynamics in the coronal region works. A final possibility, of course, is that the system may go into a time-dependent mode where it switches from one state to another on some characteristic time scale.

(2) For some ranges of $\dot{m}$ and r there may be no stable equilibrium at all. Region 4 in Figure 16.3 corresponds to this situation. What will an accreting system

that falls within region 4 do? It seems obvious that the flow will be forced into a time-dependent mode (Chen et al. 1995), perhaps oscillating between different equilibria in such a way as to satisfy the required $\dot{m}$ in the mean. The possibilities here are very rich and the time-dependent dynamics could be very interesting. Observationally, we note that most bright black hole accretors show strong variability both in their luminosity and spectrum.

16.4.2 Clues from Observations of Black Hole XRBs

Although the number of AGNs known is vastly larger than the number of black hole XRBs, the latter are likely to prove more useful for understanding black hole accretion. Because XRBs are much closer to us than AGNs they tend to be brighter and to provide better data. Also, because the time scales in XRBs are significantly shorter than those in AGNs, it is possible to obtain considerable information on variability and fluctuations of XRBs, whereas equivalent data on AGNs would require observations over prohibitively long periods of time.

Black hole XRBs exhibit four distinct states, distinguished by luminosity and spectral shape (e.g., Tanaka 1989; Grebenev et al. 1993; van der Klis 1994; Gilfanov et al. 1995). In order of increasing luminosity, these states are:

(1) *Off/Quiescent State*: This is seen predominantly in X-ray novae between outbursts. In at least two sources, V404 Cyg and A0620-00, low-level X-ray emission has been seen in quiescence, and presumably all quiescent sources have such emission. The X-ray luminosity is in the range $L_X \sim 10^{30} - 10^{34}$ ergs s^{-1} $\ll$ $L_{\rm Edd}$, where $L_{\rm Edd} = 10^{38} m$ ergs s^{-1} is the Eddington luminosity. The nature of the spectrum is not known, but according to the advection-dominated model (Fig. 16.2) the spectrum is fairly hard. Sgr A* (Fig. 16.1) may be considered an example of a galactic nucleus in the Off/Quiescent State.

(2) *Low State*: In this state, the spectrum of an X-ray binary is exceptionally hard, typically $\alpha_N \sim 1.5 - 2$. The luminosity is typically $L_X \sim 10^{36} - 10^{37.5}$ ergs s^{-1} $\lesssim 0.01 L_{\rm Edd}$. According to the model of Lasota et al. (1996) the galactic nucleus in NGC 4258 may correspond to the lower end of the Low State.

(3) *High State*: The luminosity of the X-ray binary is greater than in the Low State, $L_X \sim 0.01 - 0.9 L_{\rm EDD}$, and the spectrum is very soft. Indeed, at most about 10% of the emission is above 10 keV. Most bright AGN probably correspond to the High State or the following Very High State.

(4) *Very High State*: This state, first discovered in the galactic source GX 339-4 by Ginga observations (Miyamoto et al. 1991), corresponds to the highest luminosity, $L_X \gtrsim 0.9 L_{\rm Edd}$. The spectrum consists of an ultrasoft component which contains most of the luminosity plus a hard tail with $\alpha_N \sim 2.5$.

The existence of so many distinct states is doubtless a very important clue which we ought to be able to exploit to understand the nature of accretion flows around black holes. If we accept the advection-dominated model for the Quiescent State, an obvious question is: could the same model be applied also to the Low State? The hard spectrum of the Low State implies a very hot optically thin flow, and perhaps the only difference between this state and the Off State is the value of $\dot{m}$ (Narayan 1996). The High State appears to be a good candidate for the standard thin accretion disk model since the observed temperature is consistent with that predicted by equation (16.3), with perhaps some modifications due to electron scattering (Ebisawa 1994). Why is there an abrupt change between the Low and High States? Perhaps the boundary between the two states corresponds to the limiting $\dot{m}$ of the hot advection-dominated solution mentioned earlier (see Fig. 16.3). This proposal works quite well if $\alpha \sim 1$ (Narayan 1996; Narayan et al. 1996), and so the question is whether such a large value of α is physically reasonable. What is the nature of the Very High State? If the $\dot{m}$ in the High State is already too large to allow a hot advection-dominated flow, then the $\dot{m}$ in the Very High State certainly is much too large. Yet this state is seen to have a hard tail in the spectrum. What could produce the tail? One might speculate that the hard radiation in this case is produced by a corona. Since the typical value of α_N in the Low State is smaller than that in the Very High State (van der Klis 1994) it is reasonable to propose different origins for the two hard components. Finally, does the upper optically-thick advection-dominated solution branch (see Fig. 16.3) play a role in the Very High State?

Gilfanov et al. (1995) note that the spectral index α_N in the Low State varies considerably from one system to another but is relatively independent of the hard X-ray luminosity in any single system. What determines the value of α_N? The fact that α_N is independent of luminosity suggests that $\dot{M}$ may not be the key, though one should keep in mind that the hard X-ray luminosity is not necessarily directly proportional to $\dot{M}$. Nor is α_N likely to depend on the black hole mass, since XRBs and AGNs have similar spectra in their hard states even though they differ by orders of magnitude in mass. Therefore, there has to be a third parameter, which may be the inclination of the system or the specific angular momentum of the black hole.

Black hole XRBs tend to be variable on a wide range of time scales and some systems even display quasi-periodic oscillations. In general, it appears that the largest variability is seen in the hard spectral components, while the soft components remain relatively steady. This could well be an important clue to the nature of the hot gas in these systems. The reader is referred to van der Klis (1994), Kouveliotou (1994) and Gilfanov et al. (1995) for more details on the variability of black hole XRBs.

16.4.3 *Black Holes versus Neutron Stars*

The distinction between flows around black holes and those around weakly magnetized neutron stars is an interesting sub-topic (e.g., van der Klis 1994). (X-ray pulsars are excluded from this discussion since the neutron star magnetic field plays an important role in these systems and the spectral properties are very different.) There are two reasons why this subject is of interest. First, if the differences between neutron stars and black holes are understood well, then one could hope to develop better techniques to distinguish black hole XRBs from neutron star XRBs in the observations. This will make it easier to identify new black hole systems. Second, if the differences between the two kinds of systems can be fitted naturally into a theoretical framework, then we can have confidence in our general understanding of high energy accretion flows.

Black holes and neutron stars both have deep relativistic potentials, which is of course why the two systems look so similar when they accrete. The main distinction between the two cases is that the inner boundary conditions are different.

1. *Radiation Boundary Condition*: In the case of a black hole, the accreting matter disappears through the horizon, carrying its energy with it. In contrast, a neutron star has a surface which re-radiates whatever energy flows in. Sunyaev et al. (1991ab, see also Liang 1993) have argued that Compton-cooling of the accretion flow by stellar photons will make the spectra of neutron star XRBs steeper than those of black hole XRBs. Narayan & Yi (1995b) confirmed this proposal with detailed calculations of hot two-temperature advection-dominated flows around neutron stars. The observational situation is that hard tails do occur in neutron star systems in their low intensity states (Barret, McClintock, & Grindlay 1996; Barret et al. 1996). Fitting these tails with power laws generally yields spectral indices $\alpha_N \gtrsim 3$, though on one occasion the X-ray burster in Terzan 2 showed a very hard spectrum with $\alpha_N = 1.7$ (Barret et al. 1991). Thermal bremsstrahlung fits give temperatures $kT \sim 30 - 40$ keV (Barret & Vedrenne 1994). Therefore, on average, neutron star hard tails are cooler as well as steeper/softer than black hole tails (which have $\alpha_N \sim 1.5 - 2.5$, $kT \gtrsim 100$ keV). As further evidence that neutron star flows are cooler, we note that electron-positron annihilation features have been seen only in black hole candidates (1E1740.7-2942, Bouchet et al. 1991; Nova Muscae 1991, Goldwurm et al. 1992).

2. *Flow boundary condition*: The second difference, which has not been well explored, follows again from the fact that the black hole has a horizon whereas the neutron star has a hard surface. The accreting gas responds to these two boundaries differently. In the case of the black hole, the gas makes a sonic transition and falls into the horizon supersonically. In contrast, gas flowing onto a neutron star has to slow down, either subsonically (e.g., Popham & Narayan 1992) or following

a shock (Kluzniak & Wagoner 1985), and settle on the surface. It is conceivable that this difference leads to important changes in the gas dynamics close to the center. One might speculate for instance that the nature of outflows and jets depends very sensitively on the inner boundary condition. Many AGNs have relativistic jets (Zensus & Pearson 1987; Begelman, Blandford, & Rees 1984) and it is now becoming clear that many XRBs also have such ejections. The best examples of XRB jets are all found in black hole candidates, e.g., 1E1740.7-2942 (Mirabel et al. 1992), GRS 1915+105 (Mirabel & Rodriguez 1994a), GRO J1655-40 (Tingay et al. 1994; Hjellming & Rupen 1995), GRS 1758-258 (Mirabel & Rodriguez 1994b). The only XRBs which have jets and which might be neutron stars are SS 433 (Margon 1984) and Cyg X-3 (Molnar, Reid, & Grindlay 1988). Both of these are nonstandard XRBs and may well be accreting black holes. Relativistic jets may thus be an unambiguous signature of black holes. Alternatively, since SS 433 and Cyg X-3 have somewhat slower moving jets (relativistic $\beta \sim 0.2 - 0.3$) compared to the other cases, one could say that the speed of the jet is determined by the boundary condition in such a way that settling flows make slower jets (M. J. Rees, private communication). One point to keep in mind is that neutron star X-ray bursters probably have not been imaged in the radio to the same sensitivity level as the black hole candidates. Such observations need to be done before one can draw any firm conclusions on the differences between outflows from neutron stars and black holes.

16.5   CONCLUSION

This article has tried to show that black hole accretion is an open field with many opportunities and many unsolved problems. The field currently lacks an accepted paradigm, though the recent work on optically-thin advection-dominated accretion flows seems promising. The present time may be particularly good for an all-out assault in this area because of the wealth of observational clues pouring in from high energy space missions like CGRO and (very soon) XTE. With the AXAF and INTEGRAL missions set to be launched at the turn of the century we can anticipate much more data in the years ahead.

ACKNOWLEDGMENTS

The author is grateful to the following colleagues and collaborators for comments and instruction: Didier Barret, Josh Grindlay, Jean-Pierre Lasota, Rohan Mahadevan, Jeff McClintock, Martin Rees, and Insu Yi. Comments on the manuscript by John Bahcall, Jerry Ostriker, Fred Rasio, and Lyman Spitzer are gratefully acknowledged. This work was supported in part by grant NAG 52837 from NASA.

BIBLIOGRAPHIC NOTES

- **Abramowicz, M., Czerny, B., Lasota, J.-P., & Szuszkiewicz, E. 1988, ApJ, 332, 646.** A formal introduction to the optically thick advection-dominated accretion flows ("slim disks") relevant for moderately super-Eddington accretion flows. **Abramowicz, M., et al. 1995, ApJ, 438, L37.** The authors discuss another optically thin advection-dominated flow which is relevant at low mass accretion rates.

- **Frank, J., King, A. R., & Raine, D. 1992, Accretion Power in Astrophysics (Cambridge: Cambridge University Press).** A graduate level text on accretion disk physics in compact objects. The authors cover a variety of astrophysical systems ranging from accreting white dwarfs to active galactic nuclei with a useful list of references. Basic physical principles are discussed along with the standard phenomenological terminology.

- **Haardt, F., & Maraschi, L. 1991, ApJ, 380, L51; 1993, ApJ, 413, 507.** Important papers which describe corona models in which the temperature of the corona is obtained self-consistently by treating the disk-corona radiative transfer problem.

- **Lynden-Bell, D., & Pringle, J. E. 1974, MNRAS, 168, 603.** An important paper which provides the formal mathematical foundations of accretion disk theory. The authors' semi-analytic approach to time-dependent disk evolution is both entertaining and rewarding. Their discussion of the nature of the star-disk boundary layer is also an important contribution.

- **Narayan, R., & Yi, I. 1994, ApJ, 428, L13.** This paper describes a convenient self-similar solution of the viscous accretion equations which exhibits the basic properties of advection-dominated accretion flows.

- **Narayan, R., & Yi, I. 1995, ApJ, 452, 710.** A recent paper which discusses in detail various aspects of the energetics and radiation mechanisms of optically thin advection-dominated flows.

- **Novikov, I. D., & Thorne, K. S. 1973, in Black Holes ed. C. DeWitt & B. DeWitt (New York: Gordon and Breach).** The authors derive the relativistic version of the accretion disk equations and provide the relativistic corrections to the Newtonian disk solutions. A short but helpful summary of relevant radiation processes is also found here.

- **Pringle, J. E. 1981, ARA&A, 19, 137.** A widely referenced review on the basic principles of the standard accretion disk model with a useful list of early references.

- **Rees, M. J., Begelman, M. C., Blandford, R. D., & Phinney, E. S. 1982, Nature, 295, 17.** An early paper on optically thin hot flows (''ion tori'') which contains some useful qualitative descriptions of the basic nature of the flows and excellent order of magnitude estimates of various physical processes and their time scales.

- **Shakura, N. I., & Sunyaev, R. A. 1973, A&A, 24, 337.** A seminal paper which introduces the phenomenological α viscosity prescription. This paper derives the basic scalings of geometrically thin and optically thick accretion disks.

- **Shapiro, S. L., Lightman, A. P., & Eardley, D. M. 1976, ApJ, 204, 187.** A classic paper on hot two-temperature accretion flows. The specific solution described in this paper has been superseded by an optically thin advection-dominated solution.

- **Sunyaev, R. A., & Titarchuk, L. G. 1980, A&A, 86, 127.** An important paper which demonstrates that Comptonization of soft photons by hot electrons can account for observed power-law X-ray spectra. The derivation of the semi-analytic formulas is very helpful.

- **Svensson, R. 1994, ApJS, 92, 585.** An excellent review of hot accretion disks with non-thermal pairs. The author focuses on X-ray spectra of active galactic nuclei.

- **Tanaka, Y. 1989, Proc. 23rd ESLAB Symp. on Two Topics in X-ray Astronomy (Bologna) (ESA SP-296).** A stimulating review on X-ray observations of galactic black hole candidates based on X-ray satellite observations.

- **van der Klis, M. 1994, ApJS, 92, 511.** This review highlights the difficulties of distinguishing black hole candidates from neutron stars based on X-ray properties. The discussion of X-ray time variability is especially helpful.

- **G. Lewin, J. van Paradijs, & E. P. J. van den Heuvel, eds. 1995, X-ray Binaries (Cambridge: Cambridge University Press).** A comprehensive review of X-ray binaries, with several articles devoted to black hole candidates.

BIBLIOGRAPHY

[1] Abramowicz, M., Chen, X., Kato, S., Lasota, J. P., & Regev, O. 1995, ApJ, 438, L37.

[2] Abramowicz, M., Czerny, B., Lasota, J. P., & Szuszkiewicz, E. 1988, ApJ, 332, 646.

[3] Balbus, S. A., & Hawley, J. F. 1991, ApJ, 376, 214.

[4] Barret, D. et al. 1991, ApJ, 379, L21.

[5] Barret D. et al. 1996, Proc. Third Compton Symposium, A&A Suppl. Special Issue, in press.

[6] Barret D., McClintock, J. E., & Grindlay, J. E. 1996, ApJ, 472, in press.

[7] Barret, D., & Vedrenne, G. 1994, ApJS, 92, 505.

[8] Begelman, M. C. 1978, MNRAS, 184, 53.

[9] Begelman, M. C., Blandford, R. D., & Rees, M. J. 1984, Rev. Mod. Phys., 56, 255.

[10] Begelman, M. C., & Chiueh, T. 1988, ApJ, 332, 872.

[11] Begelman, M. C., & Meier, D. L. 1982, ApJ, 253, 873.

[12] Bouchet, L. et al. 1991, ApJ, 383, L45.

[13] Chen, X. 1995, MNRAS, 275, 641.

[14] Chen, X., Abramowicz, M., Lasota, J., Narayan, R., & Yi, I. 1995, ApJ, 443, L61.

[15] Cowley, A. 1994, in The Evolution of X-ray Binaries, eds. S. S. Holt, & C. S. Day (New York: AIP), 45.

[16] Ebisawa, K. 1994, in The Evolution of X-ray Binaries, eds. S. S. Holt, & C. S. Day (New York: AIP), 143.

[17] Elvis, M., et al. 1994a, ApJS, 95, 1.

[18] Elvis, M., Matsuoka, M., Siemiginowska, A., Fiore, F., Mihara, T., & Brinkman, W. 1994b, ApJ, 436, L55.

[19] Esin, A. A., Narayan, R., Ostriker, E., & Yi, I. 1996, ApJ, 465, 312.

[20] Fabian, A. C. 1994, ApJS, 92, 555.

[21] Fabian, A. C., & Canizares, C. R. 1988, Nature, 333, 829.

[22] Fabian, A. C., & Rees, M. J. 1995, MNRAS, 277, L55.

[23] Field, G. B., & Rogers, R. D. 1993, ApJ, 403, 94.

[24] Frank, J., King, A., & Raine, D. 1992, Accretion Power in Astrophysics (Cambridge, UK: Cambridge University Press).

[25] Genzel, R., Hollenbach, D., & Townes, C. H. 1994, Rep. Prog. Phys., 57, 417.

[26] Genzel, R., & Townes, C. H. 1987, ARA&A, 25, 377.

[27] Grebenev, S., et al. 1993, A&AS, 97, 281.

[28] Gilfanov, M., et al. 1995, NATO ASI Series, eds. A. Alpar & J. van Paradijs, vol. 450, 331.

[29] Goldwurm, A. et al. 1992, ApJ, 389, L79.

[30] Guilbert, P. W., & Rees, M. J. 1988, MNRAS, 233, 475.

[31] Haardt, F., & Maraschi, L. 1991, ApJ, 380, L51.

[32] Haardt, F., & Maraschi, L. 1993, ApJ, 413, 507.

[33] Harmon, B. A., Zhang, S. N., Wilson, C. A., Rubin, B. C., Fishman, G. J., & Paciesas, W. S. 1994, The Second Compton Symposium, eds. C. E. Fichtel, N. Gehrels, & J. P. Norris (New York: AIP), 210.

[34] Hartman, R. C., et al. 1994, The Second Compton Symposium, eds. C. E. Fichtel, N. Gehrels, & J. P. Norris (New York: AIP), 563.

[35] Hjellming, R. M., & Rupen, M. P. 1995, Nature, 375, 464.

[36] Honma, F. 1996, PASJ, 48, 77.

[37] Huang, M., & Wheeler, J. C. 1989, ApJ, 343, 229.

[38] Igumenshchev, I. V., Chen, X., & Abramowicz, M. A. 1996, MNRAS, 278, 236.

[39] Ipser, J. R., & Price, R. H. 1983, ApJ, 267, 371.

[40] Johnson, W. N., et al. 1994, The Second Compton Symposium, eds. C. E. Fichtel, N. Gehrels, & J. P. Norris (New York: AIP), 515.

[41] Kato, S., Abramowicz, M. A., & Chen, X. 1996, PASJ, 48, 67.

[42] Kato, S., & Inagaki, S. 1994, PASJ, 46, 289.

[43] Kinzer, R. L., et al. 1994, The Second Compton Symposium, eds. C. E. Fichtel, N. Gehrels, & J. P. Norris (New York: AIP), 531.

[44] Kluzniak, W., & Wagoner, R. V. 1985, ApJ, 297, 548.

[45] Kouveliotou, C. 1994, The Second Compton Symposium, eds. C. E. Fichtel, N. Gehrels, & J. P. Norris (New York: AIP), 202.

[46] Kusunose, M., & Takahara, F. 1989, PASJ, 41, 263.

[47] Kusunose, M., & Takahara, F. 1985, Prog. Theor. Phys., 73, 41.

[48] Lasota, J.-P. 1995, Invited Review at IAU Symposium 165, "Compact Stars in Binaries."

[49] Lasota, J.-P., Abramowicz, M. A., Chen, X., Krolik, J., Narayan, R., & Yi, I. 1996, ApJ, 462, 142.

[50] Liang, E. P. 1993, Proc. Compton Gamma-Ray Observatory Workshop.

[51] Lightman, A. P., & Eardley, D. N. 1974, ApJ, 187, L1.

[52] Lightman, A. P., & White, T. R. 1988, ApJ, 335, 57.

[53] Luo, C., & Liang, E. P. 1994, MNRAS, 266, 386.

[54] Lynden-Bell, D., & Pringle, J. E. 1974, MNRAS, 168, 603.

[55] Maisack, M., et al. 1993, ApJ, 407, L61.

[56] Malkan, M. A., & Sargent, W. L. W. 1982, ApJ, 254, 22.

[57] Margon, B. 1984, ARA&A, 22, 507.

[58] Matsumoto, R., Kato, S., & Fukue, J. 1985, in Theoretical Aspects on Structure, Activity, and Evolution of Galaxies III, ed. S. Aoki, M. Iye, & Y. Yoshii (Tokyo Astr. Obs: Tokyo), 102.

[59] Matsuoka, M., Piro, L., Yamauchi, M., & Murakami, T. 1990, ApJ, 361, 440.

[60] Melia, F. 1992, ApJ, 387, L25.

[61] Melia, F. 1994, ApJ, 426, 577.

[62] Melia, F., & Misra, R. 1993, ApJ, 411, 797.

[63] Meyer, F., & Meyer-Hofmeister, E. 1994, A&A, 132, 184.

[64] Mineshige, S., & Wheeler, J. C. 1989, ApJ, 343, 241.

[65] Mirabel, I. F., & Rodriguez, L. F. 1994a, Nature, 371, 46.

[66] Mirabel, I. F., & Rodriguez, L. F. 1994b, The Second Compton Symposium, eds. C. E. Fichtel, N. Gehrels, & J. P. Norris (New York: AIP), 411.

[67] Mirabel, I. F., Rodriguez, L. F., Cordier, B., Paul, J., & Lebrun, F. 1992, Nature, 358, 215.

[68] Miyamoto, S., et al. 1991, ApJ, 383, 784.

[69] Miyoshi, M., Moran, J., Herrnstein, J., Greenhill, L., Nakai, N., Diamond, P., & Inoue, M. 1995, Nature, 373, 127.

[70] Molnar, L. A., Reid, M. J., & Grindlay, J. E. 1988, ApJ, 331, 494.

[71] Narayan, R. 1992, ApJ, 394, 261.

[72] Narayan, R. 1996, ApJ, 462, 136.

[73] Narayan, R., Loeb, A., & Kumar, P. 1994, ApJ, 431, 359.

[74] Narayan, R., McClintock, J. E., & Yi, I. 1996, ApJ, 457, 821.

[75] Narayan, R., & Popham, R. 1993, Nature, 362, 820.

[76] Narayan, R., & Yi, I., 1994, ApJ, 428, L13.

[77] Narayan, R., & Yi, I., 1995a, ApJ, 444, 231.

[78] Narayan, R., & Yi, I., 1995b, ApJ, 452, 710.

[79] Narayan, R., Yi, I., & Mahadevan, R. 1995, Nature, 374, 623.

[80] Narayan, R., Yi, I., & Mahadevan, R. 1996, Proc. Third Compton Symposium, A&A Suppl. Special Issue, in press.

[81] Novikov, I. D., & Thorne, K. S. 1973, in Blackholes ed. C. DeWitt & B. DeWitt (New York: Gordon and Breach).

[82] Ostriker, J. P., McCray, R., Weaver, R., & Yahil, A. 1976, ApJ, 208, 161.

[83] Phinney, E. S. 1981, in Plasma Astrophysics, ed. T. D. Guyenne, & G. Levy (ESA SP-161), 337.

[84] Piran, T. 1978, ApJ, 221, 652.

[85] Popham, R., & Narayan, R. 1992, ApJ, 394, 255.

[86] Pounds, K. A., Nandra, K. A., Stewart, G. C., George, I. M., & Fabian, A. C. 1990, Nature, 344, 132.

[87] Rees, M. J. 1982, in The Galactic Center, ed. G. Riegler, & R. D. Blandford.

[88] Rees, M. J. 1984, ARA&A, 22, 471.

[89] Rees, M. J., Begelman, M. C., Blandford, R. D., & Phinney, E. S. 1982, Nature, 295, 17.

[90] Rybicki, G. B., & Lightman, A. P. 1979, Radiative Processes in Astrophysics (New York: Wiley-Interscience).

[91] Shakura, N. I., & Sunyaev, R. A. 1973, A&A, 24, 337.

[92] Shapiro, S. L., Lightman, A. P., & Eardley, D. M. 1976, ApJ, 204, 187 (SLE).

[93] Smak, J. 1984, PASP, 96, 5.

[94] Stepney, S., & Guilbert, P. W. 1983, MNRAS, 204, 1269.

[95] Sunyaev, R. A., et al. 1991a, in Proc. 28th Yamada Conf. on Front. X-ray Astr. (Nagoya, Japan).

[96] Sunyaev, R. A., et al. 1991b, A&A, 247, L29.

[97] Sunyaev, R. A., & Titarchuk, L. G. 1980, A&A, 86, 127.

[98] Svensson, R. 1982, ApJ, 258, 335.

[99] Svensson, R. 1984, MNRAS, 209, 175.

[100] Svensson, R. 1994, ApJS, 92, 585.

[101] Syer, D., & Narayan, R. 1993, MNRAS, 262, 749.

[102] Tanaka, Y. 1989, Proc. 23rd ESLAB Symp. on Two Topics in X-Ray Astronomy (Bologna), (ESA SP-296), 3.

[103] Tingay, S. J., et al. 1994, Nature, 374, 141.

[104] Titarchuk, L. 1994, ApJ, 434, 570.

[105] van der Klis, M. 1994, ApJS, 92, 511.

[106] van Paradijs, J., & McClintock, J. E. 1995, in X-Ray Binaries, eds. W. H. G. Lewin, J. van Paradijs, & E. P. J. van den Heuvel (Cambridge: Cambridge Univ. Press), 58.

[107] Wandel, A., & Liang, E. P. 1991, ApJ, 380, 84.

[108] White, N. E. 1994, in The Evolution of X-ray Binaries, eds. S. S. Holt, & C. S. Day (New York: AIP), 53.

[109] White, T. R., & Lightman, A. P. 1989, ApJ, 340, 1024.

[110] Williams, O. R., et al. 1992, ApJ, 389, 157.

[111] Zdziarski, A. A. 1985, ApJ, 289, 514.

[112] Zdziarski, A. A., Lightman, A. P., & Niedźwiecki, A. M. 1993, ApJ, 414, L93.

[113] Zensus, J. A., & Pearson, T. J. 1987, Editors, Superluminal Radio Sources (Cambridge: Cambridge Univ. Press).

CHAPTER 17

THE HIGHEST ENERGY COSMIC RAYS

JAMES W. CRONIN

Department of Physics and Enrico Fermi Institute,
University of Chicago, Chicago, IL

ABSTRACT

Cosmic rays with energies in excess of 10^{20}eV have been observed since 1963. They are exceedingly rare. To date, some 8 examples have been observed corresponding to a flux of about $0.5/km^2$/century. Recently, two unusual cosmic rays with energies more than 2×10^{20}eV have been reported. These events are extraordinary, there is no credible model for their acceleration. Since they are not degraded in energy by the 2.7K cosmic background radiation, these events must have originated at distances less than 100 Mpc from our galaxy. If, as is likely, these cosmic rays are extragalactic protons, they are not much deflected by the magnetic field and should point to their source. No significant astrophysical sources lie close to their hypothesized trajectories. The existence of these high-energy rays is a puzzle, the solution of which will be the discovery of new fundamental physics or astrophysics.

17.1 INTRODUCTION

Figure 17.1 shows the differential cosmic ray spectrum above 10^{14} eV, an energy beyond which direct observations of the primary cosmic rays with balloons and satellites becomes difficult. The reason very high energy cosmic rays are difficult to observe from outside the atmosphere is the lack of adequate flux to give good statistics for the apertures currently available with such techniques.

Above 10^{14} eV cosmic rays are observed by the extensive air showers they produce when they interact with the atmosphere. A cosmic ray of energy 10^{14}

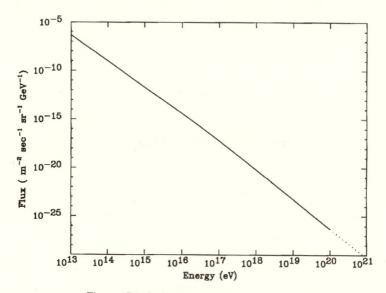

Figure 17.1: Differential cosmic ray spectrum.

eV generates a shower which spreads a detectable number of charged particles ($\geq 1/m^2$) over a diameter of 100 meters. At 10^{20}eV the shower spreads over a diameter of 5 km. Thus, arrays of charged particle detectors spread over vast areas can detect, by sampling, the presence of an extensive air shower. The energy of the shower can be inferred by the particle densities measured in the detectors.

As the shower develops longitudinally through the atmosphere, the number of charged particles at the shower maximum is about 10^{10} for a primary cosmic ray of 10^{19}eV. The shower maximum occurs at an atmospheric depth of about 800 gm/cm^2. Charged particles passing through the atmosphere excite nitrogen fluorescence. About 3 photons in the wavelength band from 300 to 400 nm are produced per meter of charged particle track. The large number of charged particles in an airshower produces an amount of light that can be detected by a photomultiplier at a distance of more than 10 km. This fluorescence technique has been successfully used to study the cosmic ray spectrum above 10^{17}.

Both the surface arrays and a fluorescence detector have been used to produce the cosmic ray spectrum displayed in Figure 17.1. The spectrum is a rather featureless power law falling as $E^{-2.7}$ up to an energy of 5×10^{15}eV and then steepening to a power law E^{-3}. This slight kink in the spectrum is the famous knee and coincides with the limit in energy predicted for supernova shock acceleration in the galaxy.

At 10^{14} eV, direct measurements show that the cosmic rays consist of atomic

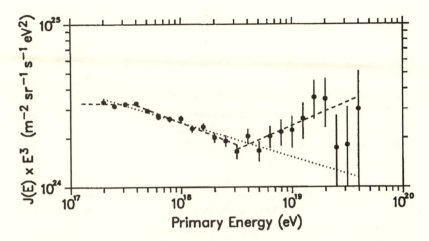

Figure 17.2: The cosmic ray spectrum as measured by the Fly's Eye experiment.

nuclei ranging from protons to iron. Beyond this energy the composition can only be measured indirectly. At the upper end of the spectrum there is some evidence that the composition has become much lighter, consistent with pure protons.

Our interest is in the very tip of the spectrum where a number of special conditions exist which present a major scientific puzzle, but where established physical phenomena can serve as analytical tools to aid in its solution. In the energy region of 10^{20}eV, the 2.7K cosmic background radiation plays a major role in the transport of the cosmic rays and limits the distance through which they can travel without severe energy loss. These cosmic rays are magnetically rigid, which suggests that their arrival directions will indicate the (projected) positions of their sources. The cosmic background radiation limits the volume from which the cosmic rays can be produced and there is no credible understanding on how such cosmic rays can be accelerated.

17.2 REVIEW OF EXISTING DATA ON THE HIGHEST ENERGY COSMIC RAYS

A thorough yet concise review of existing data prior to 1991 has been given by Watson [1]. Data on the highest energy cosmic rays come from four experiments: Haverah Park in the United Kingdom [2], and more recently, the Akeno cosmic ray observatory in Japan which now has the largest operating surface array (AGASA) covering 100 km^2[3], the Yakutsk array in north-east Siberia[4], and the Fly's Eye fluorescence detector which operated in Dugway, Utah [5]. In Figure 17.2 we show the cosmic ray energy spectrum as measured by the Fly's Eye fluorescence detector. The spectrum is multiplied by E^3 to remove the strong energy dependence and to highlight the deviations from the E^{-3} power law. Below an energy of 5×10^{18} eV, the spectrum falls faster than E^{-3}. Above 5×10^{18} eV, the spectrum flattens significantly. With the exception of one unusual event at 3×10^{20}, there are no events with energy above 10^{20} eV. This same pattern is seen in the three other experiments quoted. The Haverah Park experiment finds 4 events just slightly above 10^{20} eV. The Akeno experiment finds one event at 2.1×10^{20}eV, significantly above their next highest event which lies just below 10^{20} eV. Yakutsk finds one event just at 10^{20} eV. The first event at 10^{20}eV was found at the Volcano Ranch experiment in New Mexico in 1962[6].

As we will be discussing the two events with energy well above 10^{20} eV, it is important to ensure that the energy scales used by these experiments are accurate. In Table 17.1, we give the value of the differential cosmic ray spectrum at an energy of 10^{19} eV as reported by each of the four experiments.

Table 17.1: Comparison of differential spectrum at 10^{19}eV

Experiment	Flux at 10^{19}eV $(m^{-2}s^{-1}sr^{-1}eV^{-1})$	Energy Calibration Technique
Haverah Park	2.2×10^{-33}	Monte Carlo
Akeno	3.2×10^{-33}	Monte Carlo
Fly's Eye	2.3×10^{-33}	Fluorescence light
Yakutsk	3.4×10^{-33}	Čerenkov light

The measured fluxes agree to within 50% independent of the technique of energy measurement. Since the flux is proportional to E^{-3} a 50% variation in flux corresponds to $\sim 20\%$ in energy. The Haverah Park and Akeno spectra are derived from measurements of the density of the shower at a fixed distance of 0.6 km from the core. The constant of proportionality is calculated by Monte Carlo simulation of the shower. The Yakutsk array uses the density technique but calibrates it by the use of an array of wide angle photomultiplier detectors which observe the Čerenkov light generated by the showers on cold dark Siberian nights. The Fly's Eye obtains the energies directly from the fluorescence of the shower, which is the most direct energy calibration. Each experiment must also evaluate its efficiency as a function of the parameter which is used to determine the energy. It is remarkable that the agreement is so good at 10^{19}eV. Above that energy the data become so sparse that the important issue of the shape of the end of the spectrum cannot be addressed without the construction of much larger cosmic ray observatories. The conclusion is that each technique can accurately determine the energy of an air shower. Confidence that a surface array can accurately determine the energy is improved if it can be calibrated with optical detectors.

17.3 ACCELERATION AND TRANSPORT OF THE COSMIC RAYS $\geq 10^{19}$ EV

Cesarsky [7] has described the scientific problems concerned with acceleration and transport of these highest energy cosmic rays. For completeness we will briefly mention some of the aspects of the problem.

The familiar plot of Hillas is shown in Figure 17.3 [8]. It addresses the question of acceleration sites. For an acceleration site of size R and magnetic field B, the upper limit of energy for which a cosmic ray of charge Z can be contained is given by:

$$E(10^{18} eV) = \frac{R(kparsec) \times B(\mu gauss) \times \beta}{2Z}.$$

The factor $\beta \leq$ unity is the fraction of the maximum energy that can actually be obtained and is related to the details of the acceleration mechanism at the particular site. Lines corresponding to 10^{20} eV protons and iron nuclei are plotted. Also plotted are the size and characteristic magnetic fields of various astronomical objects. Two points can be made. First, the range of objects which might qualify as accelerators is enormous going from compact neutron stars to the entire universe. Second, very few objects have sufficient B $\times$ R to produce 10^{20} eV protons even with β equal unity. The requirement is less stringent by a factor 26 for the acceleration of iron nuclei.

That a potential acceleration site lie above the line is a necessary condition but there must also be an acceleration mechanism to boost the cosmic ray to the maximum energy. The original idea for cosmic ray acceleration was given by Fermi [9]. He imagined that the cosmic rays were scattered by clouds filled with turbulent magnetic fields. For random directions of cloud velocity and random directions of cosmic rays, there is an average fractional gain in energy on each scattering given by 4/3 β^2, where β is the mean velocity of the clouds. The maximum energy will depend on whether the magnetic fields in the clouds are strong enough to scatter the cosmic rays in a random fashion. A more efficient acceleration mechanism is provided by shock waves [7]. Here the cosmic ray bounces repeatedly back and forth across the shock wave. The bounce is produced by turbulent magnetic fields on either side of the shock. Here the fractional energy gain is given by 4/3 β, where β is the velocity of the shock wave. This mechanism is called first order Fermi

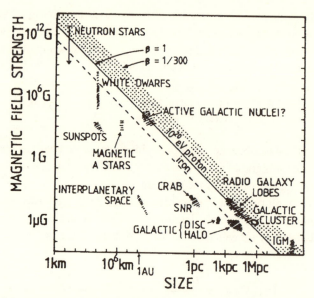

Figure 17.3: Characteristic size and magnetic field of possible acceleration sites. Objects below the solid line cannot accelerate protons to 10^{20} eV.

acceleration. The formula given above is correct for shocks where R is the radius of the shock and B is a characteristic magnetic field on either side of the shock. We refer the reader to the paper of Drury [10] for a proper discussion of cosmic ray acceleration.

The factor β is likely to be much less than unity (an extreme relativistic shock) for this mechanism. Among the most promising accelerators are the lobes at the ends of jets in the radio galaxies [11] but one has to stretch the numbers to the limit to achieve the required acceleration. There is no agreement among theoreticians that acceleration of protons to 10^{20} eV is possible in terms of what we think we know about the astrophysical objects in the universe.

These difficulties in comprehending the acceleration of the highest energy cosmic rays have stimulated the consideration of "top down" sources where topological defects from the early universe simply eject the cosmic rays at the highest energies [12]. A common source for the highest energy cosmic rays and GRB's has also been considered [13].

The transport of these highest energy cosmic rays is strongly affected by the presence of the 2.7K cosmic background radiation as was pointed out by Greisen [14] and Zatsepen and Kuz'min [15]. The title of Greisen's paper is "Is There an End to the Cosmic Ray Spectrum". Protons lose energy by photo-pion production. In the rest frame of the 10^{20} eV proton, the 2.7K photons are blue shifted by a

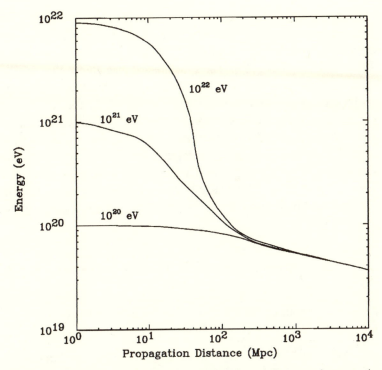

Figure 17.4: Mean energy of protons as a function of distance of propagation in the cosmic background radiation. The three curves correspond to different initial energies.

factor of 10^{11}, so that several hundred MeV is available in a collision between the proton and the 2.7K photon. We have plotted in Figure 17.4 a calculation of the mean energy as a function of distance for protons of various initial energies. One notes that for distances beyond 100 Mpc, all proton energies are less than 10^{20}eV independent of the initial energy. This is the GZK cutoff.

Photo-disintegration of nuclei by the cosmic background radiation through the giant dipole resonance is significant. An iron nucleus launched with a total energy of 10^{21} eV is totally disintegrated into its nucleon constituents after traveling 20 Mpc. The mean free path of a 3×10^{20} eV photon is only 10 Mpc in the cosmic background radiation. (For comparison, the well known Coma Cluster of galaxies is estimated to be at a distance of ~ 100 Mpc.) If a cosmic ray of 2 to 3×10^{20}eV is observed, its source must be less than $\sim$50 Mpc from our galaxy, a very short distance (or time) on a cosmological scale.

Even for the highest energy cosmic rays the effects of the galactic and extra-galactic magnetic fields cannot be ignored. Magnetic fields in our galaxy are of the order of a few μgauss, while the extragalactic fields have an upper limit of 10^{-9} gauss assuming a coherence scale of 1 Mpc. The measurements and limits on these fields are obtained by radio astronomy studies of the Faraday rotation of polarized signals from distant quasars [16]. The structure and magnitude of the magnetic field in our galaxy is roughly known so that reasonable predictions can be made about the trajectories of cosmic rays with $E/Z \geq 10^{18}$ eV [17]. The field lies in the plane of the galaxy. Parker [18] argues the the transverse thickness of the magnetic field extends between 0.6 and 2 kpc from the galactic plane. Cosmic rays with E/Z above 10^{18} eV should show a significant anisotropy if their sources follow the matter distribution of our galaxy or if their source is at the galactic center. No such anisotropy is observed [1]. In Table 17.2 we give the radius of curvature for a 3×10^{20} eV cosmic ray in magnetic fields typical of the galactic and extra galactic medium.

Table 17.2: Effect of Galactic and Extragalactic Magnetic Fields on a 3×10^{20} eV Cosmic Ray

Medium	Magnetic Field	Larmor Radius		
		Z=1	Z=8	Z=26
Galaxy	2 μG	150 kpc	19 kpc	6 kpc
Extragalactic	$\leq 1\ 10^{-9}$G	≥ 300 Mpc	≥ 40 Mpc	≥ 12 Mpc

If cosmic rays with energy $\geq 10^{20}$ are protons, they will point to their source given the distance due to the GZK cutoff. However if the cosmic rays are heavier, the effect of the galactic magnetic field must be considered. For northern hemisphere observatories the deflections are essentially perpendicular to the galactic plane as the cosmic rays enter from the galactic anticenter. A point source would be spread into a line transverse to the galactic plane.

17.4 THE BIG EVENTS

The two extraordinary events mentioned in passing can now be discussed in the light of the above remarks. These events were observed by two modern cosmic ray observatories. The first of these was observed by the Fly's Eye detector in October, 1991 at Dugway, Utah. The event is described in detail in a recent publication [19]. A thorough discussion of the possible origins of this event is given in an adjoining paper [20]. Figure 17.5 shows the longitudinal development of the shower as reconstructed from the atmospheric fluorescence. The shape of the shower devel-

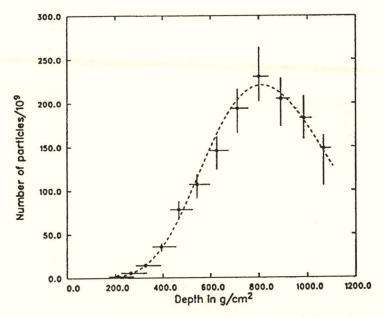

Figure 17.5: Measured number of charged particles vs. atmospheric depth for the Fly's Eye event with energy $(3.2 \pm 0.9) \times 10^{20}$ eV.

opment is consistent only with an atomic nucleus entering and interacting in the atmosphere and is consistent with a range of nuclei from proton to iron. A primary photon is probably excluded because the longitudinal development is inconsistent with a purely electromagnetic shower. A primary neutrino cannot be excluded, but its probability to have interacted in the atmosphere is $\leq 10^{-4}$. The energy of this event is $3.2 \pm 0.9 \times 10^{20}$ eV. The error includes a systematic error of 0.8×10^{20} due to the uncertainty in the absolute fluorescence efficiency of nitrogen. (The efficiency has recently been measured with much greater accuracy which does not alter the energy but reduces the error significantly [21].) The energy resolution of the Fly's Eye is 15% which excludes the possibility that the highest energy event is a fluctuation from a much lower energy. The event comes from a direction given by right ascension $(85.2 \pm 0.5)^{\circ}$ and declination of $(48 \pm 6)^{\circ}$. The next highest energy event observed by the Fly's Eye is 8×10^{19} eV.

The second highest energy event was observed by the AGASA array in December, 1993 [22]. It activated 24 of the 100 scintillators in the 100 km^2 array. Figure 17.6 shows the experimental results. The energy of this event is $2.0^{+0.6}_{-0.3} \times 10^{20}$ eV. The lateral distribution of the density of shower particles has exactly the same shape as the lower energy events and the ratio of muons to surface charged

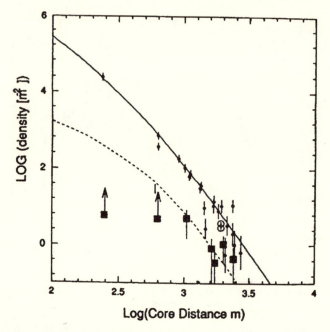

Figure 17.6: Lateral distribution of charged particles and muons for the AGASA event with energy $(2.0^{+0.6}_{-0.3}) \times 10^{20}$ eV. The solid lines are the expected lateral distributions scaled up from energies below 10^{20} eV.

particles is consistent with the lower energy events. The second highest energy event has densities a factor of two lower. This extraordinary event comes from a direction given by right ascension $(18.9 \pm 1.0)^o$ and declination $(21.1 \pm 1.0)^o$. The observed directions of these events can be significantly altered if the cosmic ray is much heavier than a proton. This is shown in Figure 17.7 where the galactic magnetic field model of Vallée [17] is used to propagate the AGASA event back out of the galaxy. The galactic latitude would have been altered by $\sim 25^o$ if this cosmic ray was an iron nucleus.

The two events just discussed are well documented and come from experimental groups with years of experience and a history of careful analysis of their results. These events, while only two, must be taken seriously.

We can ask whether these two events are unusual and are local examples of a universal distribution of cosmic ray sources sampled from a power law spectrum that extends well beyond 10^{21} eV. Figure 17.8 shows how a power law spectrum (in this case $E^{-2.5}$) is modified by the cosmic background radiation to show the GZK cutoff. At present among all the experiments there are about 50 events with energy $\geq 3 \times 10^{19}$ eV. Without the cutoff 4 events are expected with energy$\geq$

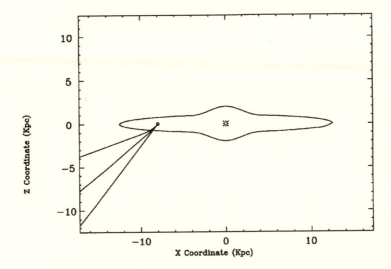

Figure 17.7: Deflection of the AGASA event by the galactic magnetic field. The nearly undeflected trajectory is for a proton. The two other trajectories correspond to an aluminum nucleus and an iron nucleus.

1.6×10^{20}eV while with the GZK cutoff only 0.1 events are expected. Two events are observed, suggesting that these two events are unusual. At the distance (50 Mpc) implied by these events, one would not expect that the assumption of a continuous distribution of sources is valid. Perhaps we are blessed with an excess of powerful sources nearby which is indeed fortunate. We have specific examples that show that sources indeed can produce energies well in excess of 10^{20} eV. At the very least, such cosmic rays should show an irregular pattern in the sky when sufficient statistics are accumulated. For distances greater than about 20 Mpc it is very unlikely that the cosmic rays can be complex nuclei since they would be disintegrated by the cosmic background radiation.

17.5 THE AUGER PROJECT

The AGASA array collects about 100 events above 10^{19} eV per year. The High Resolution Fly's Eye [23], which will be completed in less than two years, will collect about 200 events above 10^{19} eV per year. Since 1992, there has been an international effort to design and construct very large cosmic ray observatories to collect a much larger sample of events with energy $\geq 10^{19}$ eV. Following a series of workshops, an international working group was convened at Fermilab from January 30, 1995 to July 31, 1995. This group produced a conceptual de-

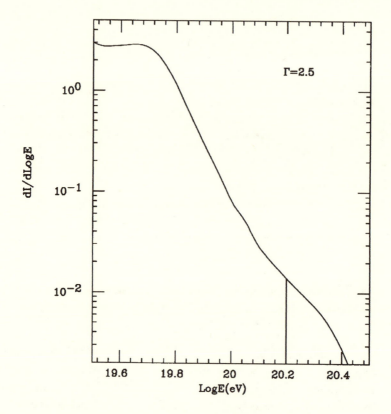

Figure 17.8: Influence of the cosmic background radiation on the cosmic ray spectrum. Sources producing a powerlaw spectrum ($E^{-2.5}$) are universally distributed. The area to the right of the vertical line corresponds to the fraction of events expected above an energy of 1.6×10^{20} eV.

sign for two observatories, each with a collecting power of 8000 km^2-sr, to be placed respectively at mid-latitude in the southern and northern hemisphere so as to observe the whole sky. It is curious that the two big events, which boost the scientific motivation, came after the project had begun. These observatories will collect a total of 8000 events above 10^{19}eV per year and $\sim$ 30 events per year $\geq 1.6 \times 10^{20}$eV if we scale from the two big events. The project has been named after Pierre Auger who first studied the extensive airshowers.

Each observatory will consist of some 1800 10 m^2 water Čerenkov detectors placed on a triangular grid with a distance between detectors of 1.5 km. The detectors will be powered by solar panels, communicate by cellular telephone techniques, and secure both fast timing and universal time from the Global Positioning Satellite system (GPS). There will be no physical connections to a station.

In addition, a fluorescence detector will be placed in the center of the array to make coincident measurement of showers that occur on dark moonless nights. These coincident events will include two independent measurements of the energy. Equally important, there will be a subset of showers for which the lateral and longitudinal distributions will be measured simultaneously. The energy resolution will be about 20% for all events. The angular resolution will be about 1° for the surface array and $\leq 0.3^o$ for the coincident events. Sensitivity to the primary composition will come from the measurement of the ratio of the muons to electromagnetic particles by the surface array and from the measurement of depth of the shower maximum and its fluctuation by the fluorescence detector. Differences in models of the highest energy hadron interactions blur somewhat the relation between the measured quantities and the primary particle. However, with the large statistics and redundant information, the primary composition can be obtained at least statistically.

In conjunction with the Fermilab working group, a site survey team has visited and evaluated sites in Argentina, South Africa, and Australia. The international Auger Project collaboration was formally constituted at a meeting at UNESCO headquarters in Paris in November 1995 and a preferred southern site, Argentina, was chosen at that time. While development work on the detectors takes place in 1996-97, financial support will be arranged. Construction will begin in 1998 and the observatories will be completed by 2001. The cost has been chosen to be no more than US 100M for the two observatories. The largest possible array will be built within that cost constraint.

17.6 What Can We Learn from Two Large Surface Arrays?

In three years of operation, one will have a sample of 2.4×10^4 events with $E \geq 10^{19}$ eV with uniform exposure over the sky. This exposure amounts to 60 events per bin of 10^0 by 10^0 with energy $\geq 10^{19}$ eV. A very natural thing to do with a sample of this size is to correlate the directions of the cosmic rays with the information available on composition obtained either by the muon to electromagnetic ratio or by the depth of shower maximum. In addition, there will be a good measurement of the time of arrival of the particles in each detector, so that the elongation rate and the fluctuation of depth of maximum of a group of showers can be inferred [24]. One might find, for example, a set of events, chosen to be rich in heavy nuclei which do correlate with the matter distribution in the galaxy. We can use our primitive knowledge of the galactic magnetic field and the absorption properties of the 2.7^0 K radiation to draw conclusions about the sources of cosmic rays.

If the cosmic rays are universal and extragalactic, one would expect an isotropic distribution and a cutoff in the spectrum. Further, because of photo disintegration, one would expect few complex nuclei.

If the cosmic rays are coming from discrete extra galactic sources, the shape of the end of the spectrum can be used as a distance measure. For a source such as the Virgo cluster at a distance of ~ 20 Mpc, one would expect a concentration of events from that direction smeared by the galactic and, perhaps, extragalactic magnetic field. One would not observe a sharp cutoff because of the relatively close distance of the source. Again, one would expect relatively few complex nuclei because of photo disintegration.

Most interesting will be the sample of roughly 100 extraordinary events with energy $\geq 2 \times 10^{20}$. These will come from nearby sources and, if protons, will point accurately to the directions of origin. They may or may not be associated with previously identified objects. If they do not, but still cluster, then they come from a source that is invisible or no longer visible. Topological defects produce a hard proton spectrum above 10^{20} eV which would be evident. The defects also produce a flux of neutrinos about 10^5 larger than the protons and might be detectable with km^2 neutrino detectors.

One may find that the extraordinary events are isotropically distributed and unclustered. Such an observation would place the highest energy cosmic rays in a category very similar to the GRB's [13]. However, even the smallest extragalactic magnetic fields and the rest mass of a proton add to the time of transit, so that a temporal correlation with a gamma ray burst can only occur if the high energy cosmic ray is a gamma ray itself. There are too many gamma ray bursts to make a convincing spatial correlation between the two phenomena.

If the extraordinary cosmic rays have a galactic source they will have a strong correlation with the galactic plane They should not show any cutoff due to the 2.7^0 K background radiation. In this case, any details of the spectrum shape will reflect the acceleration process. Indeed, such a finding would be a remarkable discovery since the known objects in our galaxy are incapable of accelerating even iron nuclei above 10^{20} eV [8]. The mean atomic number of these cosmic rays can be inferred from the muon to electromagnetic ratio and the depth of the shower maximum. For these events, this information will be quite precise because the shower will be sampled by about 20 detector stations.

17.7 SHOULD A STUDENT WORK ON THIS PROBLEM?

Research into the understanding of the highest energy cosmic rays will be a large effort with many collaborators. One might argue that a student will be lost in such a large scale effort. Nevertheless, the experiment is concerned with the exploration of a part of nature that is little understood, and the prospects for discovery are enormous. It would seem that participation by students in this effort will place them in an environment were one seeks to understand the unknown. This can be for the ambitious and self confident student an experience which will be the springboard to a successful career in science. There are many formidable technical challenges in the design and construction of the experiment. There are important measurements to be made for the development of the detector which can satisfy thesis requirements. It must be recognized that the experiment is a long term one with slow accumulation of the results. The findings of the experiment will stimulate much theoretical work which has the character of being more individual an perhaps more appropriate for a student. On the other hand we require the large scale apparatus in order to make progress. It is important that some students participate directly in the experiment. Sensitive sponsors are required so that the student becomes a genuine collaborator in the experiment. The future will require scientists who have done experimental work on large projects. In many future experiments in astrophysics, large projects will be required, and we will need the new scientists to have the strength and vision to carry them out. The student must be engaged in such a manner that the goal of understanding nature is not lost in the size and complexity of the effort. In short, the study of the highest energy cosmic rays has the intellectual strength of the best science but it is not for the "faint of heart".

17.8   FINAL REMARK

When we want to learn more about nature, exploration of extremes has consistently added to our knowledge and brought us surprises. Those of us who have worked in particle physics over the past 40 years know this all too well. I believe that a major investigation of the highest energy cosmic rays is a worthy goal on these grounds. I am confident that important new science, and fascinating phenomena, can be discovered by studying the energy extreme of the cosmic rays; this exploration can be carried out with a small fraction of the resources currently being expended in particle physics.

BIBLIOGRAPHIC NOTES

- **Gaisser, T. K. 1990, Cosmic Rays and Particle Physics (Cambridge Univ. Press). Sokolsky, P. 1988, Introduction to Ultrahigh Energy Cosmic Ray Physics (Addison Wesley).** These are two excellent introductory books on cosmic rays that the student can use to become familiar with the general subject.

- **Hillas, A. M. 1972, Cosmic Rays (Pergamon Press). Hayakawa, S. 1969, Cosmic Ray Physics (Wiley Interscience). Rossi, B. 1952, High Energy Particles (Prentice Hall).** These three books were important historically in the development of the subject.

- **Berezinskiǐ, V. S., Bulanov, S. V., Dogiel, V. A., Ginzburg, V. L., & Ptuskin, V. S. 1990, Astrophysics of Cosmic Rays (translated from the Russian) (North Holland).** This book is the classical modern treatise on cosmic rays.

- **Nucl. Phys. B (Proc. Suppl.) (1992), 28.** This journal volume contains the proceedings of the workshop held in Paris in April (1992) which launched the Auger Project.

BIBLIOGRAPHY

[1] Watson, A. A. 1991, Nucl. Phys. B (Proc Suppl.), 22B, 116.

[2] Lawrence, M. A., Reid, R. J. O., & Watson, A. A. 1991, J. Phys. G, 17, 773.

[3] Yoshida, S., et al. 1995, Astropart. Phys., 3, 105.

[4] Afanasiev, B. N., et al. 1995, Proceedings of the. 24th International Cosmic Ray Conference, Rome, Italy, 2, 756.

[5] Bird, D. J., et al. 1994, ApJ, 424, 491.

[6] Linsley, J. 1963, Phys. Rev. Lett., 10, 146.

[7] Cesarsky, C. J. 1992, Nucl. Phys. B (Proc. Suppl.), 28, 26.

[8] Hillas, A. M. 1984, ARA&A, 22, 425.

[9] Fermi, E. 1949, Phys. Rev., 75, 1169.

[10] Drury, L. O'C. 1983, Rep. Prog. Phys., 46, 973.

[11] Longair M. L. 1995, private communication; see also Biermann, P. L. 1995, Nucl. Phys. B (Proc. Suppl.) 43 221.

[12] see for example Bhattacharjee, P., Hill, C. T., & Schramm, D. N. 1992, Phys. Rev. Lett., 69, 567.

[13] Waxman, E. 1995, Phys. Rev. Lett., 75, 386; see also article by T. Piran in these proceedings.

[14] Greisen, K. 1966, Phys. Rev. Lett., 16, 748.

[15] Zatsepen, G. T., & Kuz'min, V. A. 1966, Pis'ma Zh. Eksp. Teor. Fiz., 4, 114 [Sov. Phys. JETP Lett.(Engl. Transl.) 4, 78].

[16] Kronberg, P. P. 1994, Rep. Prog. Phys., 57, 325.

[17] Vallée, J. P. 1991, ApJ, 366, 450.

[18] Parker, E. N. 1995, Proceedings of the 22nd International Cosmic Ray Conference, Dublin, Ireland, 5, 38.

[19] Bird, D. J., et al. 1995, ApJ, 441, 144.

[20] Elbert, J. A., & Sommers, P. J. 1995, ApJ, 441, 151.

[21] Loh, E., & Nagano, M. 1995, private communication.

[22] Hayashida, N., et al. 1994, Phys. Rev. Lett., 73, 3491.

[23] Bird, D. J., et al. 1995, Proceedings of the 24th International Cosmic Ray Conference, Rome, Italy, 3, 504.

[24] The elongation rate is the increase in the depth of maximum of the shower per decade increase in energy. For a discussion of the elongation rate and original references see; Cronin J. W. 1992, Nucl. Phys. B (Proc. Suppl.),28 213.

CHAPTER 18

TOWARD UNDERSTANDING GAMMA-RAY BURSTS

TSVI PIRAN

Racah Institute for Physics, The Hebrew University, Jerusalem, Israel

ABSTRACT

Gamma-ray bursts (GRBs) have puzzled astronomers since their accidental discovery in the sixties. The BATSE detector on the COMPTON-GRO satellite has been detecting one burst per day for the last four years. Its findings have revolutionized our ideas about the nature of these objects. I show, here, that the simplest, most conventional, and practically inevitable, interpretation of the observations is that GRBs result from the conversion of the kinetic energy of ultra-relativistic particles (or possibly Poynting flux) to radiation in an optically thin region. The inner "engine" that accelerates these particles is optically thick and it is hidden from direct observations. Its origin may remain mysterious for a long time.

18.1 INTRODUCTION

Gamma-ray bursts (GRBs) were discovered accidentally in the sixties by the Vela satellites. The mission of the satellites was to monitor the "outer space treaty" that forbade nuclear explosions in space. A wonderful by-product of this effort was the discovery of GRBs. Had the satellites not been needed for security purposes, it is most likely that today we would still be unaware of the existence of these mysterious bursts.

The discovery of GRBs was announced in 1973 [35]. Since then, several dedicated satellites have been launched to observe the bursts and numerous theories were put forward to explain their origin. Recently, the BATSE detector of the COMPTON-GRO has revolutionized GRB observations and consequently some of our basic ideas on the origin of GRBs. However, BATSE's observations have

343

raised as many new open questions as those they have answered. Some have even said that these observations require "new physics." I examine these questions and directions for their resolution in this lecture.

18.2   OBSERVATIONS

GRBs are short, non-thermal bursts of low energy γ-rays. After more than twenty years of GRB observations it is still difficult to summarize their basic features. This difficulty stems from the enormous variety displayed by the bursts. I will review here some features that I believe hold the key to this enigma. I refer the reader to the proceedings of the Huntsville GRB meetings [65, 23, 40] and to other recent observational reviews for a more detailed discussion [24, 22, 7, 38, 28].

Duration: The burst durations range from several microseconds to several hundred seconds, with complicated and irregular temporal structure. Several time profiles, selected from the second BATSE 2 catalog, are shown in Figure 18.1. The distribution of burst durations is bimodal [39, 43, 47, 34, 16] and can be divided into two sub-groups according to T_{90}, the time in which 90% of the burst's energy is observed: long bursts with $T_{90} > 2$sec and short bursts with $T_{90} < 2$sec. The shortest burst had a duration of 0.2 msec [2]. Some bursts are extremely long, and

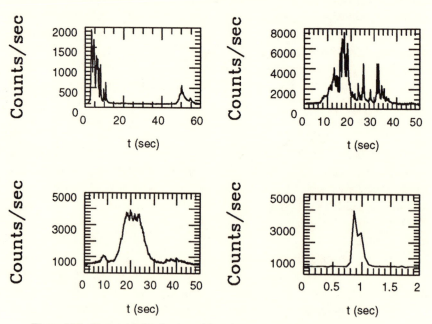

Figure 18.1: Temporal distribution of four bursts from the BATSE 2 Catalogue.

in one case high energy (GeV) photons were observed several hours after the main pulse [32]. About 3% of the bursts are preceded by a precursor with a lower peak intensity than the main burst [36].

Spectrum: The photons' energy generally peaks in the several hundred KeV range (see [6, 8] for a recent review). Presently it is not known whether this is a real feature of GRBs or it just happens that current instruments make the detection easiest in this band. While it is known that there are no softer GRBs, it is possible that there is a population of harder GRBs that emit equal power in total energy which are not observed because of selection effects [81]. It is also possible that some of the observed GRBs emit equal power per decade in higher energies. The spectrum is non-thermal. The simplest fit is a power law:

$$N(E)dE \propto E^{-\alpha}dE, \qquad (18.1)$$

with a spectral index, $\alpha \approx 2$ (see Fig. 18.2). More detailed fits include a rising power law up to some break energy and a second, generally decreasing, power law above this break [8]. Several bursts display high energy tails up to the GeV region. So far BATSE has not found any of the spectral features (absorption or emission lines) reported by earlier satellites [74].

Isotropy: The observed bursts are distributed isotropically on the sky (see Fig. 18.3). For 1005 BATSE bursts the observed dipole and quadrupole (corrected to BATSE sky exposure) relative to the galaxy are: $\langle \cos \theta \rangle = 0.017 \pm 0.018$ and $\langle \sin^2 b - 1/3 \rangle = -0.003 \pm 0.009$. These values are, respectively, 0.9σ and 0.3σ from complete isotropy [7].

Fluence and Flux Distribution: The limiting fluence observed by BATSE is $\approx 10^{-7}$ergs/cm^2. The actual fluence of the strongest bursts is larger by two

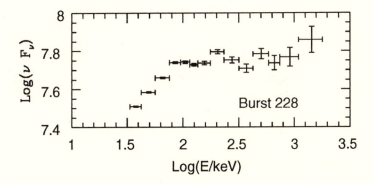

Figure 18.2: The spectrum of burst 228 from the BATSE 2 catalogue.

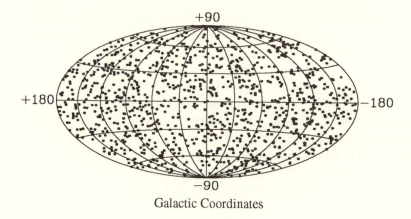

Galactic Coordinates

Figure 18.3: The Distribution of 1005 bursts on the sky.

or three orders of magnitude. A sample of 601 bursts has $\langle V/V_{max}\rangle = .328 \pm$ 0.012, which is 14σ away from the homogeneous flat space value of 0.5 [75]. Correspondingly, the peak count distribution is incompatible with a homogeneous population of sources in Euclidean space. It is compatible, however, with a cosmological distribution (see Fig. 18.4). Within the cosmological model long bursts are detected by BATSE from $0.2 \approx z_{\min} < z < z_{\max} \approx 2.1^{+1.}_{-.7}$ (assuming no source evolution). Here z_{min} and z_{max} are the minimal and maximal red shifts corresponding to the BATSE's sensitivity and observation period. Short bursts are detected from smaller distances [47, 11, 79].

Event Rate: BATSE observes about one burst per day, or ≈ 1000 bursts per year (using a detection efficiency of $\approx 30\%$). For cosmological sources, with no source evolution, this corresponds to $2.3^{-0.7}_{+1.1} \times 10^{-6}$ (long) events per galaxy per year (for $\Omega = 1$ and a galaxy density of $10^{-2}h^3$ Mpc^{-3}) [11]. The rate of short bursts is comparable. If the bursts are beamed with an opening angle θ, than the event rate should increase by a factor of $4\pi/\theta^2$ relative to this rate.

Time Dilation: Norris et al. [63, 64] found that the dimmest bursts are longer by a factor of ≈ 2.3 compared to the bright ones. This anti-correlation between the pulse's width and their intensity is compatible with the finding from the count distribution that $z_{\max} \approx 2$ and $z_{\min} \approx 0.2$ since $(1+z_{\max})/(1+z_{\min}) \approx 2.5\pm.8$. Fenimore & Bloom [19] find, on the other hand, that when the dependence of the duration on the energy band is included, this time dilation corresponds to $z_{\max} > 6$ which is incompatible with the count distribution analysis. This issue can be resolved only by a combined analysis of the count rate and of the duration using a method that avoids the issue of spectral dependence of the luminosity and the duration [12].

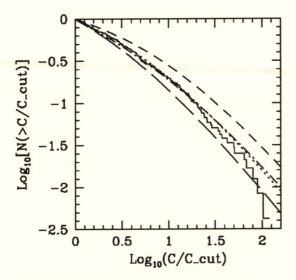

Figure 18.4: The observed long bursts' number counts distribution and three theoretical cosmological distributions with $\Omega = 1$, $\Lambda = 0$, $\alpha = -1.5$, standard candles and no source evolution: $z_{max} = 2.1$ (dotted line: best fit), $z_{max} = 1.4$ (long dashed line: lower 1% bound), $z_{max} = 3.1$ (short dashed line: upper 1% bound) and a fourth theoretical distribution with $\Omega = 0.1$ and $z_{max} = 2.1$ (dash-dot line).

Soft Gamma-Ray Repeaters (SGRs): Among more than a thousand GRBs, there is a separate group of three objects, including the famous 1979 March 5th event—the strongest GRB ever observed—that are different than all others. These unique GRBs are different in the following two ways: (i) repeated bursts are observed from the same source and (ii) the photon spectrum is softer. The three SGRs coincide with galactic SNRs (the March 5th event coincides with an SNR at the LMC). It is generally accepted that SGRs are different from regular GRBs. However [21], it has been pointed out that the spectrum of the initial part of GR 790305 is harder than the rest. This raises the possibility that GRBs and SGRs are more closely related than what was previously expected. I will not explore this possibility now and will leave it as an observational open question, a very important one, that should be resolved in the future.

Repetition: Quashnock and Lamb [84] suggested that there is evidence from the data in the BATSE 1B catalog for repetition of bursts from the same source. If true, this would severely constrain most GRB models. In particular, it would rule out the neutron star merger model [18, 79] or any other model based on a 'once in lifetime' catastrophic event. This claim has been refuted by several authors [56,

29] and most notably by the analysis of the 2B data [49]. I have mentioned it here only because of the very strong potential implications of this result, if it is true.

18.2.1   *Observational Open Questions*

There are numerous open observational questions that have not been addressed yet. Most of them deal with finding yet unknown correlations between different features of the observed data, or classifications of the bursts to sub-classes that show common characteristics. Such relations could help us distinguish between different models. In addition, there are several questions concerned with the validity of statements that have been made about the data. The best known among those are:

- *What is the relation, if any, between GRBs and SGRs?*
- *Do GRBs repeat?*
- *Are there absorption lines? Or any other spectral features?*
- *Are there bursts in other parts of the electromagnetic spectrum?*
- *What fraction of GRBs emit significant amounts of energy at the GeV range?*
- *Is the time-dilation compatible with the count distribution?*

Some believe that some of these questions have already been answered. The fact that there is not a widespread consensus qualifies them as open questions.

18.3   A Brief Summary

One could say that a fair summary of our present understanding of GRBs can be given in the form of three basic open questions:

- **What?**
- **Where?**
- **How?**

If one is more ambitious one can pose a fourth question:

- **Why?**

That this is a reasonable summary is demonstrated by the proliferation of GRB models: a recent review [59] counted more than a hundred. At the time, there were more theories than bursts! BATSE has improved this situation enormously. Even the most prolific theoreticians cannot compete with BATSE's rate of one burst per day. Today, in the post-BATSE era, the number of observed bursts exceeds the number of theories by one order of magnitude!

In the rest of this talk I will attempt to show, how the current data directs us toward some partial answers to those questions and what the new open questions are that have emerged from this understanding. An alternative open question is, of course, to find the flaws or the loopholes in this chain of arguments.

18.4 WHERE?

BATSE has revolutionized our ideas about the location of GRBs. Before BATSE, it was generally believed that GRBs originate in the galactic disk. The isotropy of the sources observed by BATSE ruled out distant galactic disk population, while the incompatibility of BATSE's count rate distribution with an Euclidean homogeneous distribution ruled out local galactic disk sources. After BATSE the only place left for GRBs in the Galaxy is at the distant parts of an extended galactic halo (with typical distances larger than 100 kpc) [27]. On the other hand, the observed distribution is compatible with a cosmological distribution which is isotropic and homogeneous, but with a count distribution that deviates from the $C^{-3/2}$ law due to cosmological effects (see e.g., [48, 77, 15, 11]). The cosmological hypothesis is supported by the fact that the predicted [77, 71] anti-correlation between the duration and intensity of the bursts was recently found [66, 63, 61, 64] (see however, [19]). The cosmological interpretation corresponds to an event rate of $2.3^{-0.7}_{+1.1} \cdot 10^{-6}$ events per galaxy per year. As we will see latter this is the best (and possibly only) direct clue that we have today to the nature of the sources. Finally, there is a preliminary evidence [10] for correlation between Abell clusters and GRBs in the 2B catalog. If confirmed this will, of course, proof, once for all the cosmological origin of GRBs.

It is tempting to enumerate the pro and con arguments for a galactic origin. However, I will not do this for two reasons. First, a *Great Debate* [73, 42] just took place on this issue and the arguments are discussed extensively there. Second, it is my personal opinion that this is no longer an open question and the observational data point clearly in the direction of an extra-galactic origin. I will, focus, therefore, on these models in the rest of my talk. I should stress, however, that current galactic models put the sources so far in the halo that the general model that present here is valid (with the appropriate numerical scaling) to such sources as well.

In addition to the classical question:

•*Extragalactic or Galactic?*
which both sides believe is not an open question, there are further questions in the context of both models. Some of those are:

•*What are the red-shifts (or distances) from which we observe GRBs?*
•*Can we rule out source evolution in the count distribution analysis?*
•*If the bursts are galactic, then where are the bursts from M31?*

18.5 HOW?

Before turning to the question of **what** can generate GRBs I shall address the question **how** this can be done. Understanding **how** might direct us towards **what**. I shall go backward from the observations toward the sources and I shall try to keep the discussion as general as possible.

The key to our discussion is the compactness problem. How can a compact source, as inferred from the rapid time variability, emit so much energy in such a short time and remain optically thin, as inferred from the observed non-thermal spectra? The only conventional resolution of this problem known today is extreme-relativistic motion of the source. All other solutions require "new physics." Once we accept the idea that the bursts involve extreme-relativistic motion, it follows that the simplest and energetically most economical way to generate the bursts is via conversion of the kinetic energy of the ultra-relativistic particles to the observed gamma-ray photons. This deflects the question of the origin of GRBs to the questions of how to produce large bursts of ultra-relativistic particles and of how to convert the kinetic energy of these particles to radiation.

18.5.1   *The Compactness Problem*

The key to understanding GRBs lies, I believe, in understanding how GRBs bypass the compactness problem. This problem was realized very early on by Schmidt [87]. At that time it was used to show that GRBs cannot originate too far from us. Now, we understand that this interpretation is false and instead we must look for ways to overcome this constraint.

The observed fluence, $F \approx 10^{-7}$ergs/cm^2, corresponds, for an isotropic source at a distance D, to a total energy release of:

$$E = 4\pi D^2 F = 10^{50} \text{ergs} \left(\frac{D}{3000 \text{ Mpc}} \right)^2 \left(\frac{F}{10^{-7}\text{ergs/cm}^2} \right). \qquad (18.2)$$

Cosmological effects change this equality by numerical factors of order unity that are not important for our discussion. The rapid temporal variability observed in some bursts (see Fig. 18.1), $\Delta T \approx 10$ msec, implies that the sources are compact with a size, R_i, smaller than $c\Delta T \approx 3000$ km (note that the shortest burst had a duration of 0.2 msec [2]). The observed spectrum (see Fig. 18.2) contains a large fraction (of the order of a few percent) of the gamma-ray photons with high energies. These photons could interact with lower energy photons and produce electron positron pairs via $\gamma\gamma \rightarrow e^+e^-$. When two photons with energies E_1 and E_2 interact, the c.m. energy is $\sim 2\sqrt{E_1 E_2}$ (up to angular factor). This energy is sufficient to produce a pair if $\sqrt{E_1 E_2} > m_e c^2$. I denote by f_p the fraction of photon pairs that satisfy this condition. The initial optical depth for this process is [26]:

$$\tau_{\gamma\gamma} = \frac{f_p \sigma_T F D^2}{R_i^2 m_e c^2},$$

or

$$\tau_{\gamma\gamma} = 10^{13} f_p \left(\frac{F}{10^{-7} \text{ergs/cm}^2} \right) \left(\frac{D}{3000 \text{ Mpc}} \right)^2 \left(\frac{\Delta T}{10 \text{ msec}} \right)^{-2}, \quad (18.3)$$

where σ_T is the Thompson cross section. This optical depth is very large. Even if there are no pairs to begin with they will form rapidly and will Compton scatter lower energy photons. The resulting huge optical depth will prevent us from observing the radiation emitted by the source. Even if the initial spectrum is non-thermal the electron-positron pairs will thermalize it and the resulting spectrum will be incompatible with the observations! This is the compactness problem.

Equation (18.3) reveals that $\tau_{\gamma\gamma} \gg 1$ even if the bursts originate in the extended galactic halo, $D \approx 100$ kpc. Thus, the following discussion is applicable to Galactic halo models as well [82].

It has been argued that the only way to avoid the compactness problem is if the sources are nearby ($D < 1$ kpc). At such distances, the total energy required is small and equation (18.3) yields $\tau_{\gamma\gamma} \lesssim 1$. Alternatively, it has been argued on the basis of this problem that "new physics" is unavoidable if GRBs are at cosmological distances. We will see, however, that it is possible to resolve this paradox within the limits of present day physics.

Compactness would not be a problem if the energy could escape from the source in some non-electromagnetic form which would be converted to electromagnetic radiation at a large distance, R_X. This radius will replace the source's size $R_i < c\Delta T$ in equation (18.3). R_X should be sufficiently large so that $\tau_{\gamma\gamma}(R_X) < 1$. A trivial solution of this kind is a weakly interacting particle, which I will call particle X, which is converted in flight to electromagnetic radiation. The only problem with this solution is that no known particle can play the role of particle X (see however [45]), and this solution requires, indeed, "new physics."
•*Can we rule out particle X or find a physical candidate?*

18.5.2 *Relativistic Motion*

Relativistic effects can fool us and, when ignored, lead to wrong conclusions. This happened 30 years ago when rapid variability implied "impossible" temperatures in extra-galactic radio sources. This puzzle was resolved when M. J. Rees suggested ultra-relativistic expansion which has been confirmed by VLBA measurements of super luminal jets. This also happened in the present case. Consider a source of radiation that is moving towards an observer at rest with a relativistic velocity characterized by a Lorentz factor, $\Gamma = 1/\sqrt{1 - v^2/c^2} \gg 1$. The observer detects photons with energy $h\nu_{obs}$. These photons have been blue shifted and their energy at the source was $\approx h\nu_{obs}/\Gamma$. Fewer photons will have sufficient energy to produce pairs, and the fraction f_p at the source is smaller by a factor $\Gamma^{-2\alpha}$ [where α is the spectral index defined in equation (18.4)] than the observed fraction. At the same time, relativistic effects allow the radius from which the radiation is emitted, $R_i < \Gamma^2 c\Delta T$ to be larger than the original estimate, $R_i < c\Delta T$, by a factor of Γ^2. We have

$$\tau_{\gamma\gamma} = \frac{f_p}{\Gamma^\alpha} \frac{\sigma_T F D^2}{R_i^2 m_e c^2} \, ,$$

or

$$\tau_{\gamma\gamma} \approx \frac{10^{13}}{\Gamma^{(4+2\alpha)}} f_p \left(\frac{F}{10^{-7} \text{ergs/cm}^2} \right) \left(\frac{D}{3000 \text{ Mpc}} \right)^2 \left(\frac{\Delta T}{10 \text{ msec}} \right)^{-2}, \quad (18.4)$$

where the relativistic limit on R_i was included in the second line. The compactness problem can be resolved if the sources are moving relativistically towards us with Lorentz factors $\Gamma > 10^{13/(4+2\alpha)} \approx 10^2$. A more detailed discussion [20, 101] gives practically the same result. Such extreme-relativistic motion ($v \approx 0.9995c$) is even faster than the ultra-relativistic motion observed in the above mentioned super luminal jets which have Lorentz factors of ~ 10. This resolution of the paradox is clearly within conventional physics, as all that it requires is special relativistic effects. But it involves extremely relativistic motions that were never met before.

The potential of relativistic motion to resolve the compactness problem was realized in the eighties by Goodman [25], Paczyński [67] and Krolik and Pier [41]. There was however a drastic difference between the first two approaches and the last one. Goodman [25] and Paczyński [67] considered relativistic motion in the dynamical context of fireballs. In this case the relativistic motion is an integral part of the burst mechanism. Krolik and Pier [41] considered, on the other hand, a kinematical solution, in which the sources move relativistically and this motion is not necessarily related to the mechanism that produces the bursts.

Is the kinematic scenario feasible? In this scenario the source moves relativistically as a whole. The radiation is beamed with an opening angle of Γ^{-1}. The total emitted energy is smaller by a factor Γ^{-3} than the isotropic estimate given in equation (18.2). The total energy required, however, is at least $(Mc^2 + 4\pi F D^2/\Gamma^3)\Gamma$, where M is the rest mass of the source (the energy would be larger by an additional amount $E_{th}\Gamma$ if an internal energy, E_{th}, remains in the source after the burst has been emitted). The whole process becomes very wasteful if the kinetic energy, $Mc^2\Gamma$, is much larger than the observed energy of the burst, $(4\pi/\Gamma^2)F D^2$.

In most cases this is exactly what happens and the total required energy is so large that the model becomes infeasible. The only exception is the most energetically-economical situation in which the kinetic energy itself is the source of the observed radiation. This is also the most conceptually economical situation, since in this case the gamma-ray emission and the relativistic motion of the source are related and are not two independent phenomena. This will be the case if GRBs result from the slowing down of ultra relativistic matter. This idea was suggested by Mészaros, & Rees [51, 52] in the context of the slowing down of fireball accelerated material [88] by the ISM and by Narayan, et al. [55] and independently by Mészaros, & Rees [85] in the context of self interaction and internal shocks within the fireball. However, it seems to be much more general, and in my mind it is an essential part of any GRB model regardless of the acceleration mechanism of the relativistic particles!

Assuming that GRBs result from the slowing down of relativistic bulk motion of massive particles, we find that the required total mass of the ultra-relativistic particles is:

$$M = \frac{\theta^2 F D^2}{\Gamma \epsilon_c}, \tag{18.5}$$

where ϵ_c is the conversion efficiency and θ is the opening angle of the emitted radiation. The relativistic motion does not imply relativistic beaming as is sometimes mistakenly asserted. θ can be as small as Γ^{-1}, the limiting relativistic beaming factor, if the matter has been accelerated along a very narrow beam. Notice that relativistic beaming requires an event rate larger by a ratio $4\pi\Gamma^2$ compared to the observed rate. With observations of about one burst per 10^{-6} year per galaxy this implies one event per year per galaxy! θ^2 can be as large as 4π as would be the case if the motion results from the relativistic expansion of a fireball [25, 67]. The opening angle can also have any intermediate value if it emerges from a beam with an opening angle $\theta > \Gamma^{-1}$, as will be the case if the source is an anisotropic fireball [78, 58] (see Fig. 18.5) or an electromagnetic accelerator with a modest beam width. In the last two cases each observer will see, indeed, radiation beamed towards him or her from a region whose width is Γ^{-1}. However, observers that are more than Γ^{-1} apart will still see a burst from the same source.

We have found particle X in the simple form of a proton. Protons escape from

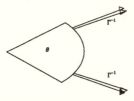

Figure 18.5: Radiation from a relativistic beam with a width θ. Each observer will detect radiation only from a very narrow beam with a width Γ^{-1}. The overall angular size of the observed phenomenon can vary, however, with $\Gamma^{-2} < \theta^2 < 4\pi$.

the source carrying the energy as kinetic energy. To produce a GRB, they should convert their kinetic energy to radiation, otherwise they are useless. The next question is, therefore, how is the energy converted?

Two variant that has been suggested recently on this theme are based on the possibility that a fraction of the energy is carried by Poynting flux [91, 93, 94]. A simple calculation reveals that this requires magnetic fields of 10^{15}Gauss or higher at the source, a large value, which might be reached in stellar collapses of highly magnetized stars. Overall the different models can be characterized by two parameters: the ratio of the kinetic energy flux to the Poynting flux and the location of the energy conversion stage ($\approx 10^{12}$ cm for internal conversion or $\approx 10^{16}$cm for external conversion). This is summarized in Table 18.1.

In the following discussion we will focus on the simplest possibility, that is of a kinetic energy flux.

18.5.3   Slowing Down of Relativistic Particles

The cross section for a direct nuclear or electromagnetic interaction between the relativistic protons and the ISM protons is too small to convert efficiently the kinetic energy to radiation. The only way that the protons can be slowed down is via some collective interaction such as a collisionless shock. The existence of supernova remnants in which the supernova ejecta is slowed down by the ISM indicates that collisionless shocks do form in similar circumstances and one can expect that they will form here as well [51].

GRBs are the relativistic analogues of supernova remnants (SNRs). In both cases the phenomenon results from the conversion of the kinetic energy of the ejected matter to radiation. Even the total energy involved is comparable. The crucial difference is the amount of ejected mass. SNRs involve several solar masses. The corresponding velocity is several thousands kilometers per second, much less

Table 18.1: General Scheme for Energy Transport

	Kinetic Energy Dominated	Kinetic Energy and Poynting Flux	Poynting Flux Dominated
Internal conversion	[55, 85]	[91]	[93]
External conversion	[51, 52]	—	[94]

than the speed of light. In GRBs, the masses are smaller by several orders of magnitude and with the same energy the matter attains ultra-relativistic velocities. The interaction between the SNR ejecta and the ISM takes place on scales of several pc and it is observed for thousands of years. The interaction of the relativistic matter in GRBs with the ISM takes place on a comparable but slightly smaller distance scale. Special relativistic effects reduce, however, the observed duration of the bursts to a few seconds! In the following sections I will discuss some of the basic features of these processes and will examine how they can be related to GRB observations. The discussion begins with a general review of the shock conditions that occur in collision between relativistic matter. Discussion of interaction with the ISM follows and I conclude, with a brief examination of the possibility that internal shocks rather than shocks with the ISM are responsible for extracting the kinetic energy of the relativistic particles [55, 85].

Shock Conditions

We consider a cold shell (whose internal energy is negligible compared to the rest mass energy) that overtakes another cold shell or moves into the cold ISM. Two shocks form: an outgoing shock that propagates into the ISM or into the external shell, and a reverse shock that propagates into the inner shell, with a contact discontinuity between the shocked material (see Fig. 18.6). Two quantities determine the shocks' structure: Γ, the Lorentz factor of the motion of the inner shell (denoted 4) relative to the external matter (denoted 1) and the ratio between the particle number densities in these regions, $f \equiv n_4/n_1$. I ignore here a third quantity, the adiabatic index of the matter, which gives rise only to factors of order unity.

In the original analysis of [51, 52, 33] it was assumed that both shocks are relativistic. In fact, this holds only if $f < \Gamma^2$. If this condition holds and if $\Gamma \gg 1$ then the shock equations between regions 1 and 2 yield: [3, 4, 78]:

$$\gamma_2 = f^{1/4}\Gamma^{1/2}/\sqrt{2} \; ; \; n_2 = 4\gamma_2 n_1 \; ; \; e \equiv e_2 = \gamma_2^2 n_1 m_p c^2, \qquad (18.6)$$

where γ_2 is the Lorentz factor of the motion of the shocked fluid relative to the rest frame of the fluid at 1 (an external observer for interaction with the ISM and the outer shell in case of internal collision). The Lorentz factor of the shock front itself is $\sqrt{2}\gamma_2$. Similar relations hold for the reverse shock:

$$\bar{\gamma}_3 = f^{-1/4}\Gamma^{1/2}/\sqrt{2} \; ; \; n_3 = 4\bar{\gamma}_3 n_4, \qquad (18.7)$$

where $\bar{\gamma}_3$ is measured relative to the rest frame of the relativistic shell. In addition we have $e_3 = e$ and $\bar{\gamma}_3 \cong (\Gamma/\gamma_2 + \gamma_2/\Gamma)/2$ which follow from the equality of pressures and velocity on the contact discontinuity.

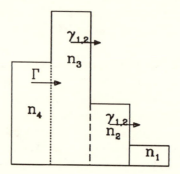

Figure 18.6: Schematic density profile in the interaction between a relativistic shell of matter (region 4) and the ISM (region 1). The shocked regions are 2, shocked ISM material, and 3, shocked shell material. The forward shock is marked by a solid line. The reverse shock is marked by a dotted line, and the contact discontinuity between regions 2 and 3 is marked by a dashed line.

If $f < \Gamma^2$ the reverse shock is non-relativistic and:

$$\gamma_2 \approx \Gamma \quad ; \quad \bar{\gamma}_3 \approx 1. \tag{18.8}$$

$$n_2 \approx 4\Gamma n_1, \quad ; \quad e \equiv e_2 = 4\Gamma^2 n_1 m_p c^2 \quad ; \quad n_3 = 7n_4, \quad ; \quad e_3 = e. \tag{18.9}$$

Comparable amounts of energy are converted to thermal energy in both shocks when both shocks are relativistic. But only a negligible amount of energy is converted to thermal energy in the reverse shock if it is Newtonian [86]. The above shock conditions follow from a planar analysis. However, numerical simulations of spherical ultra-relativistic shocks [86] show that these conditions are valid at each moment of time even for spherical systems.

Shocks with the ISM

We turn now to the scenario in which a relativistic shell interacts with the ISM. The shock structure has been described in the previous section. The critical question is where is the kinetic energy converted to thermal energy? This determines immediately the time scale of the observed bursts and other observed features such as the spectra. To answer this question we should follow the shell as it propagates outwards. Initially the shell is very dense and the density ratio between the shell and the ISM, f, is extremely large (more specifically $f > \gamma^2$). The reverse shock is a very weak Newtonian shock which converts only a small fraction of the kinetic energy to thermal energy. As the shell propagates this density ratio decreases (like R^{-2} if the width is constant) and eventually the reverse shock becomes relativistic at R_N where $f = \gamma^2$. The question where is the kinetic energy converted depends on whether the reverse shock reaches the inner edge of the shell before or after it becomes relativistic.

Like in SNRs we define the Sedov length, $l \equiv (E/n_{ism}m_p c^2)^{1/3}$. The ISM rest mass within a volume l^3 equals the energy of the GRB: E/c^2. For a canonical cosmological burst with $E \approx 10^{51}$ergs and a typical ISM density $n_{ism} = 1$particle/cm^3 we have $l \approx 10^{18}$cm. A second length scale is ΔR, the width of the relativistic shell in the observer's rest frame. The answer to our question depends on a dimensionless parameter constructed from l, ΔR and Γ [86]:

$$\xi \equiv (l/\Delta R)^{1/2}\Gamma^{-4/3} . \tag{18.10}$$

Using the expression for the velocity of the reverse shock into the shell we find that the reverse shock reaches the inner edge of the shell at R_δ [86]:

$$R_\Delta = l^{3/4}\Delta^{1/4} . \tag{18.11}$$

This should be compared with the expression for R_N

$$R_N = l^{3/2}/\Delta^{1/2}\gamma^2 = \xi^{3/2}R_\Delta . \tag{18.12}$$

If $\xi > 1$ then the reverse shock remains Newtonian or at best mildly relativistic during the whole energy extraction process. The Newtonian reverse shock reaches the inner edge of the shock at R_Δ. At this stage a reflected rarefraction wave begins to move forwards. This wave is, in turn, reflected from the contact discontinuity, between the shell's material and the ISM material, and another reverse shock begins. The overall outcome of these waves is that in this case the shell acts as a single fluid element of mass $M \approx E/\Gamma c^2$ that is interacting collectively with the ISM. A simple estimate, based on a plastic collision between a mass M moving with a Lorentz factor Γ and a mass m at rest, shows that $m = M/\Gamma$ is required to reduce Γ to $\Gamma/2$ and to convert half of the kinetic energy to thermal energy. In agreement with this analysis we find that an effective energy release occurs at R_Γ where the shell collects an ISM mass of M/Γ [51]:

$$R_\Gamma = \frac{l}{\gamma^{2/3}} = \left(\frac{M}{(4\pi/3)n_{ism}}\right)^{1/3} = \left(\frac{3FD^2}{\epsilon_c m_p c^2 n_{ism}\Gamma}\right)^{1/3} =$$

$$6 \times 10^{16} \text{ cm } \epsilon_c^{-1/3}\left(\frac{F}{10^{-7}\text{ergs/cm}^2}\right)^{1/3}\left(\frac{D}{3000 \text{ Mpc}}\right)^{2/3}\left(\frac{\Gamma}{100}\right)^{-1/3}, \tag{18.13}$$

where we have used in the second equality equation (18.5) that relates the ejected mass to the observed fluence and the distance to the source. The distance, R_Γ, is, incidentally, independent of the opening angle θ of the beam. If θ is smaller, less mass is needed, but correspondingly less mass is swept up from the interstellar medium. Substitution of R_Γ in equation (18.4) reveals that the optical depth is much smaller than one.

If the shell propagates with a constant width then $R_N/\xi = R_\Gamma = \sqrt{\xi}R_\Delta$ (see Fig. 18.7) and for $\xi > 1$ the reverse shock remains Newtonian during the energy extraction period. If there are significant variations in the particles velocity within the shell it will spread during the expansion. If the typical variation in Γ is of the same order as Γ then the shell width increases like R/Γ^2. This delays the time that the reverse shock reaches the inner edge of the shell and increases R_Δ. It also reduces the shell's density which, in turn, reduces f and leads to a decrease in R_N. The overall result is a triple coincidence $R_N/ \approx R_\Gamma \approx R_\Delta$ with a mildly relativistic reverse shock and a significant energy conversion in the reverse shock as well.

The bursts' duration is determined by the slowing down time of the shell. The emitting region moves towards the observer with a Lorentz factor γ_2. Two photons that are emitted with a time delay dt will be detected with a time delay dt/γ_2^2. Additionally, an observer detects radiation from a region with an angular size γ_2^{-1}. A photon emerging from an angle γ_2^{-1} away from the center of a region with a radius R will be detected at a time $R/\gamma_2^2 c$ after a photon that emerges from the center [33]. Thus, given a typical radius of energy conversion, R_Γ the observed

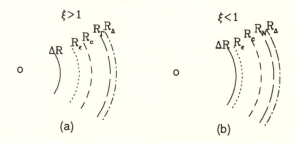

(a)   (b)

Figure 18.7: (a) Schematic description of the different radii for the case $\xi > 1$. The different distances are marked on a logarithmic scale. Beginning from the inside we have ΔR, the initial size of the shell, R_η, the radius in which a fireball becomes matter-dominated (see the following discussion), R_c, the radius where inner shells overtake each other and collide, R_Δ, where the reverse shock reaches the inner boundary of the shell, and R_Γ, where the kinetic energy of the shell is converted into thermal energy. (b) Same as (a) for $\xi < 1$. R_Γ does not appear here since it is not relevant. R_N marks the place where the reverse shock becomes relativistic.

time scale is:

$$T_{obs} \approx R_{\Gamma}/(\gamma_2^2 c) \approx R_{\Gamma}/(\Gamma^2 c) \approx 200 \text{ sec } \epsilon_c^{-1/3} \left(\frac{F}{10^{-7}\text{ergs/cm}^2}\right)^{1/3} \times$$

$$\left(\frac{D}{3000 \text{ Mpc}}\right)^{2/3} \left(\frac{n_{ism}}{1 \text{ cm}^{-3}}\right)^{-1/3} \left(\frac{\Gamma}{100}\right)^{-7/3}. \qquad (18.14)$$

This value is comparable to the duration of the long bursts. It is very sensitive to Γ. An increase of Γ by a factor of 10 will reduce the time scale by two orders of magnitudes to the transition regime between long and short bursts. Another increase by a factor of 10 in Γ is required to reach the rapid variability observed in some of the short bursts.

If Γ is higher or if ΔR is larger then $\xi < 1$. In this case the reverse shock becomes relativistic very early (see Fig. 18.7). Since $\gamma_2 \ll \Gamma$ the relativistic reverse shock converts very efficiently the kinetic energy of the shell to thermal energy. Each layer of the shell that is shocked loses effectively all its kinetic energy at once and the time scale of converting the shell's kinetic energy to thermal energy is the shell crossing time. The kinetic energy is consumed at R_Δ, where the shock reaches the inner edge of the shell. The observed time scale is now:

$$T_{obs} = R_\Delta/\gamma_2^2 \approx \Delta R/c = 1 \text{ sec } \left(\frac{\Delta R}{3 \times 10^{10}\text{cm}}\right). \qquad (18.15)$$

The observed time scale depends only on ΔR and it is independent of Γ! Spreading do not affect this estimates since for $\xi < 1$ spreading occurs very late.

An interesting feature is that γ_2 the Lorentz factor of the emitting region is in this case $\approx (l/\Delta R)^{3/8} \approx 10^3$. It is rather insensitive to the burst's parameters and in particular to the Lorentz factor γ.

Internal shocks

Internal shocks take place when an inner shell overtakes a slower outer shell. Consider an inner shell with a Lorentz factor Γ and an outer shell with a Lorentz factor $\Gamma/2$. The inner shell will overtake the outer one at:

$$R_c \approx \Gamma^2 \delta R \approx 10^{12}\text{cm}\left(\frac{\delta R}{1000\text{km}}\right)\left(\frac{\Gamma}{100}\right)^{-2} \qquad (18.16)$$

where δR is the initial separation between the shells in the observer's rest frame. The resulting shocks are mildly relativistic, with typical Lorentz factors of order $\sqrt{2}$. If the two shells have comparable masses then these shocks will convert a few percent of the kinetic energy into thermal energy.

The first question that one should ask is whether the compactness problem is resolved. Substituting R_c in equation (18.4) we find that R_c is large enough so that $\tau \approx 1$, the shock region is optically thin and the emitted photons escape freely to infinity. The observed time scale for the each collision:

$$\delta T_{obs} \approx R_c/(\Gamma^2 c) \approx \frac{\delta R}{c} \qquad (18.17)$$

is small enough to produce even the fastest observed variability if $\delta R < 10^8$cm. The overall duration of the burst is, in this scenario, the duration of the emitted pulse of relativistic particles:

$$T_{obs} = \Delta R/c = 1 \text{ sec } \left(\frac{\Delta R}{3 \times 10^{10} \text{cm}} \right). \qquad (18.18)$$

Examination of these last two equations reveals some information on the requirements from this mechanism. The observed bursts with long durations ($T_{obs} > 100$sec) require long relativistic winds that last hundred seconds, with a rapid variability on a time scale of a fraction of a second.

It was recently discovered that the situation in which an inner shell is faster than an outer shell is unstable [98]. The instability develops before the shocks form and it may drastically affect the energy conversion process. The full implications of this instability are not understood yet.

Three open questions that arise in this scenario are:
• *What could be the source of a relativistic wind satisfying these conditions?*
Another question that follows from the relatively low efficiency of this process is:
• *What happens to the bulk of the kinetic energy?*
Finally one should ask
• *What are the implications of the instability?*

Open Questions

The previous discussion contains some detailed analysis of the conditions at the place where the kinetic energy is converted to thermal energy and where the radiation is emitted. We have seen that internal shocks or shocks with the ISM can convert the kinetic energy of the relativistic baryons to radiation and that the process can take place with the right time scale. While this analysis shows that we possibly understand the main framework, it is clear that many details are missing. The basic flaw of this analysis is that there is no mechanism that anchors the resulting photons to the observed range of low-energy γ-rays. The observed energy is strongly dependent on Γ, and unless some robust process keeps different parameters, and in particular γ roughly constant, we should observe similar events with uv/x-ray photons and much harder γ-rays. Such events are not observed! The lack of observations may simply result from lack of all sky burst detectors in other parts of the spectra (recall that the soft gamma-ray band was selected because of security rather than scientific reasons). Furthermore, even in GRBs it is possible that significant amounts of energy are emitted as hard γ-rays. Still, the fact that there is no clear explanation for this feature of GRBs may cause the reader to question the whole analysis. He or she might be right, but in fairness one should recall that almost none of the GRB models that have been suggested so far provides a satisfactory answer to this question. The most puzzling question is, therefore:

• *What nails the observed spectrum to the soft gamma-ray band, or alternatively are there comparable bursts in other parts of the spectrum?*

A related question is:

• *Why there are no counterparts to GRBs events at other parts of the spectrum?*

Two other questions that are emerge from this analysis are:

• *What are the relative roles of internal vs.. ISM shocks?*

• *Can we explain the bimodal distribution of durations in terms of internal vs. ISM shocks or in terms of Newtonian vs. relativistic reverse shock?*

When we recall that both internal shocks and an interaction with the ISM can take place in the same burst we realize that the radiation from the internal shocks will arrive at a time $\approx R_\Gamma/\Gamma^2 c$ or $R_\Delta \Gamma^2 c$ before the radiation from the interaction with the ISM. With reasonable parameters this interval would be several tens of seconds which leads us to another question:

• *Can internal shocks produce the precursors observed in some bursts?*

I should now remark on the numerical values of the parameters used in this analysis. We have seen early on that equation (18.4) gave a lower limit on Γ which was of order 100. Being conservative, I have used this lower limit as the canonical value in this analysis. Historically, one would have used $\delta R \approx \Delta R \approx 1000$ km as indicated by the shortest time scale variability observed in some of the bursts. However, we have seen now that, in fact, this value is not necessarily relevant any more. Different scenarios put different constraints on the width of the shell, or, alternatively, on the size of the internal engine: Internal shocks seem to require long pulses (with a duration comparable to the observed duration) and variability on scale of $\approx 10^8$cm. ISM shocks seem to require narrow bursts, with a total width of less than 10^7cm. Lorentz factors $\Gamma \approx 10^4$ can increase, however, the allowed ΔR up to 10^{13}cm. While both models indicate the need for a compact source, the situation is not clear and the immediate question that follows is:

• *Can we determine Γ and ΔR from the current observations?*

18.5.4   *The Acceleration Mechanism?*

GRBs require a short burst of ultra-relativistic particles, with a total energy of $\approx$ $10^{50}/4\pi$ ergs per steradian and a Lorentz factor of a hundred or larger. According to the analysis presented so far, the observed γ-rays are produced when the ultra-relativistic particles are slowed down. However, there are no direct observations that can tell us about the acceleration phase. This brings us directly to the next open question:

•*What is the Acceleration Mechanism?*

There are two clear alternatives: a non-thermal, most likely electromagnetic, mechanism, or a thermal mechanism in which the particles are accelerated by thermal pressure. This second mechanism falls under the general category of fireballs.

There is little that I can say about the non-thermal acceleration mechanism. The analogy to pulsars and other steady state sources that produce high energy radiation and that accelerate particles to relativistic velocities is appealing. One has to remember, however, that the energies reached here are significantly larger than those observed in other astrophysical jets. The observed asymmetry in the temporal pulse profiles (rapid rise vs. slower fall [37]) practically rules out a "light house" (i.e. a rotating beam) model, which will be symmetric in the initial rise and the final decay, as an alternative to pulsed beams. Even with relativistic beaming the energy requirements from this accelerator are severe: $\approx 10^{46}/(\Gamma/100)^2$ ergs within a few seconds! This brings us immediately to a mysterious open question:

•*Is there a suitable non-thermal (electromagnetic) acceleration mechanism that satisfies these constraints?*

I will focus in the rest of this section on thermal acceleration that is on the fireball process.

Fireballs

GRBs involve the release of $\approx 10^{50}$ ergs of radiation into a small volume with a radius of $\approx 10^3$ km. Equation (18.3) shows that such a system will be optically thick to $\gamma\gamma \rightarrow e^+e^-$. The radiation will not be able to escape and the large optical depth will cause it to reach thermal equilibrium rapidly, with a temperature: $T_i \approx$ $1\text{MeV}(E/10^{50}\text{ergs})^{1/4}(R_i/10^3\text{km})^{-3/4}$. At this temperature there is a copious number of $e^+ - e^-$ pairs that contribute to the opacity via Compton scattering. The system turns into a fireball: a dense radiation and electron-positron pair fluid. The fluid expands under its own pressure and it cools adiabatically due to this expansion. This phase ends when the fireball becomes optically thin and stops behaving like a fluid. The location of this transition and the following phase depend critically on the fireball's constitution.

Consider, first, a pure radiation fireball. This fireball expands and cools with $T \propto R^{-1}$. The electron-positron pairs gradually annihilate and disappear. This phase ends when the local temperature is ≈ 20 KeV, and sufficiently few high energy photons are left so that $\tau \approx 1$ [25, 88]. The photons escape freely as the fireball becomes transparent. In the meantime, the fireball has been accelerated by its own pressure and the radiation-electron-positron fluid has reached extreme-relativistic velocities [25, 67] with $\Gamma \approx R/R_i \approx T_i/T$. The observed photon energy, as seen by an observer at infinity, is now blue shifted practically back to the original temperature, $T_{obs} \approx \Gamma T \approx T_i$. The resulting photon spectrum is, however, thermal and very different from those observed in GRBs.

Astrophysical fireballs include, most likely, baryonic matter in addition to radiation and $e^+ e^-$ pairs. The baryons affect the fireball in two ways: The electrons associated with the baryons increase the opacity and delay the escape of radiation. The baryons are also dragged by the accelerated leptons and this requires a conversion of the radiation energy into a kinetic energy. The second effect is more important and we will focus on it here (see [88, 83, 78, 50] for a more detailed discussion of fireball evolution). The acceleration of the baryons leads to a transition from the initial radiation dominated phase (in which most of the energy is in the form of radiation) to a matter dominated phase (in which the energy is mostly the kinetic energy of the baryons). The factor $\eta \equiv E_i/Mc^2$, the ratio of the initial radiation energy E_i to the rest energy Mc^2, determines the location of the transitions and the fate of the fireball. The transition takes place at:

$$R_\eta = 2R_i\eta = 2 \times 10^{10}\mathrm{cm} \left(\frac{R_i}{1000\mathrm{km}}\right)\left(\frac{\eta}{100}\right). \qquad (18.19)$$

The overall outcome of the fireball is the same as the outcome of a pure radiation fireball if R_η is larger than the radius in which the fireball becomes optically thin. In this case, all the initial energy is still carried away by photons, with a thermal spectrum. For most reasonable baryonic loads, R_η is, however, smaller than the radius in which the fireball becomes optically thin. Such a fireball results in a relativistic expanding shell of baryons, whose kinetic energy equals the total initial energy [88, 69]. Energy conservation dictates that

$$M = \frac{E_i}{c^2\Gamma} = 5 \cdot 10^{-7} M_\odot \left(\frac{E}{10^{50}\mathrm{ergs}}\right)\left(\frac{\Gamma}{100}\right)^{-1}, \qquad (18.20)$$

where E_i is the total *initial* energy (and not the observed energy). Comparison with the definition of η reveals that $\Gamma \approx \eta$. The width of the shell is comparable to the original size of the fireball: $\Delta R \approx R_i$ [83] . Surprisingly, we discover that the most likely outcome of a fireball is just what we need: a narrow relativistic shell of baryons with a very large Γ.

We have estimated in equation (18.20) the mass load for a spherical fireball. If the fireball is not spherical and has an angular opening θ, E_i and M are smaller by a factor $\theta^2/4\pi$. A quick glance at this mass limit reveals that the baryonic load of the fireball must be extremely small, otherwise the motion will not be relativistic. This leads us immediately to another open question:

• *How can one produce "clean" enough fireballs with sufficiently small baryonic loads?*

Some researchers have argued that this is impossible and have used this condition as an argument against the thermal acceleration mechanism. Others have argued that one can use this constraint to rule out specific models for the "engine" that generates the fireball as some engines cannot produce "clean" enough fireballs. My personal view is that this is still an open question, a very puzzling one.

Before leaving this topic, it is worth mentioning that fireballs are not necessarily spherical as their name implies. The equations governing a spherical fireball are simplest to derive. However, it has been shown that even a small fraction of a spherical shell, that is any beam whose width $\theta > \Gamma^{-1}$, behaves locally as if it were a part of a spherical shell [58]. Thus, fireballs could, in fact, be firebeams or jets if sufficient collimation takes place at the initial stages (see Fig. 18.5).

18.6   What?

We now turn to our third question, **what?** I address this question after discussing **where?** and **how?** because I hope that the previous discussion has constrained the sources that could produce GRBs. I will summarize first what have we learned so far on the nature of the "engines" and then will turn to astrophysical models.

18.6.1   *What Do We Need from the Internal Engine?*

GRBs are produced by some internal "engine" that supplies the energy for the process. This "engine" is well hidden from direct observations and it will be difficult to determine what it is from the available data. We have concluded that this "engine" should supply us with a short pulse of extreme relativistic particles. It should accelerate $4 \cdot 10^{-8}/(\Gamma/100)M_\odot$ per steradian to $\Gamma > 100$. The minimal total energy required (assuming full relativistic beaming) is $10^{46}/(\Gamma/100)^2$ ergs. The maximal mass allowed is $5 \cdot 10^{-7}M_\odot/(\Gamma/100)$ (assuming a spherical burst). The total duration of the pulse varies from several msecs to several hundred seconds. The size of the source is, most likely, less than 1000km. The acceleration can be direct, via a (unknown yet) non-thermal (most likely electromagnetic) process or indirect via a fireball phase. The source of the fireball should produce high energy photons with a total energy of $8 \cdot 10^{48}$ ergs per steradian, with no more than $4 \cdot 10^{-8} M_\odot$ per steradian within this radiation.

These are the only constraints on the sources of GRBs. These constraints are indirect and follow from our analysis. The compactness problem tells us that it is impossible to observe the sources directly, at least with electromagnetic radiation, and hence there are no direct observational constraints. The only direct observational constraint is the rate of the bursts: ≈ 1 per 10^6 years per galaxy for isotropic bursts. However, even this limit is not strict as an uncertainty in the beaming angle, θ, of the bursts leads to an uncertainty of order $4\pi/\theta^2$ in the rate. Any process that satisfies these constraints, and whose event rate is compatible with the observed event rate, is a possible model for the origin of GRBs.

18.6.2   Coincidences and Other Astronomical Hints

Before examining the origin of GRBs, it is worthwhile to consider another astronomical phenomenon, SNRs, and see how one could have reached the right conclusions there. I have chosen SNRs, which originate from the interaction of supernovae ejecta with the ISM since they seem to be the Newtonian cousins of GRBs. SNRs are observed as diffuse shells of optical and radio emission. Most SNRs are observed in a self-similar stage which is determined by two parameters: the energy of the ejected matter, which is $\approx 10^{51}$ ergs, and the density of the ISM. With so little information, how do we know that these are really supernova remnants?

The chain of arguments is very simple. Supernovae observations show the ejection of several solar masses with velocities of tens of thousands km/sec. The corresponding kinetic energy is $\approx 10^{51}$ ergs—exactly in the right range. Additionally, the birth rate of SNRs agrees with the rate of supernovae! Finally, there is a clear coincidence between pulsars that form in supernovae and SNRs.

Suppose now that we could not observe supernovae directly and that we could not see the ejected material. Could we find that the observed diffuse shells result from core collapse without these observations? Surprisingly, the answer is yes. The amount of energy needed to produce an SNR is quite large. This energy must have been released within a relatively short period of time, less than about 100 years. Only a few astronomical phenomena can do that - stellar core collapse that forms a neutron star is one. The discovery of pulsars has told us that neutron stars exist. The binding energy of a neutron star is larger than the kinetic energy required to produce an SNR, hence neutron star formation is a possible candidate for the source of SNRs. Estimates of the rate of pulsar formation and the birth rate of SNRs give a comparable rate. Thus, we have a phenomenon that can provide the energy (even though if we don't see supernovae we won't know that matter is ejected with the needed amount of energy) and it is taking place at a comparable rate. The final confirmation of the theory would come with the discovery of the Crab pulsar in the center of the remnant of the 1054 supernova. The existence of other pulsars in the centers of other SNRs will confirm that this was not a coincidence.

The situation with GRBs is remarkably similar to this conceptual toy model. At present we have very few clues to the process that causes GRBs. We know that the needed energy is $\approx 10^{50}$ ergs, which is rather close to the binding energy of a neutron star. This, in combination with the facts that the size of the source is quite likely less than 10^8cm and that GRBs are one-time events, suggests that they are related to the formation of a compact object. The only other energetically feasible alternatives that I can see are the sudden release of the total rotational energy of a millisecond pulsar or the sudden release of the total magnetic energy of a neutron star with a 10^{15} Gauss magnetic field.

No mechanism has been suggested to stop suddenly a rotating neutron star or a black hole. Thomson [91] suggested that GRBs are produced when a magnetic field of 10^{15} Gauss is suddenly destroyed, but there is no evidence that such magnetic fields exist in nature (indirect evidence that such fields exist in the soft repeater 790305 was given by [17, 72]). Woosley, suggested that GRBs occur in "failed" supernovae [102]. However, it is not known that such events take place; if they do, there is no knowledge of their event rate (the lack of observations in Kamiokande puts an upper limit of less than once per ten years per galaxy, but this limit is very weak).

We are left with binary neutron star mergers (NS^{2}Ms) [18] or, with a small variant: neutron star-black hole mergers [70]. These mergers take place because of the decay of the binary orbits due to gravitational radiation emission. The discovery of the famous binary pulsar PSR 1913+16 [31] demonstrated that this decay is taking place [89]. The discovery of other binary pulsars, and in particular of PSR 1534+12 [100], has shown that PSR 1913+16 is not unique and that such systems are common. These observations suggest that NS^{2}Ms take place at a rate of $\approx 10^{-6}$ events per year per galaxy [57, 76, 1], in amazing agreement with the GRB event rate [77, 80, 11]. Note that it has been suggested [92] that many neutron star binaries are born with very close orbits and hence with very short lifetimes. If this idea is correct, then the merger rate will be much higher. This will destroy, of course,the nice agreement between the rates of SNR and NS^{2}Ms. Consistency can be restored if we invoke beaming, which might even be advantageous for some models. The short lifetime of those systems, which is the essence of this idea, makes it impossible to confirm or rule out this speculation.

NS^{2}Ms result, most likely, in rotating black holes [14]. The process releases

$\approx 5 \times 10^{53}$ ergs [9]. Most of this energy escapes as neutrinos and gravitational radiation, but a small fraction of this energy suffices to power a GRB. The observed rate of NS^{2}Ms is similar to the observed rate of GRBs. This is not a lot - but this is more than can be said, at present, about any other GRB model. It is also remarkably similar to our conceptual SNR toy model.

How can one prove or disprove this, or any other, GRB model? Theoretical studies concerning specific details of the model can, of course, make it more or less appealing. But in view of the fact that the observed radiation emerges from a distant region which is very far from the inner "engine" I doubt if this will ever be sufficient. Again, following our conceptual toy model, it seems that the only way to confirm any GRB model will be via detecting in time-coincidence another astronomical phenomenon, whose source could be identified with certainty. This brings us directly to another open question:

•*Is there a coincidence between GRB and any other phenomenon?*

NS^{2}Ms have two accompanying signals, a neutrino signal and a gravitational radiation signal. Both signals are extremely difficult to detect, but they provide a clear prediction of coincidence that could be proved or falsified sometime in the distant future.

18.7 WHY?

The last and probably most ambitious question is **why?** That is, why were GRBs put there in the sky? Put differently, what can GRBs tell us about the Universe that we live in? It is difficult to deal with this question when we don't yet know what the origin of GRBs is and are not even certain where they are coming from. Still, it is worthwhile to speculate on the possible applications of GRBs to other branches of astrophysics and in particular to Cosmology.

If, somewhat unexpectedly, it turns out that GRBs are galactic, it will be the first indication that there is a population of stellar galactic objects that extend to distances of more than 100 kpc. At present there is no other indication that there are objects at such distances. Furthermore, the distribution of this population must differ from the halo distribution inferred from dynamical studies of the Galaxy. This might have far reaching implications to theories of galactic structure. The origin of this population is an intriguing question that might teach us a lot about the galactic halo (if the sources are born in the halo) or about stellar processes in the galactic disk (if it turns out that these objects are ejected from the disk).

If GRBs are cosmological, they seem to be a relatively homogeneous population of sources, with a narrow luminosity function (the peak luminosity of GRBs varies by less than factor of 10 [11, 30]) that is located at relatively high redshifts [77, 48, 99, 11]. The universe and our Galaxy are transparent to MeV range γ-

rays (see e.g., [103]). Hence GRBs constitute a unique homogeneous population of sources which does not suffer from any angular distortion due to absorption by the galaxy or by any other object. Could GRBs be the holy grail of Cosmology and provide us with the standard candles needed to determine the cosmological parameters H_0, Ω, and Λ? The answer is unfortunately no, at least not yet. Lacking any spectral feature, there is no indication what the redshift of individual bursts is. At present, all that we have is the number vs. peak luminosity distribution. This distribution is not sensitive enough to distinguish between different cosmological models (see Fig. 18.4), even when the sources are perfect standard candles with no source evolution [11]. Here, once more, we encounter the importance of finding counterparts to GRBs. Observation of such counterparts might provide us with additional parameters, such as the distance or the redshift, that, when combined with the GRB data, could determine the values of the cosmological parameters.

However, GRBs have other direct applications. If the bursts' source follow the matter distribution then GRBs can map the large scale structure of the Universe on scales that cannot be spanned directly otherwise. Lamb and Quashnock [44] have pointed out that a population of several thousand cosmological bursts should show angular deviations from isotropy on a scale of a few degrees. This would immediately lead to new interesting cosmological limits. So far there is no detected signal in the 1112 bursts of the BATSE 3B catalog [90]. But the potential of this population is clear and quite promising.

GRBs can also serve to explore cosmology as a background population which could be lensed by foreground objects [68]. While common gravitational lenses result in several images of the same objects. The low angular resolution of GRB detectors is insufficient to distinguish between the positions of different images of a lensed GRB. Instead a gravitationally lensed burst will appear as repeated bursts with the same time profile but different intensities from the same direction. The time delay arises due to the different times of flight along the different paths. Detection of a lensed burst will confirm once for all the cosmological origin of the bursts while lack of such events in the future will be a cause for concern for proponents of a cosmological origin. Mao [46] estimated that the probability for lensing by regular foreground galaxies is 0.04%-0.4% hence the lack of confirmed lensed event so far [62] is not problematic yet. The statistics of lensed bursts could probe the nature of the lensing objects and the dark matter of the Universe [5]. The fact that no lensed bursts where detected so far is sufficient to rule out a critical density of $10^{6.5} M_{\odot}$ to $10^{8.1} M_{\odot}$ black holes [60]. Truly, this was not the leading candidate for cosmological dark matter. Still this result is a clear demonstration of power of this technique and the potential of GRB lensing. The statistics of lensing depends on the distance to the lensed objects which is quite uncertain at present. Once more we find that a discovery of a counterpart whose distance could be measured would improve significantly this technique as well.

Finally I should mention an additional intriguing implication of the models that I have discussed so far. If GRBs are produced by ultra-relativistic particles, it is likely that some ultra-relativistic particles escape without loosing their energy. Such particles would be observed as cosmic rays [88]. With our recent understanding of energy conversion in shocks it was realized [96, 97, 53, 54, 95] that it is possible that the shocks that convert the kinetic energy to radiation also accelerate some of these particles to even higher energies [96, 97, 53, 54]. Thus the events that produce GRBs might also generate cosmic rays. This is particularly intriguing as an explanation of the three mysterious 10^{20}eV cosmic ray events [96, 97, 53] discussed by Cronin [13] at this meeting.

18.8   CONCLUSIONS

GRBs have remained an unsolved problem for more than twenty years. The analysis that I have presented suggests that GRBs are the final outcome of a complicated process in which particles are first accelerated to ultra-relativistic energies and then convert their kinetic energy, via shocks, to the observed radiation. The fact that the observed radiation emerges from a region that is far from the internal engine that accelerates the particles and supplies the energy for the burst makes it difficult to find a conclusive test that will reveal the nature of this engine. It is useful to explore the nature of the conversion mechanism of kinetic energy to radiation, possible radiation mechanisms and details of specific "engines" and acceleration processes. However, I fear that the lack of any direct observation of the source region restricts our ability to prove or falsify most models. I think we should focus on locating events that can produce the required energy and satisfy the temporal and size constraints and that are taking place at a comparable rate. Today, we have one such candidate, binary neutron star mergers. I believe that the search for coincident events in other wavelengths or other forms of emission should be the prime task of GRB research, as this will be the clearest, and perhaps the ultimate test of this and any other model.

18.9   SOME OPEN QUESTIONS

Observational questions:

General:

The 64K question:

Theoretical questions:

The Compactness problem:

Energy extraction from ultra-relativistic particles:

ACKNOWLEDGMENTS

I thank Ramesh Narayan, and Eli Waxman for many helpful discussions, Reem
Sari and Ehud Cohen for allowing me to include unpublished results in this review, and J. N. Bahcall, J. Katz, B. Paczyński and A. Ulmer for numerous critical
remarks.. This research was supported by a BRF grant to the Hebrew University
and by a NASA grant NAG5-1904 to Harvard University.

BIBLIOGRAPHIC NOTES

The following list of references provides an introduction for students to gamma-
ray bursts.

- **Klebesadel, R. W., Strong, I. B., & Olson, R. A. 1973, ApJ, 182, L85.**
  The discovery paper of GRBs.

- **Meegan, C. A. et. al . 1992, Nature, 355, 143.** The announcement of the
  BATSE team that the GRB distribution is isotropic and that there is a paucity
  of weak bursts.

- **Fishman, G. J., & Meegan,, C. A. 1995, ARA&A, 33, 415:** A recent detailed review that focuses on observations of GRBs.

- **Paciesas, W. S. & Fishman G. J., eds. AIP Conference Proceedings 265,
  Gamma-Ray Bursts, 1991, (New York: AIP);
  Fishman, G. J., Brainerd, J. J., & Hurley, K. 1994, AIP Conference Pro-
  ceedings 307, Gamma-Ray Bursts, Second Workshop, 1993 (New York:
  AIP);**

Kouveliotou, C.Briggs M.S. & G.J. Fishman eds. AIP Conference Proceedings Gamma-Ray Bursts, 1995, (New York: AIP) in press. These four volumes (one volume for each one of the first Huntsville meeting and Two volumes for the last meeting) contain numerous articles on all aspects of GRBs. The exponential increase in size of those volumes reflects the growing interest in GRBs in the last four years.

- **Paczyński, B. 1995, PASP, 107, 1167.** A summary of the arguments in favor of a cosmological origin of GRBs.

- **Lamb, D. Q. 1995, PASP, 107, 1152.** A summary of the arguments in favor of a galactic origin of GRBs.

- **Goodman, J. 1986, ApJ, 308, L47;**
  Paczyński, B. 1986, ApJ, 308, L51. The classical papers on fireballs. The first one describes fireballs in the instantaneous approximation and the second one describes them in the steady state (wind) approximation. Both papers discuss "clean" i.e. pure radiation fireballs.

- **Shemi, A., & Piran, T. 1990, ApJ, 365, L55;**
  Paczyński, B. 1990, ApJ, 363, 218. Discussion of the role of baryonic load on fireball evolution.

- **Mészaros, P., & Rees, M. J. 1992, MNRAS, 258, 41p.** The suggestion that the kinetic energy of baryons accelerated in fireballs can be recovered in collisions with the ISN.

- **Piran, T. 1994, in AIP Conference Proceedings 307, Gamma-Ray Bursts, Second Workshop, Huntsville, Alabama, 1993, eds. G. J. Fishman, J. J. Brainerd, & K. Hurley (New York: AIP), p. 495.** A detailed review of the reasons for the existence of fireballs in GRBs and of fireball evolution.

- **Eichler, D., Livio, M., Piran, T., & Schramm, D. N. 1989, Nature, 340, 126.** A discussion of Neutron Star mergers as sources of GRBs, as well as gravitational waves, neutrinos and possibly a cite for nucleosynthesis.

- **Narayan, R., Piran, T., & Shemi, A. 1991, ApJ, 379, L1.**
  Phinney, E. S. 1991, ApJ, 380, L17. Estimates of the rates of Neutron star mergers. Those were found latter to be comparable with the rate of GRBs as estimated from BATSE's observations.

- **Piran, T. 1995, Proceedings of IAU Symposium 165, eds. E. P. J. van den Heuvel, & J. van den Paradijs (Dordrecht: Kluwer).** A recent review of the status of the neutron star merger model for GRBs.

BIBLIOGRAPHY

[1] Bailes, M. 1995, to appear in Proceedings of IAU Symposium 165, eds. E. P. J. van den Heuvel, & J. van den Paradijs (Dordrecht: Kluwer).

[2] Bhat, P. N., et al. 1992, Nature, 359, 217.

[3] Blandford, R. D., & McKee, C. F. 1976, Phys. of Fluids, 19, 1130.

[4] Blandford, R. D., & McKee, C. F. 1976, MNRAS, 180, 343.

[5] Blase, O. M., & Webster, R. L. 1992, ApJ, 391, L66.

[6] Band, D., et. al., 1993, ApJ, 413, 281.

[7] Briggs, M. S. 1995, Proceedings of the 29th ESLAB symposium, eds. Bennet, K. & Winkler, C., to appear in Ap&SS.

[8] Briggs, M. S. 1995, in Proceedings of the 17th Texas Symposium, to be published in the Annals of the New York Academy of Sciences.

[9] Clark, J. P. A., & Eardley, D. 1977, ApJ, 215, 311.

[10] Cohen, U., Kollat, T., & Piran, T., 1994, Astro-ph/9406012.

[11] Cohen, U., & Piran, T. 1995, ApJ, 444, L25.

[12] Cohen, U., & Piran, T. 1995, in Gamma-Ray Bursts, Huntsville, Alabama, 1995, Kouveliotou, C. Briggs M.S. & G.J. Fishman eds. (New York: AIP) in press.

[13] Cronin, J. W. 1995, this volume.

[14] Davies, M. B., Benz, W., Piran, T., & Thielemann, F. K. 1994, ApJ, 431, 742.

[15] Dermer, C. D. 1992, Phys. Rev. Lett., 68, 1799.

[16] Dezaley, J.-P., et al. 1992, in AIP Conference Proceedings 307, Gamma-Ray Bursts, Second Workshop, Huntsville, Alabama, 1993 (New York: AIP), p. 304.

[17] Duncan, R. C., & Thompson, C. 1992, ApJ, 392, L9.

[18] Eichler, D., Livio, M., Piran, T., & Schramm, D. N. 1989, Nature, 340, 126.

[19] Fenimore E. E., & Bloom J. S. 1995, to appear in ApJ, astro-ph/9504063.

[20] Fenimore, E. E., Epstein, R. I., & Ho, C. H. 1993, A&A Supp., 97, 59.

[21] Fenimore, E. E., Klebesadel, R. W., & Laros, J. G., LAUR-95-1220, submitted to ApJ.

[22] Fishman, J. G. 1995, PASP, 107, 1145.

[23] Fishman, G. J., Brainerd, J. J., & Hurley, K. 1994, AIP Conference Proceedings 307, Gamma-Ray Bursts, Second Workshop, Huntsville, Alabama, 1993 (New York: AIP).

[24] Fishman, G. J., & Meegan,, C. A. 1995, ARA&A, 33, 415.

[25] Goodman, J. 1986, ApJ, 308, L47.

[26] Guilbert, P. W., Fabian, A. C., & Rees, M. J. 1983, MNRAS, 205, 593.

[27] Hartmann D. H. 1994, in AIP Conference Proceedings 307, Gamma-Ray Bursts, Second Workshop, Huntsville, Alabama, 1993, eds. G. J. Fishman, J. J. Brainerd, & K. Hurley (New York: AIP), p. 562.

[28] Hartmann, D. 1995, A&A Rev., 6, 225.

[29] Hartmann, D. H., et al. 1994, in AIP Conference Proceedings 307, Gamma-Ray Bursts, Second Workshop, Huntsville, Alabama, 1993, eds. G. J. Fishman, J. J. Brainerd, & K. Hurley, (New York: AIP), p. 127.

[30] Horack, J. M., & Emslie, A. G. 1994, ApJ, 428, 620.

[31] Hulse, R. A., & Taylor, J. H. 1975, ApJ, 368, 504.

[32] Hurley, K., et al. 1994, Nature, 372, 652.

[33] Katz, J. I. 1994, ApJ, 422, 248.

[34] Klebesadel, R. W. 1992, in Gamma-Ray Bursts, eds. C. Ho, R. I. Epstein, & E. E. Fenimore (Cambridge: Cambridge University Press), p. 161.

[35] Klebesadel, R. W., Strong, I. B., & Olson, R. A. 1973, ApJ, 182, L85.

[36] Koshut, T. M., et al. 1995, ApJ, in press.

[37] Kouveliotou, C. 1995, in Proceedings of the 17th Texas Symposium, to be published in the Annals of the New York Academy of Sciences.

[38] Kouveliotou, C. 1995, to appear in Proceedings of the International School of Astrophysics, D. Chalonge.

[39] Kouveliotou, C., et al. 1993, ApJ, 413, L101.

[40] Kouveliotou, C.Briggs M.S. & G.J. Fishman eds. AIP Conference Proceedings Gamma-Ray Bursts, Huntsville, Alabama, 1995, (New York: AIP) in press.

[41] Krolik, J. H., & Pier, E. A. 1991, ApJ, 373, 277.

[42] Lamb, D. Q. 1995, PASP, 107, 1152.

[43] Lamb, D. Q., Graziani, C., & Smith, I. A. 1993, ApJ, 413, L11.

[44] Lamb, D. Q., & Quashnock, J. M. 1993, ApJ, 415, L1.

[45] Loeb, A. 1993, Phys. Rev. D, 48, 3419.

[46] Mao, S. 1992, ApJ, 389, L41.

[47] Mao, S., Narayan, R., & Piran, T. 1994, ApJ, 420, 171.

[48] Mao, S., & Paczyński, B. 1992, ApJ, 388, L45.

[49] Meegan, C. A., et al. 1995, ApJ, 434, 552.

[50] Mészáros, P., Laguna, P., & Rees, M. J. 1993, ApJ, 415, 181.

[51] Mészáros, P., & Rees, M. J. 1992, MNRAS, 258, 41p.

[52] Mészáros, P., & Rees, M. J. 1993, ApJ, 405, 278.

[53] Milgrom, M., & Usov, V. 1995, ApJ, 449, L37.

[54] Milgrom, M., & Usov, V. 1995, astro-ph/9506099.

[55] Narayan, R., Paczyński, B., & Piran, T. 1992, ApJ, 395, L83.

[56] Narayan, R., & Piran, T. 1994, MNRAS, 265, 65.

[57] Narayan, R., Piran, T., & Shemi, A. 1991, ApJ, 379, L1.

[58] Narayan, R., & Piran, T. 1995, in preparation.

[59] Nemiroff, R. J. 1993, Comm. Ap., 17, 189.

[60] Nemiroff, R. J., et al. 1993, ApJ, 414, 36.

[61] Nemiroff, R. J., et al. 1994, ApJ, 435, L133.

[62] Nemiroff, R. J., et al. 1994, ApJ, 432, 478.

[63] Norris, J. P., et al. 1994, ApJ, 423, 432.

[64] Norris, J. P., et al. 1995, ApJ, 439, 542.

[65] Paciesas, W. S. & Fishman G. J., eds. AIP Conference Proceedings 265, Gamma-Ray Bursts, Huntsville, Alabama, 1991, (New York: AIP).

[66] Paciesas, W. S., et al. 1992, in AIP Conference Proceedings 265, Gamma-Ray Bursts, Huntsville, Alabama, 1991, eds. W. S. Paciesas, & G. J. Fishman (New York: AIP), p. 190.

[67] Paczyński, B. 1986, ApJ, 308, L51.

[68] Paczyński, B. 1986, ApJ, 308, L43.

[69] Paczyński, B. 1990, ApJ, 363, 218.

[70] Paczyński, B. 1991, Acta Astronomica, 41, 257.

[71] Paczyński, B. 1992, Nature, 355, 521.

[72] Paczyński, B. 1992, Acta Astronomica, 42, 145.

[73] Paczyński, B. 1995, PASP, 107, 1167.

[74] Palmer, D. M., et al. 1994, in AIP Conference Proceedings 307, Gamma-Ray Bursts, Second Workshop, Huntsville, Alabama, 1993, eds. G. J. Fishman, J. J. Brainerd, & K. Hurley (New York: AIP), p. 247.

[75] Pendleton, G. N., et al. 1995, in preparation (quoted in [7]).

[76] Phinney, E. S. 1991, ApJ, 380, L17.

[77] Piran, T. 1992, ApJ, 389, L45.

[78] Piran, T. 1994, in AIP Conference Proceedings 307, Gamma-Ray Bursts, Second Workshop, Huntsville, Alabama, 1993, eds. G. J. Fishman, J. J. Brainerd, & K. Hurley (New York: AIP), p. 495.

[79] Piran, T. 1995, to appear in the Proceedings of IAU Symposium 165, eds. E. P. J. van den Heuvel, & J. van den Paradijs (Dordrecht: Kluwer).

[80] Piran, T., Narayan, R., & Shemi, A. 1992, in AIP Conference Proceedings 265, Gamma-Ray Bursts, Huntsville, Alabama, 1991, eds. W. S. Paciesas, & G. J. Fishman (New York: AIP), p. 149.

[81] Piran, T., & Narayan, R. 1995, in Gamma-Ray Bursts, Huntsville, Alabama, 1995, Kouveliotou, C.Briggs M.S. & G.J. Fishman eds. (New York: AIP) in press.

[82] Piran, T., & Shemi, A. 1993, ApJ, 403, L67.

[83] Piran, T., Shemi, A., & Narayan, R. 1993, MNRAS, 263, 861.

[84] Quashnok, J., & Lamb, D. 1993, MNRAS, 265, L59.

[85] Rees, M. J., & Mészaros, P. 1994, ApJ, 430, L93.

[86] Sari, R., & Piran, T. 1995, ApJ in press.

[87] Schmidt, W. K. H. 1978, Nature, 271, 525.

[88] Shemi, A., & Piran, T. 1990, ApJ, 365, L55.

[89] Taylor, J. H., & Weisberg, J. M. 1982, ApJ, 253, 908.

[90] Tegmark, M., Hartmann, D. H., Briggs, M. S., & Meegan, C. A. 1995, preprint, submitted to ApJ.

[91] Thompson, C. 1994, MNRAS, 270, 480.

[92] Tutukov, A. V., & Yungelson, L. R. 1994, MNRAS, 268, 871.

[93] Usov, V. V. 1994, MNRAS, 267,1035.

[94] Usov, V. V., & Smolsky, M. V. 1995, ApJ in press.

[95] Vietri, M. 1995, preprint, astro-ph/9506081.

[96] Waxman, E. 1995, ApJ, 452, L1.

[97] Waxman, E. 1995, Phys. Rev. Lett., 75, 386.

[98] Waxman, E., & Piran, T. 1994, ApJ, 433, L85.

[99] Wickramasinghe, W. A. D. T., et al. 1993, ApJ, 411, L55.

[100] Wolszczan, A. 1991, Nature, 350, 688.

[101] Woods, E., & Loeb, A. 1995, ApJ, 383, 292..

[102] Woosley, S. E. 1993, ApJ, 405, 273.

[103] Zdziarski, A. A., & Svensson, R. 1989, ApJ, 344, 551.